D1074925

*Computer
Calculations for
Multicomponent
Vapor-Liquid and
Liquid-Liquid
Equilibria*

HOLLAND *Unsteady State Processes with Applications in Multicomponent Distillation*

KOPPEL *Introduction to Control Theory with Applications to Process Control*

LEVICH *Physiochemical Hydrodynamics*

MEISSNER *Processes and Systems in Industrial Chemistry*

MYERS AND SEIDER *Introduction to Chemical Engineering and Computer Calculations*

NEWMAN *Electrochemical Systems*

OHARA AND REID *Modeling Crystal Growth Rates from Solution*

PERLMUTTER *Stability of Chemical Reactors*

PETERSON *Chemical Reactor Analysis*

PRAUSNITZ *Molecular Thermodynamics of Fluid-Phase Equilibria*

PRAUSNITZ ET AL *Computer Calculations for Multicomponent Vapor–Liquid and Liquid–Liquid Equilibria*

PRAUSNITZ, ECKERT, ORYE, AND O'CONNELL *Computer Calculations for Multicomponent Vapor–Liquid Equilibria*

RUDD ET AL *Process Synthesis*

SCHULTZ *Polymer Materials Science*

SEINFELD AND LAPIDUS *Mathematical Methods in Chemical Engineering: Vol. III, Process Modeling, Estimation and Identification*

VILLADSEN AND MICHELSEN *Solutions of Differential Equation Models by Polynomial Approximation*

WILDE *Optimum Seeking Methods*

WILLIAMS *Polymer Science and Engineering*

WOODS *Financial Decision Making in the Process Industry*

Computer Calculations for Multicomponent Vapor-Liquid and Liquid-Liquid Equilibria

J. M. PRAUSNITZ
T. F. ANDERSON
E. A. GRENS

University of California
Berkeley

C. A. ECKERT
R. HSIEH

University of Illinois
Urbana

J. P. O'CONNELL

University of Florida
Gainesville

PRENTICE-HALL, INC.
Englewood Cliffs, New Jersey 07632

Library of Congress Cataloging in Publication Data

Main entry under title:

Computer calculations for multicomponent vapor-
   liquid and liquid-liquid equilibria.

   (Prentice-Hall international series in the
physical and chemical engineering sciences)
   Bibliography: p.
   1. Vapor–liquid equilibrium–Data processing.
2. Liquid–liquid equilibrium–Data processing.
I. Prausnitz, J. M.
TP156.E65C65   660.2′08423   79-27200
ISBN 0-13-164962-0

Production supervision by Karen Skrable
Manufacturing buyer: Gordon Osbourne

Printed in the United States of America

10 9 8 7 6 5 4 3 2 1

Prentice-Hall International, Inc., *London*
Prentice-Hall of Australia Pty. Limited, *Sydney*
Prentice-Hall of Canada, Ltd., *Toronto*
Prentice-Hall of India Private Limited, *New Delhi*
Prentice-Hall of Japan, Inc., *Tokyo*
Prentice-Hall of Southeast Asia Pte. Ltd., *Singapore*
Whitehall Books Limited, *Wellington, New Zealand*

Contents

Preface

Separation of fluid mixtures constitutes one of the main tasks of chemical engineering. While new separation methods are under active development, most large-scale separations are achieved by classical phase-contacting operations: distillation, absorption, stripping and extraction. Design of equipment for such operations requires quantitative estimates of phase equilibria.

This monograph presents a detailed discussion of a molecular-thermodynamic method for computer-implemented estimation of vapor-liquid and liquid-liquid equilibria in multicomponent systems, coupled with a minimum of experimental information. Attention is confined to nonelectrolytes, i.e. organic liquids (hydrocarbons and their derivatives, alcohols, nitriles, ketones, esters, etc.) and a few common inorganic fluids such as water and carbon dioxide. Attention is also confined to low or moderate pressures and to conditions remote from critical.

Our earlier monograph, "Computer Calculations for Multicomponent Vapor-Liquid Equilibria," published in 1967, is now out of date. The material presented here is, in a sense, an updating of the earlier work, but it is also a major extension since the present monograph, unlike the former, discusses also liquid-liquid equilibria and presents generalized iterative techniques for equilibrium calculations.

No overall statement can be made concerning the accuracy of the calculated results because that depends crucially on the accuracy of the limited experimental data on which the calculations, inevitably, must rest. When these data are accurate and when careful attention is given to all limitations specified throughout the monograph, the calculated results are likely to be reliable.

It is a pleasure to record my gratitude first, to my co-authors at the University of Illinois, Charles Eckert and Richard Hsieh,* who provided much of the material on pure-component properties and enthalpies; and to John O'Connell, University of Florida, who provided much of the material on gas-

---

*Now at Chevron Research Company, Richmond, California.

phase nonideality and on liquid mixtures containing one or more
supercritical components; second, to my Berkeley colleague,
Edward Grens, who developed all the computational procedures and
programs, and finally, to my former graduate student and co-
worker, Thomas Anderson,** who performed the illustrative calculations, prepared the figures and took care of all the many
details that are required in the preparation of a monograph.
Much of the material in Chapter 4 is taken from Tom's doctoral
dissertation. Most important, it was Tom's consistent cheerful-
ness and his willingness to undertake numerous time-consuming
tasks which sustained the momentum of our work and which enabled
us to bring it to conclusion.

Special thanks also go to Harold Null (Monsanto Company,
St. Louis, Missouri) and Carl Deal (Shell Development Company,
Houston, Texas) for their constructive review of the manuscript,
to Eldon Larsen for his generous help in attending to the details of preparing the final version, and to Diana Lorentz for
her conscientious typing services.

Molecular thermodynamics is progressing rapidly; similarly, new developments in computer science are certain to continue.
It is likely, therefore, that the techniques discussed in this
monograph will be modified, perhaps drastically so, as new under-
standing of mixtures and new computing possibilities become
available. The techniques presented here, however, reflect the
current state-of-the-art for combining thermodynamics, molecular
physics and computer science for chemical process design.

All computer subroutines and parameter compilations
(single component and binary) presented in this monograph can be
obtained on magnetic tapes; interested readers should write to
me for details. Available are 7-track tapes at 800 bpi, in
unblocked BCD with 80-character (card-image) records. It may
also be possible to furnish certain types of 9-track tapes on
special request. All subroutines are written in American
National Standard FORTRAN (FORTRAN IV), ANSI X309-1978; these
are compatible with most computer systems. The sample main
programs presented in the text require minor modifications for
use with many computer systems; these modifications, however,

---

**Now at the University of Connecticut, Storrs.

are not included on the tape.

It is my hope--and that of my co-authors--that this
monograph may find useful application by process design engin-
eers, that it may stimulate graduate students and research work-
ers toward seeking new techniques which improve upon those
presented here, and that all who use this monograph, for whatever
purpose, may derive a sense of satisfaction, as the authors have,
in combining rational thought and scientific imagination toward
the solution of practical problems.

Berkeley, California

John Prausnitz

Chapter 1

INTRODUCTION

Design of chemical processes almost always includes design
for separation operations; the most common of these are distillation, absorption, and extraction.

One of the essential ingredients for rational design of such
separation operations is a knowledge of the required phase equilibria. The purpose of the present monograph is to present a
technique, implemented for digital computers, to estimate these
equilibria from a minimum of experimental information.

While much attention has been given to the development of
computer techniques for design of distillation and absorption
columns, much less attention has been devoted to the development
of such techniques for equipment using liquid-liquid extraction.
However, regardless of the nature of the operation, few systematic attempts have been made to organize phase-equilibrium
information for direct use in chemical process design. This
monograph presents a systematic procedure for calculating multicomponent vapor-liquid and liquid-liquid equilibria for mixtures
commonly encountered in the chemical process industries. Attention is limited to systems at low or moderate pressures. Pertinent references to previous work are given at the end of this
chapter.

Need for a Thermodynamic Treatment

The possible number of liquid and vapor mixtures in tech-
nological processes is incredibly large, and it is unreasonable
to expect that experimental vapor-liquid and liquid-liquid
equilibria will ever be available for a significant fraction of
this number. Further, obtaining good experimental data requires
appreciable experimental skill, experience, and patience. It is,
therefore, an economic necessity to consider techniques for
calculating phase equilibria for multicomponent mixtures from
few experimental data. Such techniques should require only a

1

limited experimental effort and, whenever possible, should be based on a theoretical foundation to provide reliability for interpolation and extrapolation with respect to temperature, pressure, and composition.

Vapor-liquid and liquid-liquid equilibria depend on the nature of the components present, on their concentrations in both phases, and on the temperature and pressure of the system. Because of the large number of variables which determine multicomponent equilibria, it is essential to utilize an efficient organizational tool which reduces available experimental data to a small number of theoretically significant functions and parameters; these functions and parameters may then be called upon to form the building blocks upon which to construct the desired equilibria. Such an organizational tool is provided by thermodynamic analysis and synthesis. First, limited pure-component and binary data are analyzed to yield fundamental thermodynamic quantities. Second, these quantities are reduced to obtain parameters in a molecular model. That model, by synthesis, may be used to calculate the phase behavior of multicomponent liquids and vapors. In this way, it is possible to "scale up" data on binary and pure-component systems to obtain good estimates of the properties of multicomponent mixtures of a large variety of components including water, polar organic solvents such as ketones, alcohols, nitriles, etc., and paraffinic, naphthenic, and aromatic hydrocarbons.

The method proposed in this monograph has a firm thermodynamic basis. For vapor-liquid equilibria, the method may be used at low or moderate pressures commonly encountered in separation operations since vapor-phase nonidealities are taken into account. For liquid-liquid equilibria the effect of pressure is usually not important unless the pressure is very large or unless conditions are near the vapor-liquid critical region.

The detailed techniques presented here are based on particular models for the vapor phase (Hayden-O'Connell) and for the liquid phase (UNIQUAC). However, our discussion of these techniques is sufficiently general to allow the use of other models, whenever the user prefers to do so.

2

Solution of Simultaneous Thermodynamic Equations

In vapor-liquid equilibria, if one phase composition is given, there are basically four types of problems, characterized by those variables which are specified and those which are to be calculated. Let T stand for temperature, P for total pressure, x_i for the mole fraction of component i in the liquid phase, and y_i for the mole fraction of component i in the vapor phase. For a mixture containing m components, the four types can be organized in this way:

Given	Find
$P, x_1 x_2 \ldots x_m$	$T, y_1 y_2 \ldots y_m$
$T, x_1 x_2 \ldots x_m$	$P, y_1 y_2 \ldots y_m$
$P, y_1 y_2 \ldots y_m$	$T, x_1 x_2 \ldots x_m$
$T, y_1 y_2 \ldots y_m$	$P, x_1 x_2 \ldots x_m$

In each of these problems, there are m unknowns; either the pressure or the temperature is unknown and there are m − 1 unknown mole fractions.

When only the total system composition, pressure, and temperature (or enthalpy) are specified, the problem becomes a flash calculation. This type of problem requires simultaneous solution of the material balance as well as the phase-equilibrium relations.

In liquid-liquid equilibria, the total composition and temperature are known; the pressure is usually not important. This problem is similar in some ways to a vapor-liquid flash and here is referred to as a liquid-liquid flash calculation.

For vapor-liquid equilibria, the equations of equilibrium which must be satisfied are of the form

$$f_i^V = f_i^L \tag{1-1}$$

where f_i^V is the fugacity of component i in the vapor phase and f_i^L is that in the liquid phase. There are m equations of the form (1). The fugacity f_i^V is a function of $T, P, y_1 \ldots y_m$ and the fugacity f_i^L is a function of $T, P, x_1 \ldots x_m$. Once these functions

3

are established, the problem is, in principle, solved. The solution of these m simultaneous equations, however, requires tedious iterative calculations which can be effectively carried out only by an electronic computer.

In liquid-liquid equilibria, the equations of equilibrium which must be satisfied are of the form

$$f_i' = f_i'' \qquad (1-2)$$

where f_i', the fugacity of component i in the ' phase, is a function of $T, P, x_1' \ldots x_m'$ and f_i'', the fugacity of component i in the " phase, is a function of $T, P, x'' \ldots x_m''.$[†] There are m equations of the form (2). Once these functions are established, the problem is, in principle, solved. However, for multicomponent liquid-liquid equilibria, the computational problems are much more severe than those encountered in multicomponent vapor-liquid equilibria.

In vapor-liquid equilibria, it is relatively easy to start the iteration because assumption of ideal behavior (Raoult's law) provides a reasonable zeroth approximation. By contrast, there is no obvious corresponding method to start the iteration calculation for liquid-liquid equilibria. Further, when two liquid phases are present, we must calculate for each component activity coefficients in two phases; since these are often strongly nonlinear functions of compositions, liquid-liquid equilibrium calculations are highly sensitive to small changes in composition. In vapor-liquid equilibria at modest pressures, this sensitivity is lower because vapor-phase fugacity coefficients are usually close to unity and only weak functions of composition. For liquid-liquid equilibria, it is therefore more difficult to construct a numerical iteration procedure that converges both rapidly and consistently.

_____

[†]For typical conditions in the chemical industry, the effect of pressure on liquid-liquid equilibria is negligible and therefore in this monograph pressure is not considered as a variable in Equation (2).

4

In Chapter 2 we discuss briefly the thermodynamic functions whereby the abstract fugacities are related to the measurable, real quantities: temperature, pressure, and composition. This formulation is then given more completely in Chapters 3 and 4, which present detailed material on vapor-phase and liquid-phase fugacities, respectively.

Accuracy

The accuracy of our calculations is strongly dependent on the accuracy of the experimental data used to obtain the necessary parameters. While we cannot make any general quantitative statement about the accuracy of our calculations for multicomponent vapor-liquid equilibria, our experience leads us to believe that the calculated results for ternary or quarternary mixtures have an accuracy only slightly less than that of the binary data upon which the calculations are based. For multicomponent liquid-liquid equilibria, the accuracy of prediction is dependent not only upon the accuracy of the binary data, but also on the method used to obtain binary parameters. While there are always exceptions, in typical cases the technique used for binary-data reduction is of some, but not major, importance for vapor-liquid equilibria. However, for liquid-liquid equilibria, the method of data reduction plays a crucial role, as discussed in Chapters 4 and 6.

Computer Implementation

The calculation of vapor and liquid fugacities in multi-component systems has been implemented by a set of computer programs in the form of FORTRAN IV subroutines. These are applicable to systems of up to twenty components, and operate on a thermo-dynamic data base including parameters for 92 compounds.[†] The set includes subroutines for evaluation of vapor-phase fugacity

---

[†]The data base contains provisions for a simple augmentation by up to eight additional compounds or substitution of other compounds for those included. Binary interaction parameters necessary for calculation of fugacities in liquid mixtures are presently available for 180 pairs.

coefficients (See Chapter 3), liquid-phase activity coefficients and reference state fugacities (see Chapter 4), and liquid- and vapor-phase molar enthalpies (see Chapter 5). These thermodynamic subroutines can be utilized in any desired phase-equilibrium calculation program, either as replacements for thermodynamic subroutines in existing programs or by incorporation in newly developed programs. They have been developed to achieve the greatest degree of computation efficiency that is consistent with the complexity of the mathematical models employed.

The calculation of single-stage equilibrium separations in multicomponent systems is implemented by a series of FORTRAN IV subroutines described in Chapter 7. These treat bubble and dew-point calculations, isothermal and adiabatic equilibrium flash vaporizations, and liquid-liquid equilibrium "flash" separations. The treatment of multistage separation operations, which involves many additional considerations, is not considered in this monograph.

Table 1-1

Selected References for Calculation of Multicomponent Fluid-Phase Equilibria

Reference	Comments
American Petroleum Institute, Bibliographies on Hydrocarbons, Vols. 1-4, "Vapor-Liquid Equilibrium Data for Hydrocarbon Systems" (1963), "Vapor Pressure Data for Hydrocarbons" (1964), "Volumetric and Thermodynamic Data for Pure Hydrocarbons and Their Mixtures" (1964), "Vapor-Liquid Equilibrium Data for Hydrocarbon-Nonhydrocarbon Gas Systems" (1964), API, Division of Refining, Washington.	Extensive bibliography for data on hydrocarbon systems.
American Petroleum Institute Research Project 42, "Properties of Hydrocarbons of High Molecular Weight," API, Division of Science and Technology, New York, 1966.	Compilation of physical properties for 321 heavy hydrocarbons. Vapor pressures at low pressures.
American Petroleum Institute, "Technical Data Book--Petroleum Refining," 3rd ed., API, Division of Refining, Washington, 1976.	Extensive data and bibliography on hydrocarbons; a few predictive correlations are also given.
Battino, R., and H. L. Clever, Chem. Rev., 66, 395 (1966).	Thorough review and compilation of gas solubilities in liquids.
Boublik, T., V. Fried, and E. Hála: "The Vapor Pressure of Pure Substances," Elsevier, Amsterdam, 1973.	Correlation and compilation of vapor-pressure data for pure fluids. Normal and low pressure region.
Claxton, G.: "Physical and Azeotropic Data," The National Benzene and Allied Products Association, London, 1958.	Compilation of azeotropic data as well as other physical properties including melting and boiling points.
D'Ans-Lax: "Taschenbuch für Chemiker und Physiker, 3rd ed.," Vol. 1, Springer-Verlag, Berlin, 1967.	Source for liquid-liquid and vapor-liquid equilibrium data and vapor-pressure data.

Table 1-1 (continued)

Reference	Comments
Faldix, P., Fortschritte der Verfahrenstechnik, 6, 440 (1965), 7, 650 (1967), 8, 625 (1969), 9, 561 (1971).	Bibliography for vapor-liquid equilibrium data.
Francis, A. W.: "Liquid-Liquid Equilibriums," Wiley-Interscience, New York, 1963.	Phenomenological discussion of liquid-liquid equilibria with extensive data bibliography.
Francis, A. W.: "Handbook for Components in Solvent Extraction," Gordon & Breach, New York, 1972.	Comprehensive bibliography for liquid-liquid equilibrium data.
Fredenslund, A., J. Gmehling, and P. Rasmussen: "Vapor-Liquid Equilibria using UNIFAC," Elsevier, New York, 1977.	Detailed and extensive information on the UNIFAC method for estimating activity coefficients with application to vapor-liquid equilibria at moderate pressures.
Gmehling, J., and U. Onken: "Vapor-Liquid Equilibrium Data Collection," DECHEMA Chemistry Data Ser., Vol. 1 (1-10), Frankfurt, 1977.	Comprehensive data collection for more than 6000 binary and multicomponent mixtures at moderate pressures. Data correlation and consistency tests are given for each data set.
Hála, E., and others: "Vapour-Liquid Equilibrium," 2nd English ed., trans. George Standart, Pergamon, Oxford, 1967.	Excellent comprehensive survey, including a discussion of experimental methods.
Hála, E., I. Wichterle, J. Polák, and T. Boublík: "Vapor-Liquid Equilibrium Data at Normal Pressures," Pergamon, Oxford, 1968.	Compilation of experimental data for binary mixtures.
Hicks, C. P.: Bibliography of Thermodynamic Quantities for Binary Fluid Mixtures, "Chemical Thermodynamics", Vol. 2, Chap. 9, edited by M. L. McGlashan, Chemical Society, London, 1978.	Literature references for vapor-liquid equilibria, enthalpies of mixing and volume change for binary systems.

8

Table 1-1 (continued)

Reference	Comments
Hildebrand, J. H., and R. L. Scott: "Solubility of Nonelectrolytes," 3rd ed., Reinhold, New York, 1950 (reprinted by Dover, New York, 1964).	A classic in its field, giving a splendid survey of solution physical chemistry from a chemist's point of view. While seriously out of date, it nevertheless provides physical insight into how molecules "behave" in mixtures.
Hildebrand, J. H., and R. L. Scott: "Regular Solutions," Prentice-Hall, Englewood Cliffs, N.J., 1962.	Updates some of the material in Hildebrand's 1950 book; primarily for chemists.
Hildebrand, J. H., J. M. Prausnitz, and R. L. Scott: "Regular and Related Solutions," Van Nostrand Reinhold, New York, 1970.	Further updates some of the material in Hilde-brand's book; primarily for chemists.
Himmelblau, D. M., B. L. Brady, and J. J. McKetta: "Survey of Solubility Diagrams for Ternary and Quarternary Liquid Systems," Bureau of Engineering Research, Special Publ. Nr. 30, University of Texas, Austin, 1959.	Bibliography of ternary and quarternary liquid-liquid equilibrium data; temperatures are indicated.
Hirata, M., S. Ohe, and K. Nagahama: "Com-puter-aided Data Book of Vapor-Liquid Equilibria," Elsevier, Amsterdam, 1975.	Compilation of binary experimental data reduced with the Wilson equation and, for high pres-sures, with a modified Redlich-Kwong equation.
Hiza, M. J., A. J. Kidnay, and R. C. Miller: "Equilibrium Properties of Fluid Mix-tures--A Bibliography of Data on Fluids of Cryogenic Interest," NSRDS Bibliographic Series. Plenum, New York, 1975.	Extensive bibliography for data on cryogenic mixtures.
Horsley, I. H.: "Azeotropic Data," Advances in Chem. Ser., No. 6 and No. 35, Ameri-can Chem. Soc., Washington, D.C. 1952-62.	Extensive compilation of azeotropic data.

9

Table 1-1 (continued)

Reference	Comments
Jordan, T. E.: "Vapor Pressure of Organic Compounds," Interscience, New York, 1954.	Compilation of vapor-pressure data for organic compounds; data are correlated with the Antoine equation and graphs are presented.
Kehiaian, H. V. (Editor-in-Chief) and B. J. Zwolinski (Executive Officer): "International Data Series: Selected Data on Mixtures," Thermodynamics Research Center, Chemistry Department, Texas A&M University, College Station, Texas 77843 (continuing since 1973).	Presents a variety of measured thermodynamic properties of binary mixtures; these properties are often represented by empirical equations.
Kogan, V. B., and V. M. Friedman: "Handbuch der Dampf-Flüssigkeits-Gleichgewichte," VED Deutscher Verlag der Wissenschaften, Berlin, 1961.	Extensive compilation of vapor-liquid equilibrium data, particularly from Eastern Europe.
Kogan, V. B., V. M. Friedman, and V. V. Kafarov: "Vapor-Liquid Equilibria," 2 vol., Nauka, Moscow, 1966.	Extensive compilation of vapor-liquid equilibrium data, with emphasis on data from Eastern Europe (in Russian).
Landolt-Börnstein: "Zahlenwerte und Funktionen aus Physik, Chemie, Astronomie, Geophysik, und Technik," 6th ed., Vol. II (2a, 2b, 2c), Vol. IV (4b), Springer Verlag, Berlin, beginning 1950.	Vapor-liquid equilibrium data and vapor pressure data, Vol. 2 (2a) and Vol. 4 (4b); and liquid-liquid equilibrium data, Vol. 2 (2b, 2c).
Landolt-Börnstein: "Zahlenwerte und Funktionen aus Naturwissenschaft und Technik," New. Ser. Group IV, Vol. III, Springer Verlag, Berlin, 1975. Also Part 4c (Solubility of Gases in Liquids), 1976.	Updates earlier compilation of vapor-liquid equilibrium data.
Maczynski, A.: "Thermodynamic Data for Technology—Verified Vapor-Liquid Equilibrium Data," Panstwowe Wydawnictwo Naukawa, Warsaw, Volume 1, 1976; Volume 2, 1978.	Compilation of vapor-liquid equilibrium data; data are correlated with Redlich-Kister equation (in Polish). Volume 2 available in English.

Table 1-1 (continued)

Reference	Comments
Nesmeyanov, A. N.: "Vapor Pressure of the Chemical Elements," Elsevier, New York, 1963.	Compilation of vapor-pressure data for elements.
Null, H. R.: "Phase Equilibrium in Process Design," Wiley-Interscience, New York, 1970.	Useful, engineering-oriented monograph with a variety of numerical examples.
Oellrich, L. R., J. Plöcker, and H. Knapp: "Vapor-Liquid Equilibria," Technical University, Institute for Thermodynamics, Berlin, 1973.	Bibliography of vapor-liquid equilibrium data, primarily for systems at low temperatures.
Ohe, S.: "Computer Aided Data Book of Vapor Pressure," Data Book Publ. Co., Tokyo, 1976.	Vapor-pressure data correlated with the Antoine equation. Results displayed graphically.
Prausnitz, J. M., C. A. Eckert, J. P. O'Connell, and R. V. Orye: "Computer Calculations for Multicomponent Vapor-Liquid Equilibria," Prentice-Hall, Englewood Cliffs, N.J., 1967.	Discusses the thermodynamic basis for computer calculations for vapor-liquid equilibria; computer programs are given. Now out of date.
Renon, H., L. L. Asselineau, G. Cohen, and C. Raimbault: "Calcul sur ordinateur des équilibres liquide-vapeur et liquide-liquide," Editions Technip, Paris, 1971.	Discusses the thermodynamic basis for computer calculations for vapor-liquid and liquid-liquid equilibria (in French); computer programs are given. Now out of date.
Seidell, A., and W. F. Linke: "Solubilities of Inorganic and Metal-Organic Compounds," Vols. 1 and 2 and Supplement, Van Nostrand, Princeton, N.J., 1958-65.	Presents solubility data for gas-liquid, solid-solid, and liquid-liquid systems.
Selected Values of Properties of Hydrocarbons and Related Compounds, American Petroleum Institute Research Project 44, Texas A&M Research Foundation, College Station, Texas (continuing).	Vapor-pressure data and other thermodynamic properties.

Table 1-1 (continued)

Reference	Comments
Selected Values of Properties of Chemical Compounds, Thermodynamic Research Center, Texas A&M Research Foundation, College Station, Texas (continuing).	Vapor-pressure data and other thermodynamic properties.
Stage, H., Fortschritte der Verfahrenstechnik, 2, 306 (1956), 3, 364 (1958); Stage, H., and P. Faldix, Fortschritte der Verfahrenstechnik, 4, 429 (1961), 5, 515 (1962).	Presents solubility data for gas-liquid, solid-liquid, and liquid-liquid systems.
Stull, D. R., Ind. Eng. Chem., 39, 517 (1947).	Presents vapor-pressure data for a large number of substances.
Timmermans, J.: "Physical-Chemical Constants of Pure Organic Compounds," Vol. 1-2, Elsevier, Amsterdam, 1950-65.	Compilation of pure-component data including vapor pressures.
Timmermans, J.: "The Physico-Chemical Constants of Binary Systems in Concentrated Solutions," Vol. 1-4, Interscience, New York, 1959-60.	Compilation of data for binary mixtures; reports some vapor-liquid equilibrium data as well as other properties such as density and viscosity.
Treybal, R. E.: "Liquid Extraction," 2nd ed., Chapters 2 and 3, McGraw-Hill, New York, 1963.	Early chapters give good review of classical thermodynamics for liquid-liquid systems with engineering applications.
Wichterle, I., and J. Linek: "Antoine Vapor Pressure Constants of Pure Compounds," Academia, Prague, 1971.	Presents Antoine vapor-pressure constants for pure compounds for two pressure ranges.
Wichterle, I., J. Linek, and E. Hála: "Vapor-liquid Equilibrium Data Bibliography," Vols. 1 and 2, Elsevier, Amsterdam, 1973-76.	Thorough compilation of literature sources for binary and multicomponent data; includes many references to the East European literature.

12

Table 1-1 (continued)

References	Comments
Wilhelm, E., R. Battino, and R. J. Wilcock, Chem. Rev., 77, 219 (1977).	Reviews low-pressure solubilities of gases in water.
Zwolinski, B. J., and R. C. Wilhoit: "Vapor Pressures and Heats of Vaporization of Hydrocarbons and Related Compounds," Thermodynamic Research Center, Dept. of Chemistry, Texas A&M University, College Station, Texas, 1971.	Compilation of vapor pressures of organic and related compounds to one atmosphere.

13

Chapter 2

THERMODYNAMICS OF PHASE EQUILIBRIA

The thermodynamic treatment of multicomponent phase equilibria, introduced by J. W. Gibbs, is based on the concept of the
chemical potential. Two phases are in thermodynamic equilibrium
when the temperature of one phase is equal to that of the other
and when the chemical potential of each component present is the
same in both phases. For engineering purposes, the chemical
potential is an awkward quantity, devoid of any immediate sense
of physical reality. G. N. Lewis showed that a physically more
meaningful quantity, equivalent to the chemical potential, could
be obtained by a simple transformation; the result of this transformation is a quantity called the fugacity, which has units of
pressure. Physically, it is convenient to think of the fugacity
as a thermodynamic pressure since, in a mixture of ideal gases,
the fugacity of each component is equal to its partial pressure.
In real mixtures, the fugacity can be considered as a partial
pressure, corrected for nonideal behavior.

For a vapor phase (superscript V) and a liquid phase (superscript L), at the same temperature, the equation of equilibrium
for each component i is expressed in terms of the fugacity f_i:

$$f_i^V = f_i^L \tag{2-1}$$

Equation (1) is of little practical use unless the fugacities can be related to the experimentally accessible quantities
x, y, T, and P, where x stands for the composition (expressed in
mole fraction) of the liquid phase, y for the composition (also
expressed in mole fraction) of the vapor phase, T for the
absolute temperature, and P for the total pressure, assumed to
be the same for both phases. The desired relationship between
fugacities and experimentally accessible quantities is facili-
tated by two auxiliary functions which are given the symbols ϕ
and γ. The first of these, the fugacity coefficient ϕ, relates
the vapor-phase fugacity f_i^V to the mole fraction y_i and to the
total pressure P. It is defined by

$$\phi_i \equiv \frac{f_i^V}{y_i P} \tag{2-2}$$

The activity coefficient γ relates the liquid-phase fugacity

14

f_i^L to the mole fraction x_i and to a standard-state fugacity f_i^{OL}. It is defined by

$$\gamma_i \equiv \frac{f_i^L}{x_i f^{OL}} \qquad (2\text{-}3)$$

From Eqs. (1), (2), and (3) the equation of equilibrium for any component i becomes

$$\boxed{\phi_i y_i P = \gamma_i x_i f_i^{OL}} \qquad (2\text{-}4)$$

Equation (4) is the key equation for calculation of multicomponent vapor-liquid equilibria.

For a liquid phase (superscript ') in equilibrium with another liquid phase (superscript "), the equation analogous to Equation (1) is

$$f_i' = f_i'' \qquad (2\text{-}5)$$

When the same standard-state fugacity is used in both phases, Equation (5) can be rewritten

$$\boxed{(\gamma_i x_i)' = (\gamma_i x_i)''} \qquad (2\text{-}6)$$

Equation (6) is the key equation for calculation of multicomponent liquid-liquid equilibria.

The Fugacity Coefficient

A rigorous relation exists between the fugacity of a component in a vapor phase and the volumetric properties of that phase; these properties are conveniently expressed in the form of an equation of state. There are two common types of equations of state: one of these expresses the volume as a function of the temperature, pressure, and mole numbers (volume-explicit equation) while the other expresses the pressure as a function of the temperature, volume, and mole numbers (pressure-explicit equation). The latter form is more common but, since pressure (rather than volume) is a preferred independent variable, it is more convenient, whenever possible, to use a volume-explicit equation of state. At low or moderate densities, a suitable equation of state is the virial equation truncated after the second term

15

$$z = \frac{Pv}{RT} = 1 + \frac{BP}{RT} + \ldots \qquad (2\text{-}7)$$

where v is the molar volume, P is the total pressure, T is the absolute temperature and R is the gas constant; B is the second virial coefficient which depends on temperature and composition but is independent of pressure or density. For a system containing m components, the composition dependence of B is given by

$$B = \sum_{i=1}^{m} \sum_{j=1}^{m} y_i y_j B_{ij} \qquad (2\text{-}8)$$

where $B_{ij} = B_{ji}$ and y is the mole fraction. For a binary mixture,

$$B = y_1^2 B_{11} + 2y_1 y_2 B_{12} + y_2^2 B_{22} \qquad (2\text{-}8a)$$

where B_{ij} depends only on the temperature and on components i and j.

The fugacity coefficient ϕ is found from the thermodynamic relation

$$\ln \phi_i = \int_0^P \frac{\bar{z}_i - 1}{P} \, dP \qquad (2\text{-}9)$$

where

$$\bar{z}_i = \frac{P\bar{v}_i}{RT} \qquad (2\text{-}10)$$

and the partial molar volume $\bar{v}_i$ is defined by

$$\bar{v}_i = \left(\frac{\partial V}{\partial n_i}\right)_{T,P,n_j \ldots} \qquad (2\text{-}11)$$

In Equation (11), V is the total volume containing n_i moles of component i, n_j moles of component j, etc. The differentiation is carried out such that, in addition to temperature and pressure, all mole numbers (except n_i) are held constant.

When Equations (7) and (8) are substituted into Equation (9), we obtain

$$\boxed{\ln \phi_i = [2 \sum_{j}^{m} y_j B_{ij} - B]\frac{P}{RT}} \qquad (2\text{-}12)$$

Equation (12), applicable at low or moderate pressures, is used in this monograph for typical vapor mixtures. However, when the vapor phase contains a strongly dimerizing component such as carboxylic acid, Equation (7) is not applicable and

16

therefore Equation (12) cannot be used. In that event, the fugacity coefficient is calculated from a chemical theory of vapor imperfection, as discussed in Chapter 3 with details in Appendix A.

The Activity Coefficient

As indicated in Equation (3), the activity coefficient is completely defined only if the standard-state fugacity, f_i^{OL} is clearly specified. The definition of f_i^{OL}, however, is arbitrary and dictated only by convenience. It is necessary that f_i^{OL} be the fugacity of component i at the same temperature as that of the solution, at some fixed composition, and at some specified pressure; these last two variables, however, may be chosen at will.

It is strictly for convenience that certain conventions have been adopted in the choice of a standard-state fugacity. These conventions, in turn, result from two important considerations: (a) the necessity for an unambiguous thermodynamic treatment of noncondensable[†] components in liquid solutions, and (b) the relation between activity coefficients given by the Gibbs-Duhem equation. The first of these considerations leads to a normalization for activity coefficients for noncondensable components which is different from that used for condensable components, and the second leads to the definition and use of adjusted or pressure-independent activity coefficients. These considerations and their consequences are discussed in the following paragraphs.

Symmetric and Unsymmetric Convention for Normalization

When we speak of the normalization of activity coefficients, we mean a specification of the state wherein the activity coef-

---

[†]The boundary between condensable and noncondensable components is somewhat arbitrary, especially because it depends on the range of temperatures where calculations are made. In this monograph we consider only common volatile gases (e.g. H_2, N_2, O_2, CO, Ar, and CH_4) as noncondensable. Other typical fluids are considered condensable.

17

ficient is unity. For condensable components, it is customary
to normalize the activity coefficient such that

$$\gamma_i \rightarrow 1 \text{ as } x_i \rightarrow 1 \qquad (2\text{-}13)$$

For such components, as the composition of the solution approaches that of the pure liquid, the fugacity becomes equal to
the mole fraction multiplied by the standard-state fugacity. In
this case, the standard-state fugacity for component i is the
fugacity of pure liquid i at system temperature T. In many
cases all the components in a liquid mixture are condensable
and Equation (13) is therefore used for all components; in this
case, since all components are treated alike, the normalization
of activity coefficients is said to follow the symmetric convention.

However, if the liquid solution contains a noncondensable
component, the normalization shown in Equation (13) cannot be
applied to that component since a pure, supercritical liquid
is a physical impossibility. Sometimes it is convenient to
introduce the concept of a pure, hypothetical supercritical
liquid and to evaluate its properties by extrapolation; provided that the component in question is not excessively above
its critical temperature, this concept is useful, as discussed
later. We refer to those hypothetical liquids as condensable
components whenever they follow the convention of Equation (13).
However, for a highly supercritical component (e.g., H_2 or N_2 at
room temperature) the concept of a hypothetical liquid is of
little use since the extrapolation of pure-liquid properties in
this case is so excessive as to lose physical significance.

For a noncondensable component, therefore, it is convenient
to use a normalization different from that given by Equation
(13); in its place we use

$$\gamma_i^* \rightarrow 1 \text{ as } x_i \rightarrow 0 \qquad (2\text{-}14)$$

The purpose of the asterisk is to call attention to the difference in normalization.

According to Equation (14), the fugacity of component i
becomes equal to the mole fraction multiplied by the standard-

state fugacity of i in the limit as the mole fraction of component i becomes very small. The concentration region where the
activity coefficient γ_i^* is (essentially) equal to unity is
called the ideal dilute solution or Henry's-law region. In a
binary solution, the characteristic constant for the ideal
dilute solution is Henry's constant H defined by

$$H \equiv \lim_{x_i \to 0} \frac{f_i^L}{x_i} \qquad (2-15)$$

Henry's constant is the standard-state fugacity for any component i whose activity coefficient is normalized by Equation
(14).

In a binary liquid solution containing one noncondensable
and one condensable component, it is customary to refer to the
first as the solute and to the second as the solvent. Equation
(13) is used for the normalization of the solvent's activity
coefficient but Equation (14) is used for the solute. Since the
normalizations for the two components are not the same, they are
said to follow the unsymmetric convention. The standard-state
fugacity of the solvent is the fugacity of the pure liquid. The
standard-state fugacity of the solute is Henry's constant.

The use of Henry's constant for a standard-state fugacity
means that the standard-state fugacity for a noncondensable component depends not only on the temperature but also on the nature
of the solvent. It is this feature of the unsymmetric convention
which is its greatest disadvantage. As a result of this disadvantage special care must be exercised in the use of the unsymmetric convention for multicomponent solutions, as discussed
in Chapter 4.

The Gibbs-Duhem Equation

The standard-state fugacity of any component must be evaluated at the same temperature as that of the solution, regardless
of whether the symmetric or unsymmetric convention is used for
activity-coefficient normalization. But what about the pressure?
At low pressures, the effect of pressure on the thermodynamic
properties of condensed phases is negligible and under such con-

ditions the standard-state fugacities are essentially independent
of pressure. At higher pressures, however, this is no longer
the case.

The pressure at which standard-state fugacities are most
conveniently evaluated is suggested by considerations based on
the Gibbs-Duhem equation which says that at constant temperature
and pressure

$$\sum_i x_i d \ln \gamma_i = 0 \quad \text{(constant T, P)} \quad (2-16)$$

Equation (16) is a differential equation and applies equally
to activity coefficients normalized by the symmetric or unsymmetric convention. It is only in the integrated form of the Gibbs-
Duhem equation that the type of normalization enters as a boundary condition.

More general forms of the Gibbs-Duhem equation have been derived to allow for variations in temperature or pressure (or both)
but these are not useful for our purposes since they are not
easily integrated. Equation (16) is satisfied by various simple
algebraic forms relating $\ln \gamma$ to x; well-known examples are the
Margules and van Laar equations but many others exist. The particular relation used in this work, the UNIQUAC equation, while
significantly different from the equations of Margules and van
Laar, is also a solution to the Gibbs-Duhem differential equation.

If we vary the composition of a liquid mixture over all
possible composition values at constant temperature, the equilibrium pressure does not remain constant. Therefore, if integrated
forms of the Gibbs-Duhem equation [Equation (16)] are used to
correlate isothermal activity coefficient data, it is necessary
that all activity coefficients be evaluated at the same pressure.
Unfortunately, however, experimentally obtained isothermal activity coefficients are not all at the same pressure and therefore
they must be corrected from the experimental total pressure P
to the same (arbitrary) reference pressure designated P^r. This
may be done by the rigorous thermodynamic relation at constant
temperature and composition:

$$\gamma_i^{(P^r)} = \gamma_i^{(P)} \exp \int_P^{P^r} \frac{\bar{v}_i^L}{RT} \, dP \quad (2-17)$$

20

where $\bar{v}_i^L$ is the partial molar liquid volume. Equation (17) is independent of the form of normalization used for $\gamma_i^{(P)}$.

The activity coefficient $\gamma_i^{(P^r)}$ [or $\gamma_i^{*(P^r)}$] is called an adjusted activity coefficient because it has been corrected for the effect of pressure and, since it is independent of the experimental pressure P, it can be used in the isothermal and isobaric Gibbs-Duhem equation [Equation (16)] and in its various integrated forms.

Standard-State Fugacity for a Condensable Component

We can now consider the most convenient form for writing the liquid-phase fugacity of component i. First we consider a condensable component and write

$$f_i^L = \gamma_i^{(P^r)} x_i f_i^{0L} \exp \int_{P^r}^{P} \frac{\bar{v}_i^L}{RT} dP \qquad (2\text{-}18)$$

We choose this form because we want to express the fugacity f_i^L in terms of an adjusted activity coefficient $\gamma_i^{(P^r)}$ which, in turn, can be related to the composition by an integrated form of the isothermal, isobaric Gibbs-Duhem equation. The definition of f_i^{0L} now follows directly from Equation (18) by utilizing the normalization condition for a condensable component used in integrating the Gibbs-Duhem equation, viz., $\gamma_i^{(P^r)} \to 1$ as $x_i \to 1$. We find that the standard-state fugacity f_i^{0L} is the fugacity of pure liquid i at the temperature of the solution and at the reference pressure P^r.

The standard-state fugacity is given by

$$f_i^{0L} = P_i^s \phi_i^s \exp \int_{P_i^s}^{P^r} \frac{v_i^L}{RT} dP \qquad (2\text{-}19)$$

where P_i^s = saturation (vapor) pressure of pure liquid i at temperature T,

ϕ_i^s = fugacity coefficient of pure saturated vapor i at temperature T and pressure P_i^s, and

v_i^L = molar liquid volume of pure i at temperature T.

21

In this monograph, the reference pressure P^r is set equal to zero for all components.[†]

We make the simplifying assumption that both $\bar{v}_i^L$ and v_i^L are functions only of temperature, not of pressure and composition. For a condensable component it follows that at the same temperature, $\bar{v}_i^L = v_i^L$.

Standard-State Fugacity for a Noncondensable Component

For a noncondensable component we write

$$f_i^L = \gamma_i^{*(P^r)} x_i f_i^{OL} \exp \int_{P^r}^{P} \frac{\bar{v}_i^L}{RT} dP \qquad (2\text{-}20)$$

Using the normalization $\gamma_i^{*(P^r)} \to 1$ as $x_i \to 0$, we find that f_i^{OL} is Henry's constant evaluated at system temperature T and $P = P^r$.

Normally, Henry's constant for solute 2 in solvent 1 is determined experimentally at the solvent vapor pressure P_1^S. The effect of pressure on Henry's constant is given by

$$H_{2,1}^{(P^r)} = H_{2,1}^{(P_1^S)} \exp \int_{P_1^S}^{P^r} \frac{\bar{v}_2^\infty}{RT} dP \qquad (2\text{-}21)$$

where ∞ denotes infinite dilution. When $P^r = 0$, the exponential correction in Equation (21) is usually small, often negligible.

[†]When the pressure is low and mixture conditions are far from critical, activity coefficients are essentially independent of pressure. For such conditions it is common practice to set $P^r = P_i^S$ in Equations (18) and (19). Coupled with the assumption that $\bar{v}_i^L = v_i^L$, substitution gives the familiar equation

$$f_i^L = \gamma_i x_i \ (P_i^S \phi_i^S) \exp \int_{P_i^S}^{P} \frac{v_i^L}{RT} dP \qquad (2\text{-}18a)$$

In this case, there is no superscript on γ because, by assumption, γ is independent of pressure. The disadvantage of this procedure is that the reference pressure P^r is now different for each component, thereby introducing an inconsistency in the isobaric Gibbs-Duhem equation [Equation (16)]. In many, but not all, cases, this inconsistency is of no practical importance.

We know little about $\gamma_i^{*(P^r)}$ for noncondensable components because few appropriate experimental data have been obtained and little attention has been given to the molecular thermodynamics of concentrated solutions of noncondensable components dissolved in liquid solvents. Therefore, in this monograph we confine attention to the dilute region, i.e., where the mole fraction of the noncondensable component in the liquid phase is small. For that region we make the simplifying assumption

$$\gamma_i^{*(P^r)} \exp \int_{P^r}^{P} \frac{\bar{v}_i^L}{RT} \, dP = 1 \qquad (2\text{-}22)$$

In typical dilute mixtures, $\gamma_i^{*(P^r)} < 1$ and $\bar{v}_i^L > 0$. It follows therefore that Equation (22) provides a reasonable approximation whenever $x_i \ll 1$ and whenever $P - P^r$ is small, so that

$$\int_{P^r}^{P} \frac{\bar{v}_i^L \, dP}{RT} \ll 1$$

Because of the approximation given by Equation (22), we obtain a convenient method for determining f_i^{0L} for a noncondensable solute dissolved in a solvent mixture.[‡] For that case, f_i^{0L} is Henry's constant for component i in the solvent mixture, evaluated at system temperature T and reference pressure P^r.

Summary of Key Equations

For multicomponent vapor-liquid equilibria, the equation of equilibrium for every condensable component i is

$$\phi_i y_i P = \gamma_i^{(P^r)} x_i f_i^{0L} \exp \int_{P^r}^{P} \frac{\bar{v}_i^L}{RT} \, dP \qquad (2\text{-}23)$$

where the standard-state fugacity is given by

$$f_i^{0L} = P_i^s \phi_i^s \exp \int_{P_i^s}^{P^r} \frac{v_i^L}{RT} \, dP \qquad (2\text{-}24)$$

[‡]See P. M. Mathias and J. P. O'Connell, AIChE Journal (in press, 1979).

For a noncondensable component, the equation of equilibrium is limited to the dilute region ($x_i \ll 1$). It is

$$\phi_i y_i P = H_{i,M} x_i \qquad (2\text{-}25)$$

where $H_{i,M}$ is Henry's constant for solute i in the solvent mixture.

For multicomponent liquid-liquid equilibria, the equation of equilibrium for every component i is

$$\left(\gamma_i^{(P^r)} x_i \right)' = \left(\gamma_i^{(P^r)} x_i \right)'' \qquad (2\text{-}26)$$

Chapter 3 discusses calculation of fugacity coefficient ϕ. Chapter 4 discusses calculation of adjusted activity coefficient $\gamma^{(P^r)}$, fugacity of the pure liquid f_i^{0L} [Equation (24)], and Henry's constant H.

Chapter 3

THE VAPOR PHASE

In the calculation of vapor-liquid equilibria, it is necessary to calculate separately the fugacity of each component in each of the two phases. The liquid and vapor phases require different techniques; in this chapter we consider calculations for the vapor phase.

At pressures to a few bars, the vapor phase is at a relatively low density, i.e., on the average, the molecules interact with one another less strongly than do the molecules in the much denser liquid phase. It is therefore a common simplification to assume that all the nonideality in vapor-liquid systems exist in the liquid phase and that the vapor phase can be treated as an ideal gas. This leads to the simple result that the fugacity of component i is given by its partial pressure, i.e. the product of y_i, the mole fraction of i in the vapor, and P, the total pressure. A somewhat less restrictive simplification is the Lewis fugacity rule which sets the fugacity of i in the vapor mixture proportional to its mole fraction in the vapor phase; the constant of proportionality is the fugacity of pure i vapor at the temperature and pressure of the mixture. These simplifications are attractive because they make the calculation of vapor-liquid equilibria much easier; the K factors ($K_i \equiv y_i/x_i$) and relative volatilities ($\alpha_{ij} \equiv K_i/K_j$) are not functions of vapor composition. As a result, when performing bubble-point and flash calculations, no iterations are required for the vapor mole fractions.

Unfortunately, the ideal-gas assumption can sometimes lead to serious error. While errors in the Lewis rule are often less, that rule has inherent in it the problem of evaluating the fugacity of a fictitious substance since at least one of the condensable components cannot, in general, exist as pure vapor at the temperature and pressure of the mixture.

Witn the advent of electronic computers, it is no longer necessary to make drastic simplifying assumptions to reduce the calculations to manageable proportions. Under certain favorable conditions, the ideal-gas assumption and the Lewis fugacity rule can provide excellent approximations, but at even moderate pressures, the ideal-gas assumption can be in error by 5 to 10 percent for typical components, and the Lewis rule is only good for the component present in excess in the vapor phase. For a typical mixture, nonideality in the liquid phase is likely to be dominant, but in many cases, nonideality in the vapor phase is also significant.

This chapter presents a general method for estimating nonidealities in a vapor mixture containing any number of components; this method is based on the virial equation of state for ordinary substances and on the chemical theory for strongly associating species such as carboxylic acids. The method is limited to moderate pressures, as commonly encountered in typical chemical engineering equipment, and should only be used for conditions remote from the critical of the mixture.

The Fugacity Coefficient

The fugacity f_i^V of a component i in the vapor phase is related to its mole fraction y_i in the vapor phase and the total pressure P by the fugacity coefficient:

$$\phi_i \equiv \frac{f_i^V}{y_i P} \tag{3-1}$$

The fugacity coefficient is a function of temperature, total pressure, and composition of the vapor phase; it can be calculated from volumetric data for the vapor mixture. For a mixture containing m components, such data are often expressed in the form of an equation of state explicit in the pressure

$$P = P(T, V, n_1, n_2, \ldots, n_m) \tag{3-2}$$

where V is the total volume having n_1 moles of 1, n_2 moles of 2, etc.

The fugacity coefficient can be found from the equation of state using the thermodynamic relation (Beattie, 1949):

26

$$\ln \phi_i = \frac{1}{RT} \int_V^{\infty} [(\frac{\partial P}{\partial n_i})_{T,V,n_{j \neq i}} - \frac{RT}{V}] \, dV - \ln z \qquad (3-3)$$

where z is the compressibility factor of the vapor mixture,

$$z \equiv \frac{PV}{(n_1 + n_2 + \ldots + n_m) RT} \equiv \frac{Pv}{RT} \qquad (3-4)$$

Alternatively, the P-V-T-n data can be expressed in the functional form,

$$V = (n_1 + n_2 + \ldots + n_m) \, v(T, P, y_1, y_2, \ldots, y_m) \qquad (3-5)$$

The fugacity coefficient is then found from

$$\ln \phi_i = \frac{1}{RT} \int_0^P [(\frac{\partial V}{\partial n_i})_{T,P,n_{j \neq i}} - \frac{RT}{P}] \, dP \qquad (3-6)$$

If the vapor mixture contains only ideal gases, the integrals in Equations (3) and (6) are zero, z is unity for all compositions, and ϕ_i equals 1 for each component i. At low pressures, typically less than 1 bar, it is frequently a good assumption to set ϕ_i = 1, but even at moderately low pressures, say in the vicinity of 1 to 10 bars, ϕ_i is often significantly different from unity, especially if i is a polar component.

It is important to be consistent in the use of fugacity coefficients. When reducing experimental data to obtain activity coefficients, a particular method for calculating fugacity coefficients must be adopted. That same method must be employed when activity-coefficient correlations are used to generate vapor-liquid equilibria.

The Virial Equation

Numerous empirical equations of state have been proposed but the theoretically based virial equation (Mason and Spurling, 1969) is most useful for our purposes. We use this equation for systems which do not contain carboxylic acids.

The virial equation of state is a power series in the reciprocal molar volume or in the pressure:

27

$$z = 1 + \frac{B}{v} + \frac{C}{v^2} + \ldots \qquad (3\text{-}7a)$$

$$z = 1 + \frac{BP}{RT} + C'\left(\frac{P}{RT}\right)^2 + \ldots \qquad (3\text{-}7b)$$

where v is the molar volume of the gas, B is the second virial coefficient, C and C' are the third virial coefficients, etc. For a pure vapor the virial coefficients are functions only of temperature; for a mixture they are also functions of composition. An important advantage of the virial equation is that there are theoretically valid relations between the virial coefficients of a mixture and its composition. These relations are:

$$B_{mixture}(T, y_1, \ldots, y_m) = \sum_{i=1}^{m} \sum_{j=1}^{m} y_i y_j B_{ij}(T) \qquad (3\text{-}8)$$

$$C_{mixture}(T, y_1, \ldots, y_m) = \sum_{i=1}^{m} \sum_{j=1}^{m} \sum_{k=1}^{m} y_i y_j y_k C_{ijk}(T) \qquad (3\text{-}9a)$$

$$C'_{mixture} = C_{mixture} - (B_{mixture})^2 \qquad (3\text{-}9b)$$

where the individual coefficients B_{ij} and C_{ijk} are functions only of temperature. When Equations (7a), (8), and (9a) are substituted into Equation (3), we obtain

$$\ln \phi_i = \frac{2}{v} \sum_{j=1}^{m} y_j B_{ij} + \frac{3}{2v^2} \sum_{j=1}^{m} \sum_{k=1}^{m} y_j y_k C_{ijk} + \ldots - \ln z \quad (3\text{-}10a)$$

Equation (10a) is somewhat inconvenient first, because we prefer to use pressure rather than volume as our independent variable, and second, because little is known about third virial coefficients C_{ijk}. It is therefore more practical to substitute Equations (7b) and (8) into Equation (6), neglecting all third virial coefficients. We then obtain

$$\ln \phi_i = \left(2 \sum_{j=1}^{m} y_j B_{ij} - B_{mixture}\right) \frac{P}{RT} \qquad (3\text{-}10b)$$

Equation (10b) is used in this work whenever the vapor mixture does not contain one or more carboxylic acids.

Much theoretical and experimental information is available for the second virial coefficients B_{ij} (both for $i = j$ and for $i \neq j$), but relatively little is known about higher virial

28

coefficients. As a result, the virial equation may be used only for low or moderate densities; in principle, the equation holds for higher densities but at such densities a large number of virial coefficients is required to give a correct representation. Unfortunately, these coefficients are not available.

At moderate densities, Equation (3-10b) provides a very good approximation. This approximation should be used only for densities less than (about) one half the critical density. As a rough rule, the virial equation truncated after the second term is valid for the present range

$$P \leq \frac{T}{2} \frac{\sum\limits_{i=1}^{m} y_i P_{c_i}}{\sum\limits_{i=1}^{m} y_i T_{c_i}} \qquad (3\text{-}11)$$

where P_{c_i} and T_{c_i} refer to the critical pressure and temperature of component i (Prausnitz, 1957). For many industrial operations such as distillation, absorption, or flash separations at moderate pressures, Equation (11) can be satisfied.

To use Equation (10b), we require virial coefficients B_{ij} which depend on temperature. As discussed in Appendix A, these coefficients are calculated using the correlation of Hayden and O'Connell (1975). The required input parameters are, for each component: critical temperature T_c, critical pressure P_c, dipole moment μ, mean radius of gyration R_D and association parameter η; and for each binary pair, association parameter η_{ij}. These parameters are given for a large number of fluids in Appendix C.

Figure 1 shows second virial coefficients for four pure fluids as a function of temperature. Second virial coefficients for typical fluids are negative and increasingly so as the temperature falls; only at the Boyle point, when the temperature is about 2.5 times the critical, does the second virial coefficient become positive. At a given temperature below the Boyle point, the magnitude of the second virial coefficient increases with

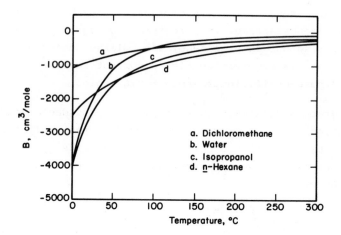

Figure 3-1. Second virial coefficients for four fluids.

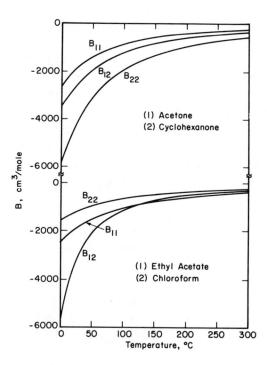

Figure 3-2.   Second virial coefficients for two binary systems.

molecular size and polarity.

Figure 2 shows second virial coefficients B_{11}, B_{22}, and B_{12} for two binary systems: acetone-cyclohexanone and ethyl acetate-chloroform. In the first binary, intermolecular forces between like molecules are similar to those between unlike molecules and therefore the curve for B_{12} lies between the curves for B_{11} and B_{22}. However, in the second binary, intermolecular forces between unlike molecules are much stronger than those between like molecules; chloroform and ethyl acetate can strongly hydrogen bond with each other but only very weakly with themselves. Therefore, the curve for B_{12} lies well below those for B_{11} and B_{22}.

Figures 3 and 4 show fugacity coefficients for two binary systems calculated with Equation (10b). Although the pressure is not large, deviations from ideality and from the Lewis rule are not negligible.

"Chemical" Theory of Vapor Nonideality for Strongly Interacting Substances (Mixtures Containing Carboxylic Acids)

The virial equation is appropriate for describing deviations from ideality in those systems where moderate attractive forces yield fugacity coefficients not far removed from unity. The systems shown in Figures 2, 3, and 4 are of this type. However, in systems containing carboxylic acids, there prevails an entirely different physical situation; since two acid molecules tend to form a pair of stable hydrogen bonds, large negative deviations from vapor ideality occur even at very low pressures. To account for dimerization,[†] expressions for fugacity coefficients were developed based on the thermodynamics of chemical equilibrium (Marek, 1955; Prausnitz, 1969; Nothnagel et al., 1973). Formally, the "chemical" theory should be consistently used in all cases, but the calculations are tedious. Since its limiting form gives an expression for the fugacity coefficient identical with that of the virial equation [Equation (10b)],

---

[†]As used here, dimerization refers to the formation of an at least moderately stable molecule from two monomers A and B, where A and B may, but need not, be identical.

31

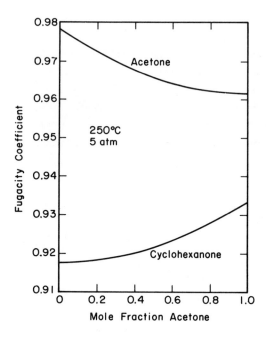

Figure 3-3. Fugacity coefficients for the system acetone-cyclohexanone.

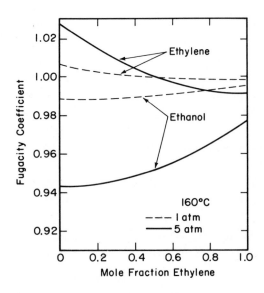

Figure 3-4. Fugacity coefficients for the system ethylene-ethanol.

there is continuity between the "chemical" and the "physical" theories. However, because the calculations are complex, the "chemical" theory is used here only when one or more carboxylic acid is in the vapor phase.

The "chemical" theory postulates that there is a dimerization equilibrium of the type

$$i + j \rightleftarrows ij$$

where i and j are two monomer molecules (that may or may not be chemically identical) which form a dimer ij. Monomer i belongs to species i and monomer j belongs to species j.

To describe this chemical equilibrium quantitatively, we use a chemical equilibrium constant

$$K_{ij} = \frac{f_{ij}}{f_i f_j} = \frac{z_{ij} \phi^{\ddagger}_{ij}}{z_i z_j \phi^{\ddagger}_i \phi^{\ddagger}_j P} \tag{3-12}$$

where f is the fugacity of the true molecular species (monomer or dimer); z is the true mole fraction; $\phi^{\ddagger}$ is the fugacity coefficient of the true species; and P is the total pressure.

As shown elsewhere (Nothnagel et al., 1973), the fugacity coefficient of component i is given by

$$\phi_i = \frac{z_i \phi^{\ddagger}_i}{y_i} \tag{3-13}$$

In Equation (13), z_i and $\phi^{\ddagger}_i$ refer to the monomer of species i while y_i is the apparent mole fraction of component i, where apparent means that dimerization has been neglected.

To use Equation (13), it is first necessary to calculate the true fugacity coefficient $\phi^{\ddagger}_i$. This calculation is achieved by utilizing the Lewis fugacity rule

$$\ln \phi^{\ddagger}_i = \frac{B^F_i P}{RT} \tag{3-14}$$

where B^F_i is the "free" contribution to the second virial coefficient of component i as discussed in Appendix A.

33

Next, and more difficult, is the calculation of the true mole fraction z_i. This calculation is achieved by simultaneous solution of material balances ($\Sigma z = 1$) and chemical equilibria [Equation (12)] for all possible dimerization reactions. Every true fugacity coefficient is calculated with the Lewis fugacity rule, i.e. with an equation analogous to Equation (14). Details are given in Appendix A.

The chemical equilibrium constant K_{ij} is found from the relation

$$K_{ij} = \frac{-B_{ij}^D (2 - \delta_{ij})}{RT} \qquad (3\text{-}15)$$

where Kronecker $\delta_{ij} = 0$ when $i \neq j$ and $\delta_{ij} = 1$ when $i = j$, and where B_{ij}^D is the dimerization contribution to the second virial coefficient as discussed in Appendix A.

Figure 5 shows fugacity coefficients for the system acetaldehyde-acetic acid at 90°C and 0.25 atm. Calculations are based on the "chemical" theory of vapor imperfection. Although the pressure is far below atmospheric, fugacity coefficients for both components are well removed from unity. Because of strong dimerization between acetic acid molecules and weak dimerization between the other possible pairs, deviations from ideality are large, much larger than one might expect at this low pressure.

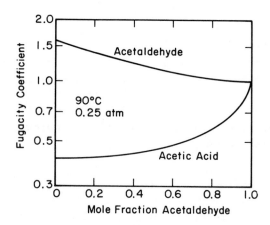

Figure 3-5. Fugacity coefficients for the system acetaldehyde-acetic acid.

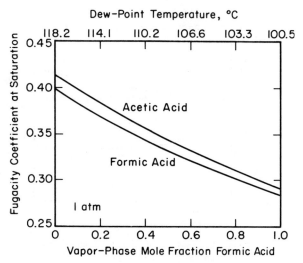

Figure 3-6. Fugacity coefficients for saturated mixtures containing two carboxylic acids: formic acid (1) and acetic acid (2). Calculations based on chemical theory show large deviations from ideal behavior.

Two additional illustrations are given in Figures 6 and 7 which show fugacity coefficients for two binary systems along the vapor-liquid saturation curve at a total pressure of 1 atm. These results are based on the chemical theory of vapor-phase imperfection and on experimental vapor-liquid equilibrium data for the binary systems. In the system formic acid (1) - acetic acid (2), ϕ_1 (for $y_1 = 1$) is lower than ϕ_2 (for $y_2 = 1$) because pure formic acid at 100.5°C has a stronger tendency to dimerize than does acetic acid at 118.2°C. Since strong dimerization occurs between all three possible pairs, ϕ_1 and ϕ_2 are not severely different from each other and, as shown, give nearly parallel curves when plotted against composition.

By contrast, in the system propionic acid (1) - methyl iso-butyl ketone (2), ϕ_1 and ϕ_2 are very much different when $y_1 \to 1$. Propionic acid has a strong tendency to dimerize with itself and only a weak tendency to dimerize with ketone; also, the ketone has only a weak tendency to dimerize with itself. At acid-rich compositions, therefore, many acid molecules have dimerized but most ketone molecules are monomers. Acid-acid dimerization lowers the fugacity of acid and thus ϕ_1 is well below unity. Because of acid-acid dimerization, the true mole fraction of ketone is signi-

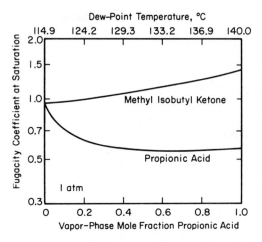

Figure 3-7. Fugacity coefficients for a saturated mixture of propionic acid (1) and methylisobutylketone (2). Calculations based on chemical method show large variations from ideal behavior.

ficantly larger than the apparent mole fraction and therefore [see Equation (13)] ϕ_2 is larger than unity. On the other hand, when y_1 is near zero, there is little dimerization and thus both ϕ_1 and ϕ_2 are close to unity.

Computational Implementation

The computation of vapor-phase fugacity coefficients is implemented by the FORTRAN subroutine PHIS. This subroutine evaluates the coefficients ϕ_i for all components, in a system of up to 20 components, at a specified temperature and pressure. This subroutine, in turn, utilizes subroutine BIJS to evaluate the second virial coefficients, B_{ij}'s. Both subroutines are described and listed in Appendix E.

Guide for Estimating Unknown Parameters

Since parameters for many fluids of interest are not given in this monograph, it may be necessary to estimate the required parameters: T_c, P_c, R_D, μ, and η.

The third edition of "Properties of Gases and Liquids" by Reid et al. (1977) lists useful group contribution methods for predicting critical properties. Contributions to the second

36

virial coefficient are most sensitive to the value of T_c, which, fortunately, is the most reliably predicted and measured.

Values of R_D, probably close to the required accuracy, can be estimated from the parachor, P'; the parachor can be calculated from a group-contribution method given by Reid et al. The approximate relation between P' and R_D is

$$P' = 50 + 7.6\ R_D + 13.75\ R_D^2 \qquad (3-16)$$

where P' is in units of $\dfrac{gm^{\frac{1}{4}}\ cm^3}{sec^{\frac{1}{2}}\ g\text{-mole}}$ and R_D is in Å.

Values of dipole moment μ are listed in the compilation by McClellan (1974).

Finally, values of η must be obtained when specific ("chemical") interactions can occur. These are difficult to estimate à priori but reasonable approximations can often be made by choosing a value (Appendix C) for a chemically similar system.

Conclusion: Effect of Independent Variables on Vapor-Phase Nonideality

A component in a vapor mixture exhibits nonideal behavior as a result of molecular interactions; only when these interactions are very weak or very infrequent is ideal behavior approached. The fugacity coefficient ϕ_i is a measure of nonideality and a departure of ϕ_i from unity is a measure of the extent to which a molecule i interacts with its neighbors. The fugacity coefficient depends on pressure, temperature, and vapor composition; this dependence, in the moderate pressure region covered by the truncated virial equation, is usually as follows:

(a) At constant T and y, an increase in P causes ϕ_i to depart further from unity, usually in the direction $\phi_i < 1$.

(b) at constant P and y, an increase in T causes ϕ_i to approach unity.

(c) at constant P and T, the effect of composition on ϕ_i is strongest when y_i is small. If y_i is near unity, the change in composition has relatively little effect on ϕ_i.

At constant P,T,y, the effect of a molecule j on ϕ_i is most pronounced whenever j is chemically much different from i. In polar mixtures, where specific interactions like hydrogen bonding may occur, the effect of composition may be very large; a well-known example is the diethyl ether-chloroform system where hydrogen bonds are formed between the dissimilar molecules but not between the similar molecules.

While vapor-phase corrections may be small for nonpolar molecules at low pressure, such corrections are usually not negligible for mixtures containing polar molecules. Vapor-phase corrections are extremely important for mixtures containing one or more carboxylic acids.

This chapter uses an equation of state which is applicable only at low or moderate pressures. Serious error may result when the truncated virial equation is used at high pressures.

Details for calculating fugacity coefficients are given in Appendix A.

References

Beattie, J. A., Chem. Rev., 44, 141 (1949).

Hayden, J. G., O'Connell, J. P., Ind. Eng. Chem., Process Des. Dev., 14, 221 (1975).

Marek, J., Coll. Czech, Chem. Commun., 20, 1490 (1955).

Mason, E. A., Spurling, T. H., The International Encyclopedia of Phys. Chem. and Chem. Phys., Topic 10, The Fluid State, Vol. 2, The Virial Equation of State, Pergamon Press, Oxford (1969).

McClellan, A. L., Tables of Experimental Dipole Moments, W. H. Freeman, San Francisco (1963-74).

Nothnagel, K. H., Abrams, D. S., Prausnitz, J. M., Ind. Eng. Chem., Process Des. Dev., 12, 25 (1973).

Prausnitz, J. M., Chem. Eng. Sci., 6, 112 (1957).

Prausnitz, J. M., "Molecular Thermodynamics of Fluid-Phase Equilibria," Prentice-Hall, Englewood Cliffs, N.J. (1969).

Reid, T. C., Prausnitz, J. M., Sherwood, T. K., "Properties of Gases and Liquids," 3rd ed., McGraw-Hill, New York (1977).

Chapter 4

THE LIQUID PHASE

To predict vapor-liquid or liquid-liquid equilibria in
multicomponent systems, we require a method for calculating the
fugacity of a component i in a liquid mixture. At system temperature T and system pressure P, this fugacity is written as a
product of three terms

$$f_i^L = \gamma_i^{(P^r)} \; x_i \; [f_i^{0(P^r)} \exp \frac{\bar{v}_i^L (P-P^r)}{RT}] \qquad (4\text{-}1)^\dagger$$

where, as discussed in Chapter 2, γ is the activity coefficient
at reference pressure P^r, x is the mole fraction, and the standard-state fugacity, in brackets, is the product of the standard-
state fugacity at P^r and the Poynting correction.[††] Equation (1)
assumes that the partial molar volume $\bar{v}_i^L$ is a function only of
temperature. It follows then that, for a condensable component,
at a fixed temperature, $\bar{v}_i^L$ can be replaced by v_i^L at the same
temperature.

For condensable components, we use the symmetric normalization $\gamma_i \to 1$ as $x_i \to 1$; therefore, the quantity in brackets is
the fugacity of pure liquid i at system temperature and pressure.

For our reference pressure, we choose $P^r = 0$. Therefore,
the quantity $f_i^{0(P^r)}$, also written $f_i^{(P0)}$ when $P^r = 0$, is the
fugacity of pure liquid i at system temperature, corrected to
zero pressure. It is found from the relation

$$f_i^{(P0)} = P_i^s \; \phi_i^s \; \exp \frac{-v_i^L P_i^s}{RT} \qquad (4\text{-}2)$$

where P_i^s is the saturation (vapor) pressure of pure liquid i and
ϕ_i^s is the fugacity coefficient of saturated pure i, both at
system temperature T. In Equation (2), the important term is

[†]f^0 is the same here as f^{0L} in Eq. (2-18).

[††]The Poynting correction is the exponential term in Equation (4-1).

P^S; the other terms provide corrections which at low or moderate pressure are close to unity. To use Equation (2), we require vapor-pressure data and liquid-density data as a function of temperature. We also require fugacity coefficients, as discussed in Chapter 3.

Using corresponding-states arguments, it is possible to derive a generalized version of Equation (2) which has the form

$$\ln \frac{f_i^{(PO)}}{P_{c_j}} = F(T/T_{c_i}, \omega_i) \qquad (4-3)$$

where the generalized function F depends on reduced temperature and acentric factor. Appendix B presents this function, in addition to a more detailed discussion on Equations (2) and (3).

Equations (2) and (3) are physically meaningful only in the temperature range bounded by the triple-point temperature and the critical temperature. Nevertheless, it is often useful to extrapolate these equations either to lower or, more often, to higher temperatures. In this monograph we have extrapolated the function F [Equation (3)] to a reduced temperature of nearly 2. We do not recommend further extrapolation. For highly supercritical components it is better to use the unsymmetric normalization for activity coefficients as indicated in Chapter 2 and as discussed further in a later section of this chapter.

The words "condensable" and "noncondensable" as used here are discussed in the footnote near Equation (13) of Chapter 2.

Activity Coefficients

In a liquid mixture, all activity coefficients are directly related to the molar excess Gibbs energy g^E which is defined by

$$g^E = RT \sum_i x_i \ln \gamma_i \qquad (4-4)^\dagger$$

$\dagger$It is understood that all activity coefficients are at reference pressure P^r. For convenience we now drop superscript (P^r).

40

A mathematical model, preferably based on molecular considerations, provides a convenient method for expressing g^E as a function of x. From this function an individual activity coefficient γ_i for component i can be calculated from g^E using the relation

$$RT \; \ln \gamma_i = (\frac{\partial n_t g^E}{\partial n_i})_{T,P,n_{j \neq i}} \tag{4-5}$$

where n_i is the number of moles $[n_t = \sum_j n_j$ and $x_i = n_i/n_t]$.

In this monograph we use for g^E the UNIQUAC model of Abrams (1975) as slightly modified by Anderson (1978).[†]

$$g^E = g^E(\text{combinatorial}) + g^E(\text{residual}) \tag{4-6}$$

For a binary mixture

$$\frac{g^E(\text{combinatorial})}{RT} = x_1 \; \ln \frac{\Phi_1}{x_1} + x_2 \; \ln \frac{\Phi_2}{x_2}$$

$$+ (\frac{z}{2}) \; [q_1 x_1 \; \ln \frac{\theta_1}{\Phi_1} + q_2 x_2 \; \ln \frac{\theta_2}{\Phi_2}] \tag{4-7}$$

$$\frac{g^E(\text{residual})}{RT} = -q_1' x_1 \; \ln(\theta_1' + \theta_2' \tau_{21})$$

$$-q_2' x_2 \; \ln(\theta_2' + \theta_1' \tau_{12}) \tag{4-8}$$

where the coordination number z is set equal to 10 and segment fraction, Φ, and area fractions, θ and θ', are given by

$$\Phi_1 = \frac{x_1 r_1}{x_1 r_1 + x_2 r_2} \qquad \Phi_2 = \frac{x_2 r_2}{x_1 r_1 + x_2 r_2}$$

$$\theta_1 = \frac{x_1 q_1}{x_1 q_1 + x_2 q_2} \qquad \theta_2 = \frac{x_2 q_2}{x_1 q_1 + x_2 q_2}$$

$$\theta_1' = \frac{x_1 q_1'}{x_1 q_1' + x_2 q_2'} \qquad \theta_2' = \frac{x_2 q_2'}{x_1 q_1' + x_2 q_2'}$$

[†]A simple derivation and critical discussion of UNIQUAC is given by G. Maurer, Fluid Phase Equilibria, 2, 91 (1978).

The parameters r, q, and q' are pure-component molecular-structure constants depending on molecular size and external surface areas. In the original formulation (Abrams and Prausnitz, 1975), q = q'. To obtain better agreement for mixtures containing water or alcohols, q' for water and alcohols has here been obtained empirically to give an optimum fit to a variety of systems containing these components. For alcohols, the surface of interaction q' is smaller than the geometric external surface q, indicating that for alcohols, intermolecular attraction is determined primarily by the OH group. Appendix C presents values of these structural parameters.

For each binary combination in a multicomponent mixture, there are two adjustable parameters, τ_{12} and τ_{21}. These, in turn, are given in terms of characteristic energies Δu_{12} and Δu_{21} by

$$\tau_{12} = \exp\left(-\frac{\Delta u_{12}}{RT}\right) = \exp\left(-\frac{a_{12}}{T}\right) \tag{4-9}$$

$$\tau_{21} = \exp\left(-\frac{\Delta u_{21}}{RT}\right) = \exp\left(-\frac{a_{21}}{T}\right) \tag{4-10}$$

Equations (9) and (10) give the primary effect of temperature on τ_{12} and τ_{21}. Characteristic energies Δu_{12} and Δu_{21} are often only weakly dependent on temperature. The activity coefficients γ_1 and γ_2 are given by

$$\ln \gamma_1 = \ln \frac{\Phi_1}{x_1} + \left(\frac{z}{2}\right)q_1 \ln \frac{\theta_1}{\phi_1} + \phi_2 \left(\ell_1 - \frac{r_1}{r_2}\ell_2\right)$$
$$- q_1' \ln(\theta_1' + \theta_2'\tau_{21}) + \theta_2' q_1' \left[\frac{\tau_{21}}{\theta_1' + \theta_2'\tau_{21}} - \frac{\tau_{12}}{\theta_2' + \theta_1'\tau_{12}}\right] \tag{4-11}$$

$$\ln \gamma_2 = \ln \frac{\Phi_2}{x_2} + \left(\frac{z}{2}\right)q_2 \ln \frac{\theta_2}{\Phi_2} + \Phi_1\left(\ell_2 - \frac{r_2}{r_1}\ell_1\right)$$
$$- q_2' \ln(\theta_2' + \theta_1'\tau_{12}) + \theta_1' q_2' \left[\frac{\tau_{12}}{\theta_2' + \theta_1'\tau_{12}} - \frac{\tau_{21}}{\theta_1' + \theta_2'\tau_{21}}\right] \tag{4-12}$$

where

$$\ell_1 = \left(\frac{z}{2}\right)(r_1 - q_1) - (r_1 - 1)$$

$$\ell_2 = \left(\frac{z}{2}\right)(r_2 - q_2) - (r_2 - 1)$$

42

Data Sources

Parameters a_{12} and a_{21} must be found from binary experimental data. Possible sources of data include

1. vapor-liquid equilibrium data (P,y,x) at constant temperature
2. total pressure data (P,x or y) at constant temperature
3. vapor-liquid equilibrium data (T,y,x) at constant pressure
4. boiling-point (or dew-point) data (T,x or y) at constant pressure
5. mutual (liquid-liquid) solubilities
6. azeotropic data
7. activity coefficients at infinite dilution

Activity-coefficient data at infinite dilution often provide an excellent method for obtaining binary parameters as shown, for example, by Eckert and Schreiber (1971) and by Nicolaides and Eckert (1978). Unfortunately, such data are rare.

When no experimental data at all are available, activity coefficients can sometimes be estimated using the UNIFAC method (Fredenslund et al., 1977a, b). However, for many real engineering problems it is often necessary to obtain new experimental data.

Since the accuracy of experimental data is frequently not high, and since experimental data are hardly ever plentiful, it is important to reduce the available data with care using a suitable statistical method and using a model for the excess Gibbs energy which contains only a minimum of binary parameters. Rarely are experimental data of sufficient quality and quantity to justify more than three binary parameters and, all too often, the data justify no more than two such parameters. When data sources (5) or (6) or (7) are used alone, it is not possible to use a three- (or more)-parameter model without making additional arbitrary assumptions. For typical engineering calculations, therefore, it is desirable to use a two-parameter model such as UNIQUAC.

Data Reduction

The most reliable estimates of the parameters are obtained from multiple measurements, usually a series of vapor-liquid equilibrium data (T, P, x and y). Because the number of data points exceeds the number of parameters to be estimated, the equilibrium equations are not exactly satisfied for all experimental measurements. Exact agreement between the model and experiment is not achieved due to random and systematic errors in the data and due to inadequacies of the model. The optimum parameters should, therefore, be found by satisfaction of some selected statistical criterion, as discussed in Chapter 6. However, regardless of statistical sophistication, there is no substitute for reliable experimental data.

Illustrative Examples

Earlier experience with the UNIQUAC equation indicated that the original formulation of UNIQUAC (q' = q) gave only fair results for highly nonideal mixtures containing water or alcohols. The empirical modification, where area parameter q has a different value in the residual contribution to g^E, has introduced an additional unicomponent parameter. However, that parameter is only adjustable to the extent that a single value is chosen after examining a representative number of binary systems. The value of q', once fixed, is independent of binary (or multi-component) data.

Figure 1 compares data reduction using the modified UNIQUAC equation with that using the original UNIQUAC equation. The data are those of Boublikova and Lu (1969) for ethanol and n-octane. The dashed line indicates results obtained with the original equation (q' = q for ethanol) and the continuous line shows results obtained with the modified equation. The original equation predicts a liquid-liquid miscibility gap, contrary to experiment. The modified UNIQUAC equation, however, represents the alcohol/n-octane system with good accuracy.

When there are sufficient data at different temperatures, the temperature dependence of the parameters is reflected in the confidence ellipses (Bryson and Ho, 1969; Draper and Smith,

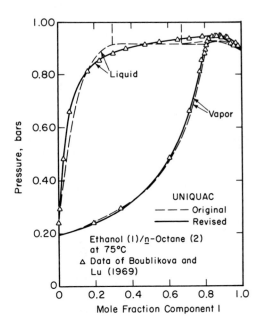

Figure 4-1. Effect of UNIQUAC equation modification for an alcohol-hydrocarbon system.

1966). To illustrate, UNIQUAC parameters were obtained for the ethanol/cyclohexane system using the extensive isothermal data of Scatchard and Satkiewicz (1964). Figure 2 shows parameters for 5, 35, and 60°C along with the confidence ellipses. These regions indicate that it is possible to choose a single value of a_{21} appropriate for all temperatures; a single value of a_{21} (e.g. 1300) can be included in all three confidence ellipses, implying that in the range 5-65°C parameter a_{21} is temperature independent. For a_{12}, however, there is no single value which can intercept all three confidence ellipses. Therefore, parameter a_{12} must be represented by a function of temperature as shown in Table 1 where the estimated variance of the fit, σ_F^L, provides a measure of how well the data are represented. The first line shows results obtained when fitting two UNIQUAC parameters, a_{12} and a_{21}, independent of temperature. The next two lines show the results of three-parameter fits: first a_{21} and then a_{12} are functions of temperature. Finally, four parameters were determined: both UNIQUAC parameters are now functions of

45

Table 4-1

Temperature Dependence of UNIQUAC Parameters for
Ethanol(1)/Cyclohexane(2) Isothermal Data
(5-65°C) of Scatchard (1964)

Number of Parameters	α_{12}	α_{21}	β_{12}	β_{21}	σ_F^2
2	-85.67	1187.36	--	--	53.33
3	-87.87	1659.25	--	9.245×10^5	24.67
3	-331.75	1304.33	7.392×10^4	--	4.44
4	-296.05	259.28	6.278×10^4	3.352×10^5	3.46

$$\sigma_F^2 = \frac{\text{Sum of squared, weighted residuals}}{\text{Number of degrees of freedom}}$$

$$a_{ij} = \alpha_{ij} + \beta_{ij}/T; \quad a_{ij} = \Delta u_{ij}/R$$

a_{ij} and T are in kelvins.

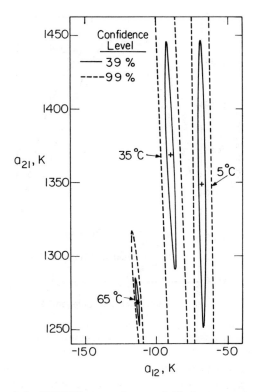

Figure 4-2. UNIQUAC parameters and their approximate confidence regions for the ethanol-cyclohexane system for three isotherms. Data of Scatchard and Satkiewicz, 1964.

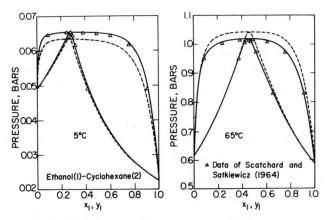

Figure 4-3. Calculated and experimental vapor-liquid equilibria.
--- Calculated with temperature-independent UNIQUAC parameters.
— Calculated with temperature-dependent UNIQUAC parameters.

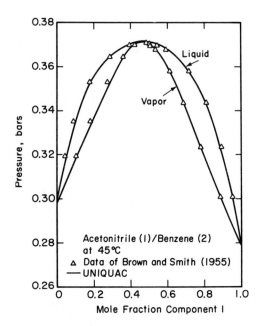

Figure 4-4. Representation of vapor-liquid equilibria for a binary system showing moderate positive deviations from Raoult's law.

temperature. Substantial improvement is obtained when a_{12} is temperature dependent, while little improvement is obtained when a_{21} is temperature dependent, supporting the conclusions drawn from Figure 2. Finally, Figure 3 shows agreement between observed and calculated equilibria when the three-parameter fit is used.

An adequate prediction of multicomponent vapor-liquid equilibria requires an accurate description of the phase equilibria for the binary systems. We have reduced a large body of binary data including a variety of systems containing, for example, alcohols, ethers, ketones, organic acids, water, and hydrocarbons with the UNIQUAC equation. Experience has shown it to do as well as any of the other common models. When all types of mixtures are considered, including partially miscible systems, the modified UNIQUAC equation performs as well as, or better than, any other two-parameter equation for excess Gibbs energy g^E.

Figure 4 shows experimental and predicted phase equilibria for the acetonitrile/benzene system at 45°C. This system exhibits moderate positive deviations from Raoult's law. The high-quality data of Brown and Smith (1955) are very well represented by the UNIQUAC equation.

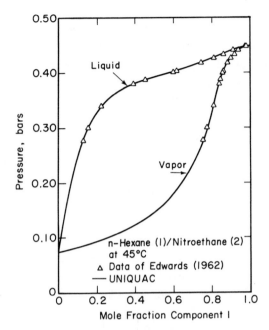

Figure 4-5. Representation of vapor-liquid equilibria for a binary system showing strong positive deviations from Raoult's law.

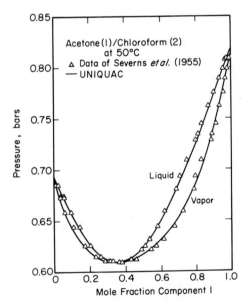

Figure 4-6. Representation of vapor-liquid equilibria for a binary system showing strong negative deviations from Raoult's law.

Figure 5 shows the isothermal data of Edwards (1962) for
n-hexane and nitroethane. This system also exhibits positive
deviations from Raoult's law; however, these deviations are much
larger than those shown in Figure 4. At 45°C the mixture shown
in Figure 5 is only 15° above its critical solution temperature.
Again, representation with the UNIQUAC equation is excellent.

The acetone/chloroform system, shown in Figure 6, exhibits
strong negative deviations from Raoult's law because of hydrogen
bonding between the single hydrogen atom of chloroform and the
carbonyl oxygen of acetone. The data of Severns et al. (1955)
are well represented by the UNIQUAC equation.

Figure 7 shows a fit of the UNIQUAC equation to the isobaric data of Nakanishi et al. (1967) for the methanol-diethyl-
amine system; this system also exhibits strong negative deviations from Raoult's law. The UNIQUAC equation correctly re-

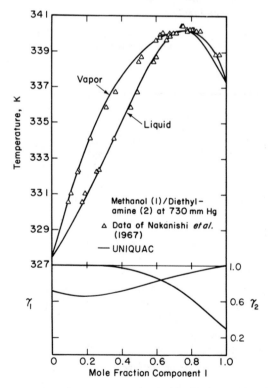

Figure 4-7. Vapor-liquid equilibria and activity coefficients
in a binary system showing a weak minimum in the activity
coefficient of methanol.

produces a weak minimum in the activity coefficient of methanol.
Agreement with experiment is not as good as that in previous examples because the data are badly scattered, particularly near
the azeotrope.

As discussed in Chapter 3, at moderate pressures, vapor-
phase nonideality is usually small in comparison to liquid-phase
nonideality. However, when associating carboxylic acids are
present, vapor-phase nonideality may dominate. These acids dimerize appreciably in the vapor phase even at low pressures;
fugacity coefficients are well removed from unity. To illustrate,
Figures 8 and 9 show observed and calculated vapor-liquid
equilibria for two systems containing an associating component.
In the first, both components strongly associate with themselves
and with each other. In the second, only one of the components
associates strongly. For both systems, representation of the
data is very good. However, the interesting quality of these
systems is that whereas the fugacity coefficients are significantly remote from unity, the activity coefficients show only
minor deviations from ideal-solution behavior. Figures 6 and 7
in Chapter 3 indicate that the fugacity coefficients show marked
departure from ideality. In these systems, the major contribution to nonideality occurs in the vapor phase. Failure to take
into account these strong vapor-phase nonidealities would result
in erroneous activity-coefficient parameters, a_{12} and a_{21}, leading to poor prediction of multicomponent equilibria.

Multicomponent Systems

Equations (4) and (5) are not limited to binary systems;
they are applicable to systems containing any number of components.

The UNIQUAC equation for excess Gibbs energy g^E is

$$\frac{g^E \text{(combinatorial)}}{RT} = \sum_i x_i \, \ell n \, \frac{\Phi_i}{x_i} + \frac{z}{2} \sum_i q_i x_i \, \ell n \, \frac{\theta_i}{\Phi_i} \qquad (4-13)$$

$$\frac{g^E \text{(residual)}}{RT} = - \sum_i q_i' x_i \, \ell n \, (\sum_j \theta_j' \tau_{ji}) \qquad (4-14)$$

where segment fraction Φ and area fractions θ and θ' are given by

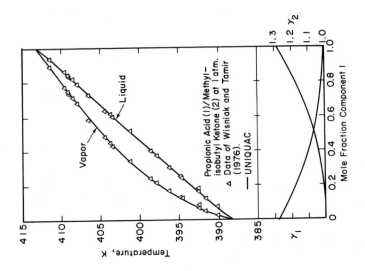

Figure 4-9. Vapor-liquid equilibria for a binary system where one component dimerizes in the vapor phase. Activity coefficients show only small deviations from liquid-phase ideality.

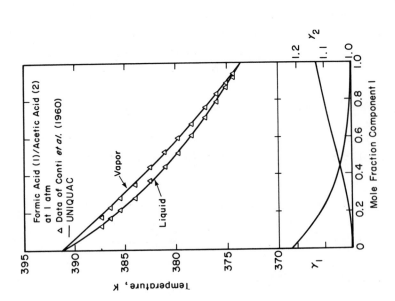

Figure 4-8. Vapor-liquid equilibria for a binary system where both components solvate and associate strongly in the vapor phase.

$$\Phi_i = \frac{r_i x_i}{\sum_j r_j x_j}$$

$$\theta_i = \frac{q_i x_i}{\sum_j q_j x_j}$$

$$\theta_i' = \frac{q_i' x_i}{\sum_j q_j' x_j}$$

For any component i, the activity coefficient is given by

$$\ln \gamma_i = \ln \frac{\Phi_i}{x_i} + (\frac{z}{2})q_i \ln \frac{\theta_i}{\Phi_i} + \ell_i - \frac{\Phi_i}{x_i}\sum_j x_j \ell_j$$

$$- q_i' \ln \left(\sum_j \theta_j' \tau_{ji}\right) + q_i' - q_i'\sum_j \frac{\theta_j' \tau_{ij}}{\sum_k \theta_k' \tau_{kj}} \qquad (4\text{-}15)$$

where

$$\ell_j = \frac{z}{2}(r_j - q_j) - (r_j - 1)$$

Equation (15) requires only pure-component and binary parameters.[†]

Using UNIQUAC, Table 2 summarizes vapor-liquid equilibrium predictions for several representative ternary mixtures and one quaternary mixture. Agreement is good between calculated and experimental pressures (or temperatures) and vapor-phase compositions.

The largest errors in predicted compositions occur for the systems acetic acid-formic acid-water and acetone-acetonitrile-water where experimental uncertainties are significantly greater than those for other systems.

Moderate errors in the total pressure calculations occur for the systems chloroform-ethanol-n-heptane and chloroform-acetone-methanol. Here strong hydrogen bonding between chloroform and alcohol creates unusual deviations from ideality; for both alcohol-chloroform systems, the activity coefficients show

---

[†]Coordination number $z = 10$.

Table 4-2

Prediction of Multicomponent Vapor-Liquid Equilibria

System and no. of data points[†]		Press. mm Hg/ Temp °C	Dev in temp or % dev in press Avg (Max)	Dev in vapor comp, mole % Avg (Max)	Data Source
Mthylcycptn Ethanol Benzene	48	760 60-71	0.25(0.31)°C	0.51(-3.03) 0.55(2.99) 0.35(-1.25)	(1)
Acetic acid Formic acid Water	40	760 102-110	0.55(-1.80)°C	1.00(-2.08) 1.60(3.77) 2.18(-5.36)	(2)
Acetone Acetonitrile Water	30	762 63-92	1.13(-3.67)°C	1.22(3.24) 1.27(-3.45) 1.53(-4.68)	(3)
Methanol Crbntetraclrd Benzene	8	665-717 55	0.11(-0.27)%	0.44(0.99) 0.39(-0.89) 0.09(0.17)	(4)
Mek n-Heptane Toluene	39	760 77-103	0.17(-0.63)°C	0.79(2.00) 0.52(-1.31) 0.38(-1.18)	(5)
Chloroform Ethanol n-Heptane	92	262-501 50	1.57(-3.03)%	*	(6)
Chloroform Acetone Methanol	29	463-645 50	1.10(-3.12)%	0.86(1.03) 0.77(2.68) 0.81(1.03)	(7)
Chloroform Methanol Ethyl acetate	72	760 57-72	0.36(1.77)°C	0.74(2.06) 1.11(2.40) 0.80(2.47)	(8)
n-Hexane Mthylcycptn Ethanol Benzene	10	760 60-65	0.38(-0.45)°C	0.31(0.60) 0.44(0.95) 0.55(-1.13) 0.44(0.96)	(1)

*P-T-x data only

Data sources: (1) Sinor, 1960; (2) Conti, 1960; (3) Pratt, 1947; (4) Scatchard, 1952; (5) Steinhauser, 1949; (6) Abbott, 1975; (7) Severns, 1955; (8) Nagata, 1962.

[†]Mthylcycptn = Methylcyclopentane
Crbntetraclrd = Carbontetrachloride
Mek = Methylethylketone

well-defined extrema. Since extrema are generally not well re-
produced by the UNIQUAC equation, these binaries are not well
represented.[†] The overall ternary deviations are of the order
of those for the worst fitted binaries, methanol-chloroform and
ethanol-chloroform. In spite of the relatively large deviations
in calculated pressure, the predicted vapor compositions agree
well with the experimental data of Severns (1955).

Predictions for the other isobaric systems (experimental
data of Sinor, Steinhauser, and Nagata) show good agreement.
Excellent agreement is obtained for the system carbon tetrachlor-
ide-methanol-benzene, where the binary data are of superior
quality.

The results shown in Table 2 indicate that UNIQUAC can be
used with confidence for multicomponent vapor-liquid equilibria
including those that exhibit large deviations from ideality.

Noncondensable Components

As discussed in Chapter 2, for noncondensable components,
the unsymmetric convention is used to normalize activity coefficients. For a noncondensable component i in a multicomponent
mixture, we write the fugacity in the liquid phase

$$f_i^L = \gamma_i^*{}^{(P^r)} x_i H_{i,M}^{(P^r)} \exp \int_{P^r}^{P} \frac{\bar{v}_i^L dP}{RT} \qquad (4-16)$$

where $H_{i,M}$ is Henry's constant for component i in the mixture
evaluated at reference pressure P^r. Since we restrict attention
to the dilute region ($x_i \ll 1$), we use the simplifying assumption

$$\gamma_i^*{}^{(P^r)} \exp \int_{P^r}^{P} \frac{\bar{v}_i^L dP}{RT} = 1 \qquad (4-17)$$

[†]Null (1970) discusses some alternate models for the excess
Gibbs energy which appear to be well suited for systems whose
activity coefficients show extrema.

In subsequent discussion, therefore, we drop the superscript (P^r) on H. Further, we designate a noncondensable component "solute" and a condensable component "solvent."

For solute i in solvent j, we designate Henry's constant by $H_{i,j}$. For a given binary pair, $H_{i,j}$ is taken to be a function only of temperature.

To estimate Henry's constant for solute i in a mixed solvent, we use the approximation

$$\ln H_{i,M} = \sum_j \theta_j \ln H_{i,j} \qquad (4\text{-}18)$$

where the summation is over all solvents j. The surface fraction θ_j is taken on a solute-free basis where solute-free refers to all solutes. This basis is automatically achieved by setting $q_i = q'_i$ equal to zero for every solute.

For convenience, we express $H_{i,j}$ as a product of two functions

$$H_{i,j} = (\gamma^\infty_{i,j}) \, f^{0L}_i \qquad (4\text{-}19)$$

where superscript ∞ denotes infinite dilution. The standard-state fugacity f^{0L}_i is a hypothetical quantity: the fugacity of pure liquid i at system temperature T. For noncondensable components we choose an essentially arbitrary function

$$\ln \frac{f^{0L}_i}{P_{c_i}} = 7.224 - 7.534 \left(\frac{T}{T_{c_i}}\right)^{-1} - 2.598 \ln \left(\frac{T}{T_{c_i}}\right) \qquad (4\text{-}20)$$

where P_{c_i} and T_{c_i} are, respectively, the critical pressure and critical temperature of component i. A plot of Equation (20) is shown in Figure 10.

Equation (20) gives a rough first approximation for the effect of temperature on Henry's constant. Activity coefficient $\gamma^\infty_{i,j}$ is often only a weak function of temperature provided that the temperature range is not large and provided the system temperature T is not near T_{c_j}, the critical temperature of solvent j. For our purposes here, we give the effect of temperature on $\gamma^\infty_{i,j}$ by the empirical equation

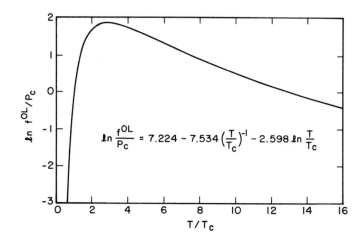

Figure 4-10. Generalized fugacities for noncondensable fluids.

$$\ln \gamma_{i,j}^{\infty} = \delta_{ij}^{(0)} + \delta_{ij}^{(1)} T^{-1} \tag{4-21}$$

where $\delta_{ij}^{(0)}$ and $\delta_{ij}^{(1)}$ are binary parameters. Figure 11 shows $\ln \gamma_{i,j}^{\infty}$ for a few binary systems.

For a mixed solvent, it follows from Equation (18) that

$$\ln \gamma_i^{\infty} = \sum_j \theta_j \ln \gamma_{i,j}^{\infty} \tag{4-22}$$

where θ_j, again on an (all) solute-free basis, is the surface fraction of solvent j.

When a condensable[‡] solute is present, the activity coefficient of a solvent is given by Equation (15) provided that all composition variables (x, θ, and Φ) are taken on an (all) solute-free basis. Composition variables θ and Φ are automatically on a solute-free basis by setting $q_i = q_i' = r_i = 0$ for every solute.

Consistent with our restriction to solutions where all solute mole fractions are small, we set UNIQUAC parameters $\tau_{ij} = \tau_{ji} = 1$ whenever i is a solute and j is either a solvent or a solute.

[‡]See footnote near Equation 13 of Chapter 2.

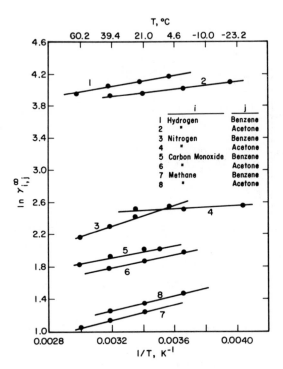

Figure 4-11. Activity coefficients for noncondensable solutes at infinite dilution.

Vapor-Liquid Equilibria for Mixtures Containing One or More Noncondensable Components

In some cases, the temperature of the system may be larger than the critical temperature of one (or more) of the components, i.e., system temperature T may exceed T_{c_i}. In that event, component i is a supercritical component, one that cannot exist as a pure liquid at temperature T. For this component, it is still possible to use symmetric normalization of the activity coefficient ($\gamma_i \to 1$ as $x_i \to 1$) provided that some method of extrapolation is used to evaluate the standard-state fugacity which, in this case, is the fugacity of pure liquid i at system temperature T. For highly supercritical components ($T \gg T_{c_i}$), such extrapolation is extremely arbitrary; as a result, we have no assurance that when experimental data are reduced, the activity coefficient tends to obey the necessary boundary condition $\gamma_i \to 1$ as $x_i \to 1$. For highly supercritical components, therefore, unsymmetric normalization is used: $\gamma_i^* \to 1$ as $x_i \to 0$.

There is no sharp dividing line between "slightly" super-critical and "highly" supercritical. Experience has shown, how-ever, that for most practical purposes the dividing line is in the region $T/T_{c_i} \sim 1.8$. Using this criterion as a guide, it follows that at ordinary temperatures, hydrogen, helium, nitro-gen, argon, oxygen, and carbon monoxide should be designated non-condensable components, which means that their activity coeffi-cients are normalized by the unsymmetric convention. On the other hand, again at ordinary temperatures, ethane, ethylene, carbon dioxide, propane, and propylene should be designated condensable components, which means that their activity coefficients are normalized symmetrically. For these condensable components, at temperatures near or above the critical, the standard-state fugacity is obtained by extrapolation as discussed in Appendix B.

The critical temperature of methane is 191°K. At 25°C, therefore, the reduced temperature is 1.56. If the dividing line is taken at $T/T_{c_i} = 1.8$, methane should be considered condensable at temperatures below (about) 70°C and noncondensable at higher temperatures. However, in process design calculations, it is often inconvenient to switch from one method of normalization to the other. In this monograph, since we consider only equilibria at low or moderate pressures in the region 200-600°K, we elect to consider methane as a noncondensable component.

To illustrate calculations for a binary system containing a supercritical, condensable component, Figure 12 shows isobaric equilibria for ethane-n-heptane. Using the virial equation for vapor-phase fugacity coefficients, and the UNIQUAC equation for liquid-phase activity coefficients, calculated results give an excellent representation of the data of Kay (1938). In this case, the total pressure is not large and therefore, the mixture is at all times remote from critical conditions. For this binary system, the particular method of calculation used here would not be successful at appreciably higher pressures.

Figure 13 presents results for a binary where one of the components is a supercritical, noncondensable component. Vapor-phase fugacity coefficients were calculated with the virial

59

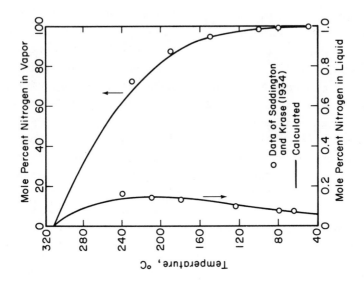

Figure 4-13. Vapor-liquid equilibria for the system water-nitrogen at 100 atm.

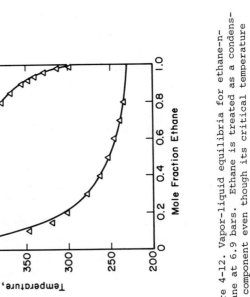

Figure 4-12. Vapor-liquid equilibria for ethane-n-heptane at 6.9 bars. Ethane is treated as a condensable component even though its critical temperature is 305.4 K.

equation. For the liquid phase, the fugacity of nitrogen was
calculated using Henry's constant for nitrogen in water, coupled
with Equation (17); consistent with that equation, the liquid-
phase mole fraction of nitrogen is small. The activity coefficient for water is unity. The experimental data are represented
well because, once again, the conditions shown in Figure 13 are
remote from the mixture's critical.

Table 3 shows results obtained from a five-component, isothermal flash calculation. In this system there are two condensable components (acetone and benzene) and three noncondensable
components (hydrogen, carbon monoxide, and methane). Henry's
constants for each of the noncondensables were obtained from
Equations (18-22); the simplifying assumption for dilute solutions [Equation (17)] was also used for each of the noncondensables. Activity coefficients for both condensable components
were calculated with the UNIQUAC equation. For that calculation,
all liquid-phase composition variables are on a solute-free
basis; the only required binary parameters are those for the
acetone-benzene system. While no experimental data are available for comparison, the calculated results are probably reliable because all simplifying assumptions are reasonable: the
pressure is moderate, the liquid phase is dilute with respect to
the noncondensable components, and conditions are remote from
the mixture's critical.

Our experience with multicomponent vapor-liquid equilibria
suggests that for system temperatures well below the critical
of every component, good multicomponent results are usually obtained, especially where binary parameters are chosen with care.
However, when the system temperature is near or above the critical of one (or more) of the components, multicomponent predictions may be in error, even though all binary pairs are fit well.

Liquid-Liquid Equilibria

In multicomponent liquid-liquid equilibria, the equation of
equilibrium, for every component i, is

$$(\gamma_i x_i)' = (\gamma_i x_i)'' \qquad (4-23)$$

61

Table 4-3

Isothermal Flash Calculations for
Mixtures Containing Condensable
and Noncondensable Components

System: Hydrogen(1)-Carbon Monoxide(2)-Acetone(3)-
Benzene(4)-Methane(5) at 25°C and 30 atm

| Mixture | Component | Composition, Mole Percent | | | Split$(V/F)^{\dagger}$ |
		Feed	Liquid	Vapor	
I	1	10	0.28	34.78	
	2	10	0.67	33.80	
	3	0	0.00	0.00	0.282
	4	70	97.21	0.60	
	5	10	1.84	30.82	
II	1	10	0.29	34.62	
	2	10	0.69	33.61	
	3	20	27.70	0.46	0.283
	4	50	69.53	0.45	
	5	10	1.78	30.86	
III	1	10	0.30	34.54	
	2	10	0.71	33.49	
	3	35	48.55	0.72	0.283
	4	35	48.70	0.35	
	5	10	1.74	30.90	
IV	1	10	0.31	34.46	
	2	10	0.74	33.39	
	3	50	69.42	0.97	0.284
	4	20	27.83	0.22	
	5	10	1.70	30.96	
V	1	10	0.32	34.38	
	2	10	0.77	33.25	
	3	70	97.26	1.33	0.284
	4	0	0.00	0.00	
	5	10	1.65	31.04	

†Split $(\frac{V}{F})$ = $\frac{\text{moles of vapor}}{\text{mole of feed}}$.

where ' and " refer to the two equilibrated liquid phases.

Equation (23) holds only when, for every component i, the same standard-state fugacity is used in both liquid phases.

Since we make the simplifying assumption that the partial molar volumes are functions only of temperature, we assume that, for our purposes, pressure has no effect on liquid-liquid equilibria. Therefore, in Equation (23), pressure is not a variable. The activity coefficients depend only on temperature and composition. As for vapor-liquid equilibria, the activity coefficients used here are given by the UNIQUAC equation, Equation (15).

Liquid-liquid equilibria are much more sensitive than vapor-liquid equilibria to small changes in the effect of composition on activity coefficients. Therefore, calculations for liquid-liquid equilibria should be based, whenever possible, at least in part, on experimental liquid-liquid data.

In the next three sections we discuss calculation of liquid-liquid equilibria (LLE) for ternary systems and then conclude the chapter with a discussion of LLE for systems containing more than three components.

Ternary Systems

In ternary systems, we distinguish between two common types. In type II, two binaries are partially miscible and the third binary is completely miscible; in type I, only one binary is partially miscible. (A third type, where all three binaries are only partially miscible, is relatively rare and not considered here.)

For systems of type II, if the mutual binary solubility (LLE) data are known for the two partially miscible pairs, and if reasonable vapor-liquid equilibrium (VLE) data are known for the miscible pair, it is relatively simple to predict the ternary equilibria. For systems of type I, which has a plait point, reliable calculations are much more difficult. However, sometimes useful quantitative predictions can be obtained for type I systems with binary data alone provided that

63

1. reliable, preferably isothermal, VLE data are used for the two miscible binaries;

2. a suitable model is chosen which accurately represents the excess Gibbs energy of each constituent binary;

3. mutual solubility data are used for the partially miscible binary pairs. In addition, if other data are available for these pairs, those data may also be useful (Heinrich, 1975).

Best ternary predictions are usually obtained for mixtures having a very broad two-phase region, i.e., where the two partially miscible liquids have only small mutual solubilities. Fortunately, this is the type of ternary that is most often used in commercial liquid-liquid extraction.

For all calculations reported here, binary parameters from VLE data were obtained using the principle of maximum likelihood as discussed in Chapter 6. Binary parameters for partially miscible pairs were obtained from mutual-solubility data alone.

To illustrate, predictions were first made for a ternary system of type II, using binary data only. Figure 14 compares calculated and experimental phase behavior for the system 2,2,4-trimethylpentane-furfural-cyclohexane. UNIQUAC parameters are given in Table 4. As expected for a type II system, agreement is good.

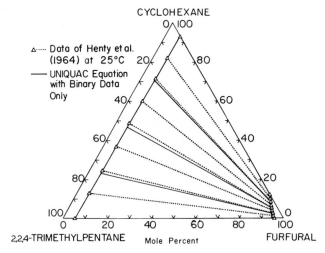

Figure 4-14. Predicted liquid-liquid equilibria for a typical type-II system shows good agreement with experimental data, using parameters estimated from binary data alone.

64

Table 4-4

Parameters for Ternary Liquid-Liquid Systems

System	Type	Temp °C	UNIQUAC Parameters, K						Data Reference
			a_{12}	a_{21}	a_{13}	a_{31}	a_{23}	a_{32}	
A	I	60	155.78	471.21	122.02	122.07	183.65	-142.35	Volpicelli, 1968
B	I	28	1557.23	131.36	330.03	-124.12	196.41	-173.64	Smith, 1941
C	I	32.8	2.49	1419.32	-86.89	1284.21	122.21	-55.81	Wittrig, 1977
D	I	25	2057.42	115.13	1131.13	-149.34	573.61	-163.72	Bancroft, 1942
E	I	45	23.71	545.71	60.28	89.57	245.42	-135.93	Palmer, 1972
F	I	25	517.19	105.01	335.25	-173.60	73.79	82.20	Weck, 1954
G	II	25	410.08	-4.98	141.01	-112.66	41.17	354.83	Henty, 1964
H	II	25	283.76	34.82	-138.84	162.13	54.36	228.71	Darwent, 1943
I	I	25	-4.98	410.08	71.00	12.00	80.91	-27.13	Henty, 1964
J	I	25	354.83	41.17	-32.57	88.26	71.00	12.00	Henty, 1964

Systems: A = Water(1)-Acrylonitrile(2)-Acetonitrile(3)

B = Vinyl Acetate(1)-Water(2)-Acetic Acid(3)

C = Methanol(1)-n-Heptane(2)-Benzene(3)

D = Benzene(1)-Water(2)-Ethanol(3)

E = Acetonitrile(1)-n-Heptane(2)-Benzene(3)

F = Cyclohexane(1)-Nitromethane(2)-Benzene(3)

G = 2,2,4-Trimethylpentane(1)-Furfural(2)-Cyclohexane(3)

H = n-Hexane(1)-Aniline(2)-Methylcyclopentane(3)

I = Furfural(1)-2,2,4-Trimethylpentane(2)-Benzene(3)

J = Cyclohexane(1)-Furfural(2)-Benzene(3)

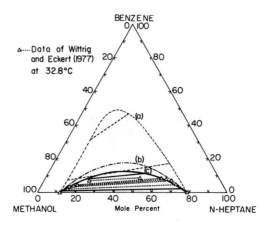

Figure 4-15. Effect of UNIQUAC equation modification on prediction of liquid-liquid equilibria for a type-I system containing
an alcohol. --- Original UNIQUAC equation. —— Modified
UNIQUAC equation with binary parameters estimated by supple-
menting binary VLE data with ternary tie-line data.

Figure 15 shows results for a difficult type I system:
methanol-n-heptane-benzene. In this example, the two-phase
region is extremely small. The dashed line (a) shows predictions using the original UNIQUAC equation with q = q'. This form
of the UNIQUAC equation does not adequately fit the binary vapor-
liquid equilibrium data for the methanol-benzene system and
therefore the ternary predictions are grossly in error. The
ternary prediction is much improved with the modified UNIQUAC
equation (b) since this equation fits the methanol-benzene system much better. Further improvement (c) is obtained when a few
ternary data are used to fix the binary parameters.

Unfortunately, good binary data are often not available,
and no model, including the modified UNIQUAC equation, is entirely adequate. Therefore, we require a calculation method
which allows utilization of some ternary data in the parameter
estimation such that the ternary system is well represented. A
method toward that end is described in the next section.

Data Reduction Using Ternary Information

For type I systems, we cannot consistently obtain good

66

representation of ternary LLE using only binary data because
these are insufficient to determine binary model parameters. In
most cases, good representation can only be obtained by incorporating into the data-reduction scheme a limited quantity of ternary
information. Since tie-line data are the most common and most
readily obtainable form of ternary data, we use one or two experimental ternary tie lines to fix the optimum binary parameters.
Reduction of binary data, coupled with limited ternary tie-line
data, is achieved by using the principle of maximum likelihood.

To illustrate the criterion for parameter estimation, let
1, 2, and 3 represent the three components in a mixture. Components 1 and 2 are only partially miscible; components 1 and 3,
as well as components 2 and 3 are totally miscible. The two
binary parameters for the 1-2 binary are determined from
mutual-solubility data and remain fixed. Initial estimates of
the four binary parameters for the two completely miscible
binaries, 1-3 and 2-3, are determined from sets of binary vapor-
liquid equilibrium (VLE) data. The final values of these parameters are then obtained by fitting both sets of binary vapor-
liquid equilibrium data simultaneously with the limited ternary
tie-line data.

As indicated in Chapter 6, and discussed in detail by
Anderson et al. (1978), optimum parameters, based on the maximum-
likelihood principle, are those which minimize the objective
function

$$
S = \sum_{i=1}^{L} \left\{ \frac{(P_i^0 - P_i^e)^2}{\sigma_{P_i}^2} + \frac{(T_i^0 - T_i^e)^2}{\sigma_{T_i}^2} + \frac{(x_{1i}^0 - x_{1i}^e)^2}{\sigma_{x_{1i}}^2} + \frac{(y_{1i}^0 - y_{1i}^e)^2}{\sigma_{y_{1i}}^2} \right\}
$$

$$
+ \sum_{i=L+1}^{M} \left\{ \frac{(P_i^0 - P_i^e)^2}{\sigma_{P_i}^2} + \frac{(T_i^0 - T_i^e)^2}{\sigma_{T_i}^2} + \frac{(x_{2i}^0 - x_{2i}^e)^2}{\sigma_{x_{2i}}^2} + \frac{(y_{2i}^0 - y_{2i}^e)^2}{\sigma_{y_{2i}}^2} \right\}
$$

$$
+ \sum_{i=M+1}^{N} \left\{ \frac{(T_i^0 - T_i^e)^2}{\sigma_{T_i}^2} + \frac{(x_{1i}^{0\alpha} - x_{1i}^{e\alpha})^2}{\sigma_{x_{1i}}^2} + \frac{(x_{3i}^{0\alpha} - x_{3i}^{e\alpha})^2}{\sigma_{x_{3i}}^2} + \frac{(x_{2i}^{0\beta} - x_{2i}^{e\beta})^2}{\sigma_{x_{2i}}^2} + \frac{(x_{3i}^{0\beta} - x_{3i}^{e\beta})^2}{\sigma_{x_{3i}}^2} \right\}
$$

$$(4-24)$$

The total number of experimental data points is N. Data points
1 through L and L+1 through M refer to VLE measurements (P, T,
x, y) for the 1-2 binary and 2-3 binary, respectively. Data
points M+1 through N are ternary liquid-liquid equilibrium
measurements (T, x_1^α, x_3^α, x_2^β, x_3^β). The 1-rich phase is indicated
by superscript α and the 2-rich phase by β. Superscript e indi-
cates an experimentally measured value, and superscript 0 indicates an estimated true value corresponding to each measured
point.

The estimated true values must satisfy the appropriate
equilibrium constraints. For points 1 through L, there are two
constraints given by Equation (2-4); one each for components 1
and 2. For points L+1 through M the same equilibrium relations
apply; however, now they apply to components 2 and 3. The constraints for the tie-line points, M+1 through N, are given by
Equation (2-6), applied to each of the three components.

In Equation (24), σ is the estimated standard deviation
for each of the measured variables, i.e. pressure, temperature,
and liquid-phase and vapor-phase compositions. The values assigned to σ determine the relative weighting between the tie-
line data and the vapor-liquid equilibrium data; this weighting
determines how well the ternary system is represented. This
weighting depends first, on the estimated accuracy of the ternary data, relative to that of the binary vapor-liquid data; and
second, on how remote the temperature of the binary data is from
that of the ternary data; and finally, on how important in a
design the liquid-liquid equilibria are relative to the vapor-
liquid equilibria. Typical values which we use in data reduc-
tion are σ_P = 1 mm Hg, σ_T = 0.05°C, σ_x = 0.001, and σ_y = 0.003
for the binary VLE data; and σ_T = 0.05°C and σ_x = 0.0005 for all
tie-line measurements.

In most cases only a single tie line is required. When
several are available, the choice of which one to use is somewhat
arbitrary. However, our experience has shown that tie lines
which are near the middle of the two-phase region are most useful for estimating the parameters. Tie lines close to the plait
point are less useful, since no common models for the excess
Gibbs energy can adequately describe the flat region near the

the plait point.

The method described here is based on the high degree of correlation of model parameters, in this case, UNIQUAC parameters. Thus, although a certain set of binary parameters may be best for VLE data, we are able to find other sets of binary parameters for the miscible binaries which significantly improve ternary LLE prediction while only slightly decreasing accuracy of representation of the binary VLE. Fitting ternary LLE data only, may yield unrealistic parameters that predict grossly erroneous results when used in regions not identical to those employed in data reduction. By contrast, fitting ternary LLE data simultaneously with binary VLE data, effectively provides constraints on the binary parameters, preventing them from attaining arbitrary values of little physical significance. Determination of a single set of parameters which can adequately represent both VLE and LLE is particularly important in three-phase distillation.

Results

Using the method outlined above, calculations were performed for ten ternary systems. All binary parameters are shown in Table 4. Some typical results are shown in Figures 16 to 19.

Figure 16 shows observed and calculated VLE and LLE for the system benzene-water-ethanol. In this unusually fortunate case, predictions based on the binary data alone (dashed line) are in good agreement with the experimental ternary data. Several factors contribute to this good agreement: VLE data for the miscible binaries are of high quality and they are well represented by the (modified) UNIQUAC equation; the temperatures of the two isothermal VLE data sets are not far removed from that of the ternary data; and finally, the two-phase region is fairly large.

The continuous line in Figure 16 shows results from fitting a single tie line in addition to the binary data. Only slight improvement is obtained in prediction of the two-phase region; more important, however, prediction of solute distribution is improved. Incorporation of the single ternary tie line into the method of data reduction produces only a small loss of accuracy in the representation of VLE for the two binary systems.

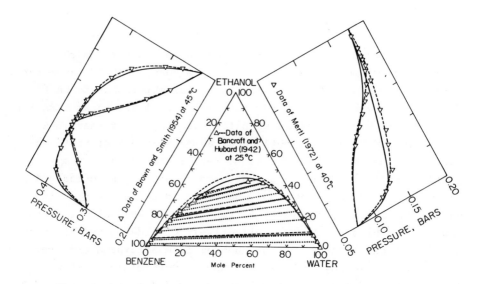

Figure 4-16. Representation of ternary liquid-liquid equilibria using the UNIQUAC equation is improved by incorporating ternary tie-line data into binary-parameter estimation. Representation of binary VLE shows small loss of accuracy. --- Binary data only. —— One tie-line plus binary data.

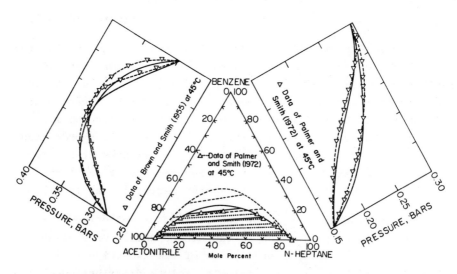

Figure 4-17. Representation of ternary liquid-liquid equilibria using the UNIQUAC equation is improved by incorporating ternary tie-line data into binary-parameter estimation. Representation of binary VLE shows some loss of accuracy. --- Binary data only. —— One tie-line plus binary data.

Figure 17 shows results for the acetonitrile-n-heptane-benzene system. Here, however, the two-phase region is somewhat smaller; ternary equilibrium calculations using binary data alone considerably overestimate the two-phase region. Upon including a single ternary tie line, satisfactory ternary representation is obtained. Unfortunately, there is some loss of accuracy in the representation of the binary VLE (particularly for the acetonitrile-benzene system where the shift of the aceotrope is evident) but the loss is not severe.

On triangular diagrams, comparisons of calculated and experimental results can be deceiving. A more realistic representation is provided by Figure 18, comparing experimental solute distributions with those calculated from the UNIQUAC equation for four ternary systems. For three of these systems, calculations were made using the parameters determined from binary data plus one ternary tie line; however, for the 2,2,4-trimethylpentane-furfural-cyclohexane system, parameters were obtained from binary data alone. With the exception of the region very near the plait point, calculated distributions are good. Fortunately, commercial extractions are almost never conducted near the plait point since the small density difference in the plait-point region causes hydrodynamic difficulties (flooding).

Two further examples of type I ternary systems are shown in Figure 19 which presents calculated and observed selectivities. For successful extraction, selectivity is often a more important index than the distribution coefficient. Calculations are shown for the case where binary data alone are used and where binary data are used together with a single ternary tie line. It is evident that calculated selectivities are substantially improved by including limited ternary tie-line data in data reduction.

Liquid-Liquid Equilibria for Four
     (or More) Components

For systems containing four components, most previous attempts for calculating LLE use geometrical correlations of ternary data (Branckner, 1940), interpolation of ternary data (Chang and Moulton, 1953), or empirical correlations of ternary data (Prince, 1954; Henty, 1964). These methods all have two

71

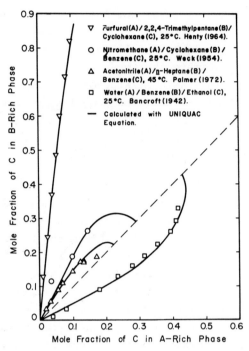

Figure 4-18. Calculated distribution for ternary liquid-liquid systems show good agreement with experiment except very near the plait point.

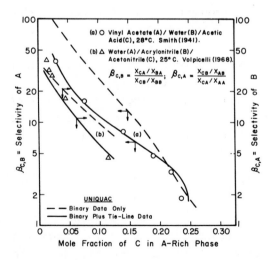

Figure 4-19. Calculated selectivities in two ternary systems show large improvements when tie-line data are used to supplement binary VLE data for estimating binary parameters.

problems in common: they are not readily generalizable to more than four components and they are not easily adaptable for computer implementation.

Guffey and Wehe (1972) used excess Gibbs energy equations proposed by Renon (1968a, 1968b) and Black (1959) to calculate multicomponent LLE. They concluded that prediction of ternary data from binary data is not reliable, but that quarternary LLE can be predicted from accurate ternary representations. Here, we carry these results a step further; we outline a systematic procedure for determining binary parameters which are suitable for multicomponent LLE.

We consider three types of m-component liquid-liquid systems. Each system requires slightly different data reduction and different quantities of ternary data. Figure 20 shows quarternary examples of each type.

Type A. Component 1 is only partially miscible with components 2 through m, but components 2 through m are completely miscible with each other. Binary data only are required for this type of system:
    1. mutual solubility data for m minus 1 binaries, e.g. 1-2, 1-3, . . . , 1-m;
    2. vapor-liquid equilibrium data for all the remaining binaries.

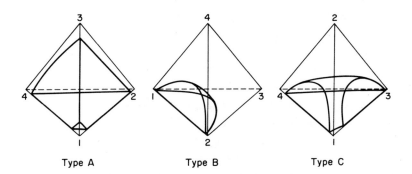

Figure 4-20. Quarternary examples of three types of multicomponent liquid-liquid mixtures.

Type B. Components 1 and 2 are only partly miscible with each other. Both 1 and 2 are completely miscible with all other components in the system (3 through m). Components 3 through m are also miscible in all proportions. Both binary and ternary data are needed for a reliable description of the multicomponent LLE:

1. mutual solubility data for the single partially miscible binary, 1-2;

2. binary VLE data for all miscible pairs;

3. ternary tie-line data are used with the binary VLE data to determine the binary parameters for each of the constituent ternaries 1-2-i for i = 3 ... m, as discussed in the previous section.

Type C. Component 1 is only partially miscible with components 3 through m, but it is totally miscible with component 2. Components 2 through m are miscible with each other in all proportions. Again, both binary data and ternary tie-line data are needed:

1. mutual-solubility data for all partially miscible binaries: 1-3, 1-4, ... , 1-m;

2. binary VLE data for all miscible pairs;

3. ternary tie-line data for all ternary systems containing two liquid phases: 1-2-i for i = 3 ... m.

Using the ternary tie-line data and the binary VLE data for the miscible binary pairs, the optimum binary parameters are obtained for each ternary of the type 1-2-i for i = 3 ... m. This results in multiple sets of the parameters for the 1-2 binary, since this binary occurs in each of the ternaries containing two liquid phases. To determine a single set of parameters to represent the 1-2 binary system, the values obtained from initial data reduction of each of the ternary systems are plotted with their approximate confidence ellipses. We choose a single optimum set from the intersection of the confidence ellipses. Finally, with the parameters for the 1-2 binary set at their optimum value, the parameters are adjusted for the remaining miscible binary in each ternary, i.e. the parameters for the 2-i binary system in each ternary of the type 1-2-i for i = 3 ... m. This adjustment is made, again, using the ternary tie-line data and binary VLE data.

Often in multicomponent mixtures represented by types B and C, a particular solute is not expected to occur in substantial amounts. In this case, it is permissible to omit ternary tie-line data for ternaries containing that component.

Type C requires the most complex data analysis. To illustrate, we have reduced the data of Henty (1964) for the system furfural-benzene-cyclohexane-2,2,4-trimethylpentane. VLE data were used in conjunction with one ternary tie line for each ternary to determine optimum binary parameters for each of the two type-I ternaries: cyclohexane-furfural-benzene and 2,2,4-trimethylpentane-furfural-benzene. Two sets of binary parameters were obtained for the furfural-benzene binary; these are shown in Figure 21 along with confidence ellipses. Also shown are binary parameters obtained from VLE data alone.

The optimum parameters for furfural-benzene are chosen in the region of the overlapping 39% confidence ellipses. The ternary tie-line data were then refit with the optimum furfural-benzene parameters; final values of binary parameters were thus obtained for benzene-cyclohexane and for benzene-2,2,4-trimethylpentane. Table 4 gives all optimum binary parameters for this quarternary system.

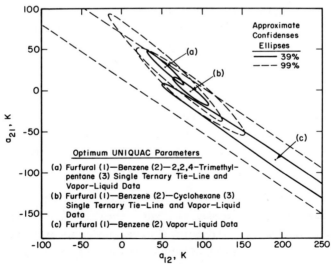

Figure 4-21. Parameters obtained for the furfural-benzene binary are different for the two ternary systems. An optimum set of these parameters is chosen from the overlapping confidence regions, capable of representing both ternaries equally well.

75

The ternary diagrams shown in Figure 22 and the selectivities and distribution coefficients shown in Figure 23 indicate very good correlation of the ternary data with the UNIQUAC equation. More important, however, Table 5 shows calculated and experimental quarternary tie-line compositions for five of Henty's twenty measurements. The root-mean-squared deviations for all twenty measurements show excellent agreement between calculated and predicted quarternary equilibria.

Computational Implementation

The computation of pure-component liquid fugacities, f_i^{OL} (or hypothetical pure-component fugacities for noncondensable components), at system temperature and pressure is implemented by the FORTRAN subroutine PURF, which evaluates the fugacities for all components, in a system of up to 20 components. As part of this computation, PURF calculates the zero-pressure, pure-component fugacities, $f_i^{(PO)}$, at system temperature.

Evaluation of the activity coefficients, γ_i (or γ_i^{∞} for noncondensable components), is implemented by the FORTRAN subroutine GAMMA, which finds simultaneously the coefficients for all components. This subroutine references subroutine TAUS to obtain the binary parameters, τ_{ij}, at system temperature. These subroutines are described and listed in Appendix E.

Conclusion

A liquid-phase model for the excess Gibbs energy provides the most practical method for correlating and predicting VLE and LLE provided that care is taken in obtaining optimum binary parameters. Ternary parameters are not required.[†]

For multicomponent VLE, the method used for obtaining binary parameters is of some, but not crucial, importance. Provided that the binary parameters give good representation of the binary data, good multicomponent results are usually obtained

---

[†] However, some ternary data may be useful for fixing binary parameters.

76

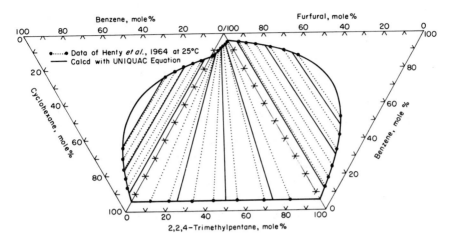

Figure 4-22. Calculated equilibria for the three ternaries in a quarternary of type C.

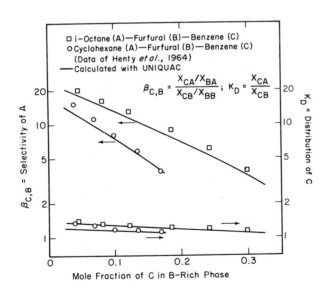

Figure 4-23. Calculated and experimental selectivities and distribution coefficients for the type-I ternaries in the 2,2,4-trimethyl pentane-cyclohexane-furfural-benzene system.

Table 4-5

Calculated and Experimental Liquid-Liquid Equilibria
for a Quarternary System at 25°C
2,2,4-trimethylpentane(1)-furfural(2)-cyclohexane(3)-benzene(4)

	Phase α				Phase β			
	x_1	x_2	x_3	x_4	x_1	x_2	x_3	x_4
measured	0.1206	0.1897	0.4587	0.2310	0.0396	0.5538	0.1940	0.2126
calculated	0.1232	0.1863	0.4603	0.2302	0.0375	0.5552	0.1939	0.2134
deviation	-0.0025	0.0034	-0.0016	0.0007	0.0021	-0.0014	0.0001	-0.0008
measured	0.2543	0.1749	0.3085	0.2623	0.0651	0.5777	0.1281	0.2291
calculated	0.2498	0.1817	0.3120	0.2564	0.0696	0.5707	0.1247	0.2350
deviation	0.0045	-0.0068	-0.0035	0.0058	-0.0046	0.0070	0.0034	-0.0059
measured	0.4164	0.1426	0.1717	0.2693	0.0810	0.6320	0.0541	0.2329
calculated	0.4074	0.1591	0.1682	0.2652	0.0875	0.6189	0.0568	0.2368
deviation	0.0089	-0.0165	0.0036	0.0041	-0.0065	0.0130	-0.0027	-0.0038
measured	0.1781	0.0824	0.6451	0.0944	0.0160	0.7863	0.1270	0.0707
calculated	0.1753	0.0887	0.6462	0.0897	0.0185	0.7812	0.1249	0.0753
deviation	0.0028	-0.0063	-0.0012	0.0047	-0.0025	0.0051	0.0021	-0.0046
measured	0.6141	0.0654	0.2574	0.0631	0.0393	0.8785	0.0388	0.0433
calculated	0.6115	0.0701	0.2584	0.0599	0.0407	0.8754	0.0374	0.0464
deviation	0.0026	-0.0047	-0.0010	0.0032	-0.0014	0.0031	0.0015	-0.0031

root-mean-square deviations for 20 experimental points

	x_1	x_2	x_3	x_4	x_1	x_2	x_3	x_4
	0.0049	0.0097	0.0026	0.0042	0.0035	0.0073	0.0022	0.0041

even though the binary parameters are not unique. For multi-component LLE equilibria, however, much care must be exercised in choosing optimum binary parameters.

Many well-known models can predict ternary LLE for type-II systems, using parameters estimated from binary data alone. Unfortunately, similar predictions for type-I LLE systems are not nearly as good. In most cases, representation of type-I systems requires that some ternary information be used in determining optimum binary parameter.

As illustrated here with the UNIQUAC equation, an optimum set of binary parameters can be obtained using simultaneously binary VLE data, binary LLE data, and one (or more) ternary tie-line data. The maximum-likelihood principle described in Chapter 6 provides the basis for parameter estimation. The parameters obtained give good representation of ternary data for a wide variety of systems. More important, however, as outlined here, calculations based on a model for the excess Gibbs energy provide a systematic procedure for predicting VLE and LLE for systems containing more than three components.

References

Abbott, M. M., Van Ness, H. C., A.I.Ch.E. J., 21, 62 (1975).

Abrams, D. S., Prausnitz, J. M., A.I.Ch.E. J., 21, 116 (1975).

Anderson, T. F., Abrams, D. S., Grens, E. A., A.I.Ch.E. J., 24, 20 (1978).

Anderson, T. F., Prausnitz, J. M., Ind. Eng. Chem., Process Des. Dev., 17, 552, 561 (1978).

Bancroft, W. D., Hubard, S. S., J. Am. Chem. Soc., 64, 347 (1942).

Black, C., Ind. Eng. Chem., 50, 403 (1958).

Boublikova, L., Lu, B. C. Y., J. Appl. Chem. (London), 19 (3), 89 (1969).

Brancker, A. V., Hunter, T. G., Nash, A. W., J. Phys. Chem., 44, 683 (1940).

Brown, I., Smith, F., Austr. J. Chem., 7, 264 (1954).

Brown, I., Smith, F., Austr. J. Chem., 8, 62 (1955).

Bryson, A. E., Ho, Y. B., "Applied Optimal Control; Optimization, Estimation and Control," Blaisdell Publishing, Waltham, Mass. (1969).

Chang, Y., Moulton, R. W., Ind. Eng. Chem., 45, 2350 (1953).

Conti, J. J., Othmer, P. F., Gilmont, R., J. Chem. Eng. Data, 5, 301 (1960).

Darwent, B. de B., Winkler, C. A., J. Phys. Chem., 47, 442 (1943).

Draper, N. R., Smith, H. "Applied Regression Analysis," John Wiley, New York (1966).

Edwards, J. B., Ph.D. Dissertation, Geortia Inst. Tech. (1962).

Fredenslund, Aa., Gmehling, J., Michelsen, M., Rasmussen, P., Prausnitz, J. M., Ind. Eng. Chem. Process Des. Dev., 16 450 (1977a).

Fredenslund, Aa., Gmehling, J., Rasmussen, P., "Vapor-Liquid Equilibria Using UNIFAC, a Group-Contribution Method," Elsevier, Amsterdam (1977b).

Guffey, C. G., Wehe, A. H., A.I.Ch.E. J., 18, 913 (1972).

Heinrich, J., Coll. Czech. Chem. Commun., 40, 2221 (1975).

Henty, C. J., McManamey, W. J., Prince, R. G. H., J. Appl. Chem., 14, 148 (1964).

Kay, W. B., Ind. Eng. Chem., 30, 459 (1938).

Mertl, I., Coll. Czech. Chem. Commun., 37, 366 (1972).

Nagata, I., Hayashida, H., J. Chem. Eng. Japan, 3, 161 (1970).

Nakaniski, K., Shirai, H., Minamiyama, T., J. Chem. Eng. Data, 12, 591 (1967).

Nicolaides, G. L., Eckert, C. A., Ind. Eng. Chem. Fund., 17, 331 (1978).

Null, H. R., "Phase Equilibrium in Process Design," John Wiley, New York (1970).

Palmer, D. A., Smith, B. D., J. Chem. Eng. Data, 17, 71 (1972).

Prince, R. G. H., Chem. Eng. Sci., 8, 175 (1954).

Renon, H., Prausnitz, J. M., A.I.Ch.E. J., 14, 135 (1968a).

Renon, H., Prausnitz, J. M., Ind. Eng. Chem. Process Des. Dev., 7, 220 (1968b).

Pratt, H. R. C., Trans. Inst. Chem. Engrs. (London), 25, 43 (1947).

Saddington, A. W., Krase, N. W., J. Am. Chem. Soc., 56, 353 (1934).

Scatchard, G., Satkiewicz, F. G., J. Am. Chem. Soc., 86, 130 (1964).

Schreiber, L. B., Eckert, C. A., Ind. Eng. Chem., Process Des. Dev., 10, 572 (1971).

Severns, W. H., Sesonske, A., Perry, R. H., Pigford, R. L., A.I.Ch.E. J., 1, 401 (1955).

Sinor, J. E., Weber, J. H., J. Chem. Eng. Data, 5, 243 (1960).

Smith, J. C., J. Phys. Chem., 45, 1301 (1941).

Steinhauser, H. H., White, R. R., Ind. Eng. Chem., 41, 2912 (1949).

Volpicelli, G., J. Chem. Eng. Data, 13, 150 (1968).

Weck, H. I., Hunt, H., Ind. Eng. Chem., 46, 2521 (1954).

Wisniak, J., Tamir, A., J. Chem. Eng. Data, 21, 88 (1976).

Wittrig, T. S., "The Prediction of Liquid-Liquid Equilibria by the UNIQUAC Equation," B.S. Degree Thesis, University of Illinois, Urbana (1977).

Chapter 5

ENTHALPIES

In modern separation design, a significant part of many phase-equilibrium calculations is the mathematical representation of pure-component and mixture enthalpies. Enthalpy estimates are important not only for determination of heat loads, but also for adiabatic flash and distillation computations. Further, mixture enthalpy data, when available, are useful for extending vapor-liquid equilibria to higher (or lower) temperatures, through the Gibbs-Helmholtz equation.[†]

Enthalpies are referred to the ideal vapor. The enthalpy of the real vapor is found from zero-pressure heat capacities and from the virial equation of state for non-associated species or, for vapors containing highly dimerized vapors (e.g. organic acids), from the chemical theory of vapor imperfections, as discussed in Chapter 3. For pure components, liquid-phase enthalpies (relative to the ideal vapor) are found from differentiation of the zero-pressure standard-state fugacities; these, in turn, are determined from vapor-pressure data, from vapor-phase corrections and liquid-phase densities. If good experimental data are used to determine the standard-state fugacity, the derivative gives enthalpies of liquids to nearly the same precision as that obtained with calorimetric data, and provides reliable heats of vaporization.

[†]The general form of the Gibbs-Helmholtz equation is

$$\left(\frac{\partial\ g/T}{\partial\ 1/T}\right)_{P,x} = h$$

where x refers to the composition. If we apply this equation to a liquid mixture, it is convenient to rewrite the equation in the form

$$\left(\frac{\partial\ g^E/T}{\partial\ 1/T}\right)_{P,x} = h^E$$

where g^E is the molar excess Gibbs energy of the mixture and h^E is the molar excess enthalpy, also called the enthalpy of mixing.

The enthalpy of mixing for a liquid solution cannot be calculated accurately without specific data for that solution. Although it is always possible to differentiate an expression for the excess Gibbs energy with respect to temperature to find the enthalpy of mixing, such a procedure is often not reliable because the temperature dependence of the binary-mixture parameter of the liquid-phase model is usually not known. Far better results are achieved if vapor-liquid-equilibrium data are available over a range of temperatures or if some calorimetric mixing data can be used to fix temperature-dependent parameters.[‡] Fortunately, the enthalpy of mixing provides only a minor contribution to the total enthalpy of nonelectrolyte liquid mixtures. For engineering purposes it is therefore usually sufficient to estimate the enthalpy of mixing crudely or to neglect it entirely.

Vapor Phase

The enthalpy of a vapor mixture is obtained first, from zero-pressure heat capacities of the pure components and second, from corrections for the effects of mixing and pressure.

The heat capacity of an ideal vapor is a monotonic function of temperature; in this work it is expressed by the empirical relation

$$c_{P_i}^0 = D_{1i} + D_{2i}/T + D_{3i}T + D_{4i} \, \ln T \tag{5-1}$$

where $c_{P_i}^0$ is expressed in joules/mole-K and T in K. The form of Equation (1) appears to be the best available for representing within experimental error heat-capacity data for typical vapors over the temperature range 200-600°K. Appendix C gives the sets of constants D_{ij} for each pure component j. Equation (1) should never be extrapolated outside the range of fit to avoid what may well be serious error.

[‡] See, for example, G. L. Nicolaides and C. A. Eckert, Ind. Eng. Chem. Fund., 17, 331 (1978).

The enthalpy of a pure ideal vapor at temperature T, relative to some reference temperature T_0, is given by

$$h_i^I = \int_{T_0}^{T} c_{P_i}^0 \, dT \tag{5-2}$$

Substitution of Equation (1) gives

$$h_i^I = D_{1i}(T-T_0) + D_{2i} \ln(T/T_0) + D_{3i}(T^2-T_0^2)/2$$

$$+ D_{4i}(T \ln T - T_0 \ln T_0 + T_0 - T) \tag{5-3}$$

For an ideal vapor mixture of m components, there is no enthalpy of mixing. The enthalpy of such a mixture is then

$$h^I = \sum_{i=1}^{m} y_i h_i^I \tag{5-4}$$

where y is the mole fraction.

Substitution of Equation (3) yields:

$$h^I = \sum_{i=1}^{m} y_i [D_{1i}(T-T_0) + D_{2i} \ln(T/T_0) + D_{3i}(T^2 - T_0^2)/2$$

$$+ D_{4i}(T \ln T - T_0 \ln T_0 + T_0 - T)] \tag{5-5}$$

For a real vapor mixture, there is a deviation from the ideal enthalpy that can be calculated from an equation of state. The enthalpy of the real vapor is given by

$$h^V = h^I + \Delta h \tag{5-6}$$

where Δh is the molar enthalpy correction for a vapor at temperature T and pressure P relative to the ideal vapor at the same T and composition. This correction is given exactly by

$$\Delta h = \int_{0}^{P} [v - T(\frac{\partial v}{\partial T})_{P,y}] dP \tag{5-7}$$

where v is the vapor-phase molar volume. To evaluate the integral in Equation (7), for non-associating mixtures, the virial equation is used, truncated after the second coefficient, as discussed in Chapter 3.

The result is

$$\Delta h = JP \sum_{i=1}^{m} \sum_{j=1}^{m} y_i y_j [B_{ij} - T \frac{dB_{ij}}{dT}] \qquad (5-8)$$

where J is the appropriate conversion factor (0.0988 joules/cm^3-bar).

For an associating vapor mixture, where strong dimerization occurs (i.e., either $\eta_{ij} > 4.5$ or $\eta_{ii} > 4.5$), the molar enthalpy, based on one stoichiometric (apparent) mole of the vapor is

$$h^V = h^I + \Delta h_{assoc} \qquad (5-9)$$

where Δh_{assoc} is the enthalpy correction, per stoichiometric (apparent) mole, of the vapor at temperature T and pressure P, relative to the ideal vapor at the same T and composition.

For such associating vapor mixtures, the "chemical" theory is used, as discussed in Chapter 3. The derivation of Δh_{assoc} is given in Appendix A. The result, based on Equation (7), is

$$\Delta h_{assoc} = \Delta h^D + \Delta h^F \qquad (5-10)$$

where

$$\Delta h^D = -(\frac{n_t}{n_t^a}) RT \{ \sum_{i=1}^{m} \sum_{j=1}^{i} z_{ij} [1 - \frac{T}{B_{ij}^D} (\frac{dB_{ij}^D}{dT})] \qquad (5-11a)$$

$$\Delta h^F = (\frac{n_t}{n_t^a}) P \{ \sum_{i=1}^{m} z_i [B_{ii}^F - T(\frac{dB_{ii}^F}{dT})]$$

$$+ \sum_{i=1}^{m} \sum_{j=1}^{i} z_{ij} [B_{ij}^F - T(\frac{dB_{ij}^F}{dT})] \qquad (5-11b)$$

where n_t/n_t^a is the ratio of the true number of moles to the stoichiometric (apparent) number of moles, as discussed in detail in Appendix A.

Liquid Phase

The enthalpy of a liquid mixture is given by

$$h^L = h^I + \sum_{i=1}^{m} x_i \overline{\Delta h}_i \qquad (5-12)$$

85

where h^I is given by Equation (5) with y's replaced by x's. Here $\Delta \bar{h}_i$ is the partial molar enthalpy of component i, at temperature T, pressure P, and composition x, relative to the molar enthalpy of pure, ideal vapor i at the same temperature.

The partial molar enthalpy for every component i is found from an appropriate form of the Gibbs-Helmholtz equation

$$\left(\frac{\partial \ln f_i^L}{\partial T}\right)_{P,x} = -\frac{\overline{\Delta h}_i}{RT^2} \tag{5-13}$$

where, in the liquid mixture, f_i^L is the fugacity of component i.

As discussed in Chapter 4,

$$f_i^L = \gamma_i^{(P0)} x_i \, f_i^{(P0)} \exp\left(\frac{\bar{v}_i P}{RT}\right) \tag{5-14}$$

For condensable components, $f_i^{(P0)}$ is the fugacity of pure liquid i at temperature T corrected to zero pressure.

Substitution of Equation (14) into Equation (12) yields

$$h^L = h^I - \sum_{i=1}^{m} x_i RT^2 \left(\frac{\partial \ln \gamma_i^{(P0)}}{\partial T}\right)_{P,x}$$

$$- \sum_{i=1}^{m} x_i RT^2 \left(\frac{d \ln f_i^{(P0)}}{dT}\right)_{pure} + \sum_{i=1}^{m} x_i P \left(\bar{v}_i - T \frac{d\bar{v}_i}{dT}\right) \tag{5-15}$$

Since attention is here confined to moderate pressures, the last term in Equation (15) can be neglected. The first term in Equation (15) is given by Equation (5), with x's replacing y's.

In Equation (15), the third term is much more important than the second term. The third term gives the enthalpy of the ideal liquid mixture (corrected to zero pressure) relative to that of the ideal vapor at the same temperature and composition. The second term gives the excess enthalpy, i.e. the liquid-phase enthalpy of mixing; often little basis exists for evaluation of this term, but fortunately its contribution to total liquid enthalpy is usually not large.

For pure liquids the standard-state fugacity is represented by

$$\ln f_i^{(P0)} = C_{1i} + C_{2i}T^{-1} + C_{3i}T + C_{4i}\ln T + C_{5i}T^2 \qquad (5\text{-}16)$$

where the C_i's are constants for liquid i. These constants are determined primarily from vapor-pressure data and secondarily, from vapor-phase and liquid-phase volumetric data. Appendix B gives details. Appendix C gives constants for 92 fluids.

Differentiation of Equation (16) gives

$$\sum_{i=1}^{m} x_i RT^2 \left(\frac{d \ln f_i^{(P0)}}{dT}\right)$$

$$= R \sum_{i=1}^{m} x_i (-C_{2i}T + C_{3i}T^2 + C_{4i}T + 2C_{5i}T^3) \qquad (5\text{-}17)$$

Equation (17) provides the third term in Equation (15).

The molar excess enthalpy h^E is related to the derivatives of the activity coefficients with respect to temperature according to

$$h^E = -RT^2 \sum_{i=1}^{m} x_i \left(\frac{\partial \ln \gamma_i^{(P0)}}{\partial T}\right)_{P,x} \qquad (5\text{-}18)$$

When the UNIQUAC equation (Chapter 4) is substituted into Equation (16), assuming all parameters a_{ij} and a_{ji} to be independent of temperature, we obtain

$$h^E = R \sum_{i=1}^{m} \left[\frac{q_i' x_i}{\sum_{j=1}^{m} \theta' \tau_{ji}} \left(\sum_{j=1}^{m} \theta_j' \tau_{ji} a_{ji} \right) \right] \qquad (5\text{-}19)$$

For many liquid mixtures, Equation (19) can be used to provide a crude estimate of excess enthalpy. A much better estimate is obtained if the UNIQUAC parameters are considered temperature-dependent. For example, suppose Equations (4-9) and (4-10) are modified to

$$\tau_{ji} = \exp\left[-\frac{a_{ji} + b_{ji}/T}{T}\right] \qquad (5\text{-}20)$$

In that event, Equation (19) must be modified to

$$h^E = R \sum_{i=1}^{m} \left[\frac{q_i' x_i}{\sum_{j=1}^{m} \theta_j' \tau_{ji}} \left[\sum_{j=1}^{m} \theta_j' \tau_{ji} (a_{ji} + 2b_{ji}/T) \right] \right] \qquad (5\text{-}19a)$$

In typical situations, we do not have the necessary experimental data to find constants b_{ij}. To obtain these constants,
we need experimental vapor-liquid equilibria (i.e. activity coefficients) as a function of temperature.

Liquid Mixtures Containing Noncondensable Components

For liquid mixtures containing both condensable and noncondensable components, Equation (15) is applicable. However it
is now convenient to rewrite that equation. Neglecting, as before, the last term in Equation (15), we obtain:

$$h^L = h^I - \sum_{i=1}^{m} x_i RT^2 \left(\underbrace{\frac{\partial \ln \gamma_i^{(PO)}}{\partial T}}_{} \right)_{P,x} - \sum_{i=1}^{m} x_i RT^2 \left(\frac{d \ln f_i^{(PO)}}{dT} \right)_{pure}$$

$$\text{all condensables}$$

$$- \sum_{i=1}^{m} x_i RT^2 \left(\frac{\partial \ln \gamma_i^{\infty}}{\partial T} \right)_x - \sum_{i=1}^{m} x_i RT^2 \left(\frac{d \ln f_i^{OL}}{dT} \right) \qquad (5\text{-}15a)$$

$$\text{all noncondensables}$$

In Equation (15a), the first term h^I is found from Equation (5)
where x's replace y's and the fugacities of the pure condensables
are found as a function of temperature from Equation (16). The
last two terms in Equation (15a) are found as functions of temperature from Equations (4-20) and (4-21). Note that Equation
(4-20) has the same form as that of Equation (16) when the constants C_{3i} and C_{5i} are set equal to zero.

When Equations (4-21) and (4-22) are used,

$$\sum_{i=1}^{m} x_i RT^2 \left(\frac{\partial \ln \gamma_i^{\infty}}{\partial T} \right)_x = -R \sum_{i=1}^{m} \left[x_i \left(\sum_{j=1}^{m} \theta_j \delta_{ij}^{(1)} \right) \right] \qquad (5\text{-}21)$$

where index i is summed over all noncondensable components and
index j is summed over all condensable components. The value of
gas constant R used here is 8.31439 joules/mole, K.

The second term in Equation (15a) gives the enthalpy of
mixing of the condensable components. It is difficult to estimate that enthalpy but fortunately it makes only a small contri-

bution compared to that of the third term. In many cases, it is permissible to neglect the second term entirely. An estimate for that term may be obtained from the UNIQUAC equation as indicated in the previous section. Equations (18), (19) and (19a) are applicable also for liquid mixtures containing noncondensable components within the restrictions that were set in Chapter 4 [Equations (4-17)-(4-22) and the two paragraphs following Equation (4-22)].

Examples

To illustrate the enthalpy calculations outlined above, Figures 1, 2, and 3 present calculated enthalpies for three binary systems.

Figure 1 gives an enthalpy-concentration diagram for ethanol(1)-water(2) at 1 atm. (The reference enthalpy is defined as that of the pure liquid at 0°C and 1 atm.) In this case both components are condensables. The liquid-phase enthalpy of mixing

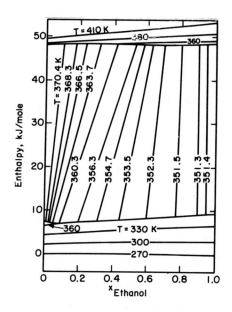

Figure 5-1. Enthalpy concentration diagram for ethanol-water at 1.013 bar.

89

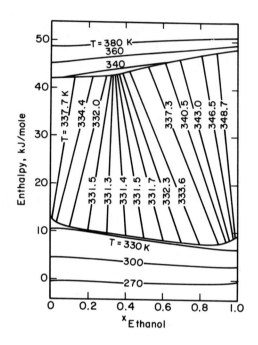

Figure 5-2. Enthalpy concentration diagram for ethanol-n-hexane at 1.013 bar.

was estimated from the UNIQUAC equation using temperature-independent parameters [Equation (17)]. The straight slanted lines are tie lines relating vapor compositions to liquid compositions at equilibrium. The nearly horizontal lines at the bottom of the figure represent isotherms for the subcooled liquid. The nearly horizontal lines at the top represent isotherms for the superheated vapor. Calculated results are in good agreement with experimental data as quoted by Hougen et al. (1954).

Figure 2 shows similar results for ethanol(1)-n-hexane(2) at 1 atm. The liquid-phase enthalpy of mixing was again estimated from UNIQUAC using temperature-independent parameters.

The enthalpies of vaporization for the pure components are in excellent agreement with experiment, as is the composition of the azeotrope. The enthalpy of the saturated vapor is also in

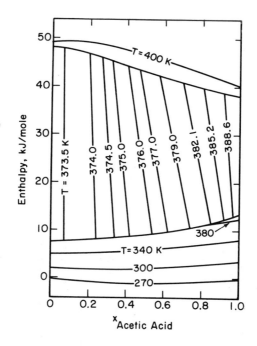

Figure 5-3. Enthalpy concentration diagram for acetic acid-water at 1.013 bar.

very good agreement with calorimetric data assuming that the vapor-phase is ideal.[†]

Table 1 indicates that the enthalpy of mixing in the liquid phase is not important when calculating enthalpies of vaporization, even though for this system, the enthalpy of mixing is large (Brown, 1964) when compared to other enthalpies of mixing for typical mixtures of nonelectrolytes.

Figure 3 presents results for acetic acid(1)-water(2) at 1 atm. In this case deviations from ideality are important for the vapor phase as well as the liquid phase. For the vapor phase, calculations are based on the chemical theory of vapor-phase imperfections, as discussed in Chapter 3. Calculated results are in good agreement with similar calculations reported by Lemlich et al. (1957).

[†]There is a significant difference between the results shown in Figure 2 and calculated results given in Brit. Chem. Eng. Proc. Tech. 16 1036 (1971). We believe the latter to be in error.

Table 5-1

Significance of Enthalpy of Mixing
for the System Ethanol(1)-n-Hexane(2)

x_1	T,K	Enthalpy of Saturated Liquid,* kJ/mole			Integral Enthalpy of Vaporization, kJ/mole		
		for ideal liquid mixture	With h^E estimated with UNIQUAC	With experimental $h^{E\dagger}$	for ideal liquid mixture	With h^E estimated with UNIQUAC	With experimental $h^{E\dagger}$
0.1	332.0	10.6	10.8	11.3	30.6	30.3	29.9
0.3	331.3	9.6	9.7	10.5	32.8	32.8	31.9
0.5	331.5	8.8	8.6	9.6	35.0	35.1	34.2
0.7	332.3	8.0	7.8	8.5	37.1	37.3	36.6
0.9	337.3	7.9	7.7	8.1	38.9	39.1	38.7

*Enthalpies are referred to zero enthalpies of the pure liquids at 1.013 bars and 273.2 K.

†Experimental values were interpolated from the data of Brown et al. (1964).

Finally, Table 2 shows enthalpy calculations for the system
nitrogen-water at 100 atm. in the range 313.5-584.7°K. [See also
Figure (4-13).] The mole fraction of nitrogen in the liquid
phase is small throughout, but that in the vapor phase varies
from essentially unity at the low-temperature end to zero at the
high-temperature end. In the liquid phase, the enthalpy is
determined primarily by the temperature, but in the vapor phase
it is determined by both temperature and composition.

Computational Implementation

The computation of pure-component and mixture enthalpies is
implemented by FORTRAN IV subroutine ENTH, which evaluates the
liquid- or vapor-phase molar enthalpy for a system of up to 20
components at specified temperature, pressure, and composition.
The enthalpies calculated are in J/mol referred to the ideal gas
at 300°K. Liquid enthalpies can be determined either with
neglect of liquid excess enthalpy (h^E) or with approximate excess enthalpy based upon temperature-independent UNIQUAC parameters. If experimental excess enthalpy data are available for
a particular system, appropriate prediction of excess enthalpy
can be incorporated into ENTH when it is used with that system.
If values for the binary parameters b_{ij} in Equation (20) are
available, they can be added to the data base and used in ENTH,
which now includes this parameter, however, set to zero.

Conclusion

This chapter presents quantitative methods for calculation
of enthalpies of vapor-phase and liquid-phase mixtures. These
methods rely primarily on pure-component data, in particular
ideal-vapor heat capacities and vapor-pressure data, both as
functions of temperature. Vapor-phase corrections for nonideality
are usually relatively small. Liquid-phase excess enthalpies are
also usually not important. As indicated in Chapter 4, for mixtures containing noncondensable components, we restrict attention
to liquid solutions which are dilute with respect to all noncondensable components.

Table 5-2

Vapor and Liquid Enthalpies at Saturation
for the Nitrogen(1)-Water(2) System
at 100 Atmospheres

Temperature K	Equilibrium Mole Fractions		Saturated Enthalpy[†] K Joule/mole	
	$10^4 x_1$	y_1	Liquid	Vapor
584.7	0	0	-23.3	6.83
575.0	3.4	0.1	-24.0	6.68
565.0	6.3	0.2	-24.7	6.53
554.4	8.7	0.3	-25.4	6.37
543.1	10.8	0.4	-26.2	6.19
530.6	12.5	0.5	-27.1	5.97
516.5	13.8	0.6	-28.1	5.69
499.6	14.5	0.7	-29.3	5.31
478.5	14.7	0.8	-30.8	4.79
446.5	13.7	0.9	-33.1	3.92
426.1	12.6	0.94	-34.6	3.32
387.6	10.3	0.98	-37.4	2.14
367.2	9.0	0.99	-38.5	1.49
313.5	5.9	0.999	-42.6	-0.25

[†]Enthalpies are zero for the ideal vapors at 300 K.

The accuracy of the calculations depends directly on the reliability of the experimental data. The correlated data presented in the Appendices were taken from standard literature sources; while these data are probably reliable for most fluids, it is not possible to be certain that they are reliable for all.

References

Brown, I., Fock, W., Smith, F., Aust. J. Chem., $\underline{17}$, 1106 (1964).

Hougan, O. H., Watson, K. M., Ragatz, R. A., "Chemical Process Principles, 2nd ed.", John Wiley & Sons, Inc., New York (1954).

Lemlich, R., Gottschlich, C., Hoke, R., Ind. Eng. Chem., $\underline{2}$, 32 (1957).

Chapter 6

PARAMETER ESTIMATION

Two generally accepted models for the vapor phase were discussed in Chapter 3 and one particular model for the liquid phase
(UNIQUAC) was discussed in Chapter 4. Unfortunately, these, and
all other presently available models, are only approximate when
used to calculate equilibrium properties of dense fluid mixtures.
Therefore, any such model must contain a number of adjustable
parameters, which can only be obtained from experimental measurements. The predictions of the model may be sensitive to the
values selected for model parameters, and the data available may
contain significant measurement errors. Thus, it is of major
importance that serious consideration be given to the proper
treatment of experimental measurements for mixtures to obtain the
most appropriate values for parameters in models such as UNIQUAC.

While many methods for parameter estimation have been proposed, experience has shown some to be more effective than others.
Since most phenomenological models are nonlinear in their adjustable parameters, the best estimates of these parameters can be
obtained from a formalized method which properly treats the
statistical behavior of the errors associated with all experimental observations. For reliable process-design calculations,
we require not only estimates of the parameters but also a measure
of the errors in the parameters and an indication of the accuracy of the data.

There are two types of measurement errors, systematic and
random. The former are due to an inherent bias in the measurement procedure, resulting in a consistent deviation of the experimental measurement from its true value. An experimenter's
skill and experience provide the only means of consistently detecting and avoiding systematic errors. By contrast, random or
statistical errors are assumed to result from a large number of
small disturbances. Such errors tend to have simple distributions subject to statistical characterization.

A major consequence of these random errors is the corresponding presence of errors or uncertainties in the estimated parameters. Because of these errors in the data, and also because of inaccuracies in the model, it is not possible for a model to represent the experimental data exactly. However, a method of parameter estimation which properly utilizes all the pertinent information available, can give a "best fit" of the model to the data and minimize parameter uncertainty.

Unfortunately, many commonly used methods for parameter estimation give only estimates for the parameters and no measures of their uncertainty. This is usually accomplished by calculation of the dependent variable at each experimental point, summation of the squared differences between the calculated and measured values, and adjustment of parameters to minimize this sum. Such methods routinely ignore errors in the measured independent variables. For example, in vapor-liquid equilibrium data reduction, errors in the liquid-phase mole fraction and temperature measurements are often assumed to be absent. The total pressure is calculated as a function of the estimated parameters, the measured temperature, and the measured liquid-phase mole fraction. The sum of the squared differences between calculated and measures pressures is minimized as a function of model parameters. This method, often called Barker's method (Barker, 1953), ignores information contained in vapor-phase mole fraction measurements; such information is normally only used for consistency tests, as discussed by Van Ness et al. (1973). Nevertheless, when high-quality experimental data are available, Barker's method often gives excellent results (Abbott and Van Ness, 1975).

The method used here is based on a general application of the maximum-likelihood principle. A rigorous discussion is given by Bard (1974) on nonlinear-parameter estimation based on the maximum-likelihood principle. The most important feature of this method is that it attempts properly to account for all measurement errors. A discussion of the background of this method and details of its implementation are given by Anderson et al. (1978).

Maximum-Likelihood Principle

In the maximum-likelihood analysis, it is assumed that all

measured data are subject to random errors. If each experiment were replicated, the average value for each replicated experimental point would approach some true value. Usually the distribution of a measured variable about its true value is approximated by the normal distribution, characterized by an associated variance. If there is any coupling between the measurement method (e.g. measurements of overlapping peaks on a chromatograph), then there are also associated covariances between these measured variables. These variances and covariances must be known or estimated, although covariances are almost always assumed to be negligible. The variances are ideally obtained from replicated experiments, but they may be estimated from experience associated with a particular type of experimental apparatus. It is customary to assume that the random errors in different experiments are uncorrelated.

For each experiment, the true values of the measured variables are related by one or more constraints. Because the number of data points exceeds the number of parameters to be estimated, all constraint equations are not exactly satisfied for all experimental measurements. Exact agreement between theory and experiment is not achieved due to random and systematic errors in the data and to "lack of fit" of the model to the data. Optimum parameters and true values corresponding to the experimental measurements must be found by satisfaction of an appropriate statistical criterion.

If this criterion is based on the maximum-likelihood principle, it leads to those parameter values that make the experimental observations appear most likely when taken as a whole. The likelihood function is defined as the joint probability of the observed values of the variables for any set of true values of the variables, model parameters, and error variances. The best estimates of the model parameters and of the true values of the measured variables are those which maximize this likelihood function with a normal distribution assumed for the experimental errors.

For binary vapor-liquid equilibrium measurements, the parameters sought are those that minimize the objective function

$$S = \sum_{i=1}^{M} \left\{ \frac{(P_i^0 - P_i^e)^2}{\sigma_{P_i}^2} + \frac{(T_i^0 - T_i^e)^2}{\sigma_{T_i}^2} + \frac{(x_{1i}^0 - x_{1i}^e)^2}{\sigma_{x_{1i}}^2} + \frac{(y_{1i}^0 - y_{1i}^e)^2}{\sigma_{y_{1i}}^2} \right\} \qquad (6\text{-}1)$$

subject to the two equilibrium constraints given by Equation (2-1) for each component. The summation is over all **M** data points (P,T,x,y). Superscript e designates a measured value and superscript 0 designates the estimated true value corresponding to each measured point. In Equation (6-1), σ^2 is the estimated variance of each of the measured variables, i.e. pressure, temperature, liquid-phase and vapor-phase mole fractions. The important feature is that all experimental data are used. The "true" value of each measured variable is also found in the course of parameter estimation.

The algorithm employed in the estimation process linearizes the constraint equations at each iterative step at current estimates of the true values for the variables and parameters. This reduces the calculation at each step to solution of a set of linear equations. The program description and listing are given in Appendix H.

Application to Parameter Estimation from VLE Data

Application of the algorithm for analysis of vapor-liquid equilibrium data can be illustrated with the isobaric data of Othmer (1928) for the system acetone(1)-methanol(2). For simplicity, the van Laar equations are used here to express the activity coefficients.

$$\ln \gamma_1 = \frac{A_{12}}{\left[1 + \dfrac{A_{12} x_1}{A_{21} x_2}\right]^2} \qquad (6\text{-}2)$$

$$\ln \gamma_2 = \frac{A_{21}}{\left[1 + \dfrac{A_{21} x_2}{A_{12} x_1}\right]^2} \qquad (6\text{-}3)$$

Substitution of Equations (2) and (3) into the equilibrium relations dictated by Equation (2-1), yields two constraints which are used to estimate the two adjustable parameters, A_{12} and A_{21}.

Standard deviations in the measured variables are taken as: σ_P = 2.0 mm Hg, σ_T = 0.2°C, σ_x = 0.005, and σ_y = 0.01. With initial parameter estimates of A_{12} = 1.0 and A_{21} = 1.0, convergence was achieved in three iterations giving A_{12} = 0.857 and A_{21} = 0.681. Table 1 gives the measured data, estimates of the true values corresponding to the measurements, and deviations of the measured values from model predictions. Figure 1 shows the phase diagram corresponding to these parameters, together with the measured data.

When there is significant random error in all the variables, as in this example, the maximum-likelihood method can lead to better parameter estimates than those obtained by other methods. When Barker's method was used to estimate the van Laar parameters for the acetone-methanol system from these data, it was estimated that A_{12} = 0.960 and A_{21} = 0.633, compared with A_{12} = 0.857 and A_{21} = 0.681 using the method of maximum likelihood. Barker's method uses only the P-T-x data and assumes that the T and x measurements are error free.

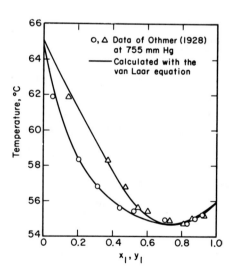

Figure 6-1. Calculated phase diagram for the acetone(1)-methanol(2) system.

Table 6-1

Measured Variables and Estimates of Their True Values
for Acetone(1)/Methanol(2) System (Othmer, 1928)

	Pressure, mm Hg			Temperature, °C			Liquid composition, x_1			Vapor composition, y_1		
	Meas	Calc	Dev	Meas	Calc	Dev	Meas	Calc	Dev	Meas	Calc	Dev
1	755.0	753.8	1.2	61.90	62.22	-0.32	0.0553	0.0599	-0.0046	0.1442	0.1502	-0.0060
2	755.0	754.9	0.1	58.30	58.32	-0.02	0.2046	0.2077	-0.0031	0.3722	0.3606	0.0116
3	755.0	755.0	0.0	56.80	56.79	0.01	0.3127	0.3162	-0.0035	0.4737	0.4543	0.0194
4	755.0	754.7	0.3	55.60	55.67	-0.07	0.4405	0.4413	-0.0008	0.5431	0.5399	0.0032
5	755.0	755.7	-0.7	55.40	55.20	0.20	0.5283	0.5292	-0.0008	0.6048	0.5953	0.0095
6	755.0	755.7	-0.7	54.90	54.70	0.20	0.7071	0.7109	-0.0030	0.7335	0.7161	0.0174
7	755.0	754.6	0.4	54.70	54.79	-0.09	0.8308	0.8304	0.0004	0.8124	0.8134	-0.0010
8	755.0	755.0	-0.0	55.00	55.00	0.00	0.8783	0.8796	-0.0013	0.8654	0.8600	0.0054
9	755.0	754.6	0.4	55.20	55.28	-0.08	0.9264	0.9804	-0.0040	0.9298	0.9140	0.0158
Standard Deviations			.58			.16			.0030			.0124

Dev = Meas - Calc

Parameter Significance, Uniqueness, and Error

The primary purpose for expressing experimental data through model equations is to obtain a representation that can be used confidently for systematic interpolations and extrapolations, especially to multicomponent systems. The confidence placed in the calculations depends on the confidence placed in the data and in the model. Therefore, the method of parameter estimation should also provide measures of reliability for the calculated results. This reliability depends on the uncertainties in the parameters, which, with the statistical method of data reduction used here, are estimated from the parameter variance-covariance matrix. This matrix is obtained as a last step in the iterative calculation of the parameters.

The diagonal elements of this matrix approximate the variances of the corresponding parameters. The square roots of these variances are estimates of the standard errors in the parameters and, in effect, are a measure of the uncertainties of those parameters.

The off-diagonal elements of the variance-covariance matrix represent the covariances between different parameters. From the covariances and variances, correlation coefficients between parameters can be calculated. When the parameters are completely independent, the correlation coefficient is zero. As the parameters become more correlated, the correlation coefficient approaches a value of +1 or -1.

When two parameters are determined, it is useful to plot the confidence ellipses of the estimated parameters. These are obtained from the eigenvalues and eigenvectors of the variance-covariance matrix of the parameters (Beck and Arnold, 1977; Bryson and Ho, 1969). The variance-covariance matrix is given in Table 2 for the parameters determined from the acetone-methanol data of Othmer. For this example, the eigenvalues are 0.00630 and 0.00085 and the major axis of the ellipse is given by the eigenvector $A_{21} = -0.481 \, A_{12}$. Confidence ellipses for this example are shown in Figure 2. The regions shown, represent areas within which the parameter values can be expected to lie at the confidence level associated with the ellipse (here 0.87 and 0.99).

Table 6-2

Vapor-Liquid Equilibrium Data Reduction for
Acetone(1)-Methanol(2) System (Othmer, 1928)

Calculated parameters for van Laar equation

$$A_{12} = 0.857$$
$$A_{21} = 0.681$$

Covariance matrix

$$\begin{bmatrix} 0.00528 & -0.00213 \\ -0.00213 & 0.00187 \end{bmatrix}$$

Correlation coefficient matrix

$$\begin{bmatrix} 1.000 & -0.678 \\ -0.678 & 1.000 \end{bmatrix}$$

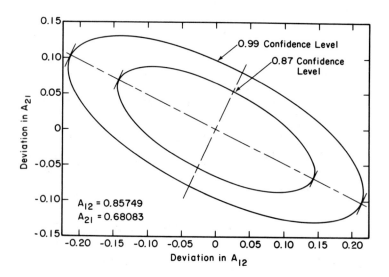

Figure 6-2. Confidence ellipses for van Laar parameters.
Acetone(1)-methanol(2) system at 755 mm Hg (Othmer, 1928).

Large confidence regions are obtained for the parameters because
of the random error in the data. For a "correct" model, the
regions become vanishingly small as the random error becomes
very small or as the number of experimental measurements becomes
very large.

For the acetone-methanol data of Othmer, the correlation coefficient is -0.678, indicating a moderate degree of correlation
between the two van Laar parameters. The elongated confidence
ellipses shown in Figure 2 further emphasize this correlation.
If the parameters were to become increasingly correlated, the
confidence ellipses would approach a 45° line and it would become
impossible to determine a unique set of parameters. As discussed by Fabries and Renon (1975), strong correlation is common
for nearly ideal solutions whenever the two adjustable parameters
represent energy differences.

A high degree of correlation may be beneficial. When the parameters are strongly related, some linear combination of the two
parameters may represent the data as well as do the individual parameters. In that case a method similar to that of Bruin and Praus-

nitz (1971) and Abrams (1975) may be used
the parameters, yielding a single-paramete[r]
binary vapor-liquid equilibrium data.

Analysis of Residuals

In the maximum-likelihood method used
of each measured variable is also found in t
estimation. The differences between these "
corresponding experimentally measured values
(also called deviations). When there are ma[ny]
residuals can be analyzed by standard statist
and Smith, 1966). If, however, there are onl[y]
examination of the residuals for trends, when
system variables, may provide valuable inform
plots can indicate at a glance excessive expe[rimental]
systematic error, or "lack of fit." Data poi[nts]
viously bad can also be readily detected. If
able and if there are no systematic errors, su
the residuals randomly distributed with zero m[ean]
ior is shown in Figure 3 for the ethyl-acetate[-]
of Murti and Van Winkle (1958), fitted with the

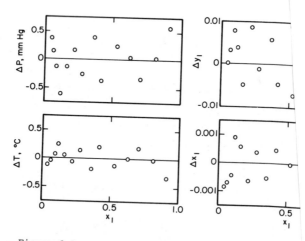

Figure 6-3. Residuals for the system ethyl acetate
propanol(2) at 60°C. Data of Murti and van Winkle

Standar[d]
$\sigma_P = 0.$
illustr
no syst
ted err
The poi
random]

Effects

sureme[nt]
accura
One of
varian

depend
ratios
wrong
behavi
deviat

Syste[m]

and i
data
If su
the [r]
some

idea
negl
tren
syst
vapo
sen[t]

105

d deviations were assumed for the experimental data:
5 mm Hg, σ_T = 0.1°C, σ_x = 0.001, and σ_y = 0.005.   Figure 3
ates the expected behavior of the residuals when there is
ematic error, the model is nearly correct, and the estima-
ors are nearly equal to the experimental random errors.
nts are within the stated standard deviation and are
y distributed about the zero axis.

of Estimated Variances

he need for good estimates of the variances of the mea-
ts cannot be overemphasized.   Unfortunately, these can be
ely estimated only by completely replicated experiments.
the limitations of most phase-equilibrium data is that
ces of experimental measurements are seldom known.

Fortunately, however, the technique used here does not
on the magnitude of the variances, but only on their
.   If estimates of the magnitudes of the variances are
but the ratios are correct, the residuals display the random
or shown in Figure 3.   However, the magnitudes of these
ions are then not consistent with the estimated variances.

atic Error

If there is sufficient flexibility in the choice of model
f the number of parameters is large, it is possible to fit
to within the experimental uncertainties of the measurements.
ch a fit is not obtained, there is either a shortcoming of
odel, greater random measurement errors than expected, or
systematic error in the measurements.

At low pressures,it is often permissible to neglect non-
ities of the vapor phase.   If these nonidealities are not
gible, they can have the effect of introducing a nonrandom
into the plotted residuals similar to that introduced by
ematic error.   Experience here has shown that application of
r-phase corrections for nonidealities gives a better repre-
ation of the data by the model, even when these corrections

must be estimated. For this reason, all data reduced in this work are corrected for vapor-phase imperfections as discussed in Chapter 3.

An apparent systematic error may be due to an erroneous value of one or both of the pure-component vapor pressures as discussed by several authors (Van Ness et al., 1973; Fabries and Renon, 1975; Abbott and Van Ness, 1977). In some cases, highly inaccurate estimates of binary parameters may occur. Fabries and Renon recommend that when no pure-component vapor-pressure data are given, or if the given values appear to be of doubtful validity, then the unknown vapor pressure should be included as one of the adjustable parameters. If, after making these corrections, the residuals again display a nonrandom pattern, then it is likely that there is systematic error present in the measurements.

An additional advantage derived from plotting the residuals is that it can aid in detecting a bad data point. If one of the points noticeably deviates from the trend line, it is probably due to a mistake in sampling, analysis, or reporting. The best action would be to repeat the measurement. However, this is often impractical. The alternative is to reject the datum if its occurrence is so improbable that it would not reasonably be expected to occur in the given set of experiments.

Model Selection

In many process-design calculations it is not necessary to fit the data to within the experimental uncertainty. Here, economics dictates that a minimum number of adjustable parameters be fitted to scarce data with the best accuracy possible. This compromise between "goodness of fit" and number of parameters requires some method of discriminating between models. One way is to compare the uncertainties in the calculated parameters. An alternative method consists of examination of the residuals for trends and excessive errors when plotted versus other system variables (Draper and Smith, 1966). A more useful quantity for comparison is obtained from the sum of the weighted squared residuals given by Equation (1).

This sum, when divided by the number of data points minus the number of degrees of freedom, approximates the overall variance of errors. It is a measure of the overall fit of the equation to the data. Thus, two different models with the same number of adjustable parameters yield different values for this variance when fit to the same data with the same estimated standard errors in the measured variables. Similarly, the same model, fit to different sets of data, yields different values for the overall variance. The differences in these variances are the basis for many standard statistical tests for model and data comparison. Such statistical tests are discussed in detail by Crow et al. (1960) and Brownlee (1965).

The maximum-likelihood method is not limited to phase equilibrium data. It is applicable to any type of data for which a model can be postulated and for which there are known random measurement errors in the variables. P-V-T data, enthalpy data, solid-liquid adsorption data, etc., can all be reduced by this method. The advantages indicated here for vapor-liquid equilibrium data apply also to other data.

The maximum-likelihood method, like any statistical tool, is useful for correlating and critically examining experimental information. However, it can never be a substitute for that information. While a statistical tool is useful for minimizing the required experimental effort, reliable calculated phase equilibria can only be obtained if at least some pertinent and reliable experimental data are at hand.

References

Abbott, M. M., Van Ness, H. C. A.I.Ch.E. J., 21, 62 (1975).

Abbott, M. M., Van Ness, H. C., Fluid-Phase Equilibria, 1, 3 (1977).

Abrams, D. S., Prausnitz, J. M., A.I.Ch.E. J., 21, 116 (1975).

Anderson, T. F., Abrams, D. S., Grens, E. A., A.I.Ch.E. J., 24, 20 (1978).

Bard, Y., Nonlinear Parameter Estimation, Academic Press, New York (1974).

Barker, J. A., Austr. J. Chem., 6, 207 (1953).

Beck, J. V., Arnold, K. J., "Parameter Estimation in Engineering and Science," John Wiley & Sons, New York (1977).

Britt, H. I., Luecke, R. H., Technometrics, 15 (2), 233 (1973).

Brownlee, K. A., "Statistical Theory and Methodology in Science and Engineering," 2nd ed., John Wiley and Sons, New York (1965).

Bruin, S., Prausnitz, J. M., Ind. Eng. Chem. Process Des. Dev., 10, 562 (1971).

Bryson, A. E., Ho, Y. B., "Applied Optimal Control; Optimization, Estimation, and Control," Blaisdell Publishing, Waltham, Mass. (1969).

Clifford, A. A., "Multivariate Error Analysis," Halsted Press, Division of John Wiley & Sons, New York (1973).

Crow, E. L., Davis, F. A., Maxfield, M. W., "Statistics Manual," Dover Publications, New York (1960).

Deming, E. W., "Statistical Adjustment of Data," John Wiley & Sons, New York (1943).

Draper, N. R., Smith, H., "Applied Regression Analysis," John Wiley & Sons, New York (1966).

Fabries, J., Renon, H., A.I.Ch.E. J., 21, 735 (1975).

Fisher, F. A., Phil. Trans. Roy. Soc. London (A), 222, 309 (1922).

Kreyszig, E., "Introductory Mathematical Statistics," John Wiley & Sons, New York (1970).

Murti, P. S., Van Winkle, M., Chem. Eng. Data Series, 3, 73 (1958).

Othmer, D. F. Ind. Eng. Chem., 20, 743 (1928).

Powell, P. R., MacDonald, J. R., Computer Journal, 15, 148 (1972).

Southwell, W. H., Computer Journal, 19, 69 (1976).

Van Ness, H. C., Byer, S. M., Gibbs, R. E., A.I.Ch.E. J., 19, 238 (1973).

Chapter 7

CALCULATION OF EQUILIBRIUM SEPARATIONS
IN MULTICOMPONENT SYSTEMS

The most frequent application of phase-equilibrium calculations in chemical process design and analysis is probably in
treatment of equilibrium separations. In these operations, often
called flash processes, a feed stream (or several feed streams)
enters a separation stage where it is split into two streams of
different composition that are in equilibrium with each other.
The product streams can be a vapor and a liquid or two immis-
cible liquids. The process may consist of a single equilibrium
stage, as with a flash drum or single mixer-settler, or of a
cascade of equilibrium stages, as with staged distillation or
extraction. Multistage separations are usually arranged with
countercurrent flow of the phases between the stages. The calculation of all such equilibrium separations is based on enthalpy
and component material balances over the separation stage, in
combination with the requirements that, for each component, the
fugacities be equal in the two streams exiting the stage.

In this chapter we present efficient calculation procedures
for single-stage equilibrium separations; subroutines implement-
ing these procedures are given in Appendices F and G. While
we recognize the great importance of multistage separations, it
must be realized that the efficient computation of such processes
requires very careful resolution of the large number of simultaneous equilibrium stages involved in a countercurrent cascade.
The dominant consideration in such multistage computation procedures is usually the technique used to achieve this simultaneous solution rather than the equilibrium treatment of the
stages themselves. (Goldstein and Stanfield, 1970; Holland,
1975; King, 1971; Naphtali and Sandholm, 1971; Newman, 1963;
and Tomich, 1970). Moreover the choice of appropriate computation procedures for distillation, absorption, and extraction is
highly dependent on the system being separated, the conditions of
separation, and the specifications to be satisfied (Friday and
Smith, 1964; Seppala and Luus, 1972). The thermodynamic methods
presented in Chapters 3, 4, and 5, particularly when combined to

110

yield equilibrium ratios as discussed later in this chapter, can
be incorporated into many of the well-documented calculational
procedures for distillation columns and absorber-strippers (e.g.
Naphtali and Sandholm, 1971; Ricker and Grens, 1974). The
situation for extraction is not so well in hand, but the liquid-
liquid equilibrium separation calculation procedures offered here
should be adaptable to many multistage extraction problems.

The single-stage separations for which we present computational procedures are the incipient separations (one product
phase present in very small amount) represented by bubble and
dew-point calculations, vapor-liquid equilibrium separations at
fixed pressure under isothermal or adiabatic conditions, and
liquid-liquid equilibrium separations at fixed pressure and temperature. These calculations are implemented by FORTRAN IV subroutines designed to minimize the number of vapor and liquid-
phase fugacity evaluations necessary to achieve satisfactory
solutions. This criterion for efficiency of the algorithms is
based on the recognition that, with relatively rigorous thermodynamic methods such as those used here, most of the computation
effort in any separation calculation is devoted to evaluation of
thermodynamic equilibrium functions. It is important
to avoid unnecessary calculations of fugacities or fugacity
(activity) coefficients in computer programs used in chemical
engineering practice.

Equilibrium Separation Calculations

In an equilibrium separation, a feed stream containing m
components at given composition, pressure, and enthalpy (or
temperature if in a single phase) is split into two streams in
equilibrium, here taken to be a vapor and a liquid. The flow
rates of the feed, vapor, and liquid streams are, respectively,
F, V, and L moles/unit time and their mole-fraction compositions
are, respectively, w_i, y_i, and x_i.

Component balances over the separation yield

$$Fw_i = Vy_i + Lx_i \qquad i = 1,\ldots,m \qquad (7-1)$$

and

$$F = V + L \qquad (7-2)$$

The enthalpy balance for the process is represented by

$$Fh^F + Q = Vh^V + Lh^L \tag{7-3}$$

where h^F, h^V, and h^L are the molal enthalpies of the feed, vapor, and liquid streams and Q is the rate of external heat (enthalpy) addition to the separation stage. For an adiabatic separation $Q = 0$ while for an isothermal separation it takes on the value necessary to achieve a specified separation or flash temperature, T.

Phase equilibrium between the vapor and liquid streams requires that $T^V = T^L = T$, $P^V = P^L = P$, and that, for each component

$$f_i^V = f_i^L \qquad i = 1, \ldots, m. \tag{2-1}$$

In the terms developed in Chapter 2, this latter requirement becomes

$$\phi_i y_i P = \gamma_i x_i f_i^{OL}. \tag{2-4}$$

For separation calculations, phase equilibrium is most conveniently represented in terms of the equilibrium ratios K_i defined by

$$K_i \equiv \frac{y_i}{x_i} \tag{7-4}$$

These ratios can be derived from Equation (2-4):

$$K_i = \frac{\gamma_i f_i^{OL}}{\phi_i P} \tag{7-5}$$

and are functions of liquid-phase composition (through γ_i), vapor-phase composition (through ϕ_i), temperature and pressure. We expect, however, that their dependence on vapor composition often is weak and that their temperature dependence frequently is dominated by an approximately exponential relationship of f_i^{OL} with $1/T$.

If the m Equations (7-1) are considered independent, additional restrictions must be imposed as

$$\sum_{i=1}^{m} x_i = 1 \tag{7-6}$$

112

$$\sum_{i=1}^{m} y_i = 1. \tag{7-7}$$

This leads, for a specified pressure, to a system with $2m+3$ unknown variables (x_i, y_i, V, L, and T or Q) described by $2m + 3$ equations ((7-1), (7-3), (7-4), (7-6), (7-7)).

Many different manipulations of these equations have been used to obtain solutions. As discussed by King (1971), many of the older approaches work in terms of V/L, which has the disadvantage of being unbounded and which, in the classical implementation, leads to poorly convergent iterative calculations. A preferable arrangement of this equation system for solution is based on the ratio V/F, which must lie between 0 and 1. If we substitute in Equation (7-1) for L from Equation (7-2) and for y_i from Equation (7-4), and then divide by F, we obtain

$$w_i = \frac{V}{F} K_i x_i + (1 - \frac{V}{F}) x_i.$$

Solution for x_i gives

$$x_i = \frac{w_i}{(K_i - 1)\alpha + 1} \tag{7-8}$$

where, for convenience, α has been used to denote the fractional vaporization V/F. Since $y_i = K_i x_i$, it follows that

$$y_i = \frac{K_i w_i}{(K_i - 1)\alpha + 1} \tag{7-9}$$

If the equilibrium ratios K_i were known or specified, Equation (7-8) could be substituted in Equation (7-6) or Equation (7-9) in (7-7) to give implicit relations for α:

$$\sum_{i=1}^{m} \frac{w_i}{(K_i - 1)\alpha + 1} = 1 \quad \text{or} \quad \sum_{i=1}^{m} \frac{K_i w_i}{(K_i - 1)\alpha + 1} = 1 \tag{7-10}$$

However, each of these forms possesses a spurious root and has other characteristics (maxima or minima) that often give rise to convergence problems with common iterative-solution techniques. A much preferable form of Equation (7-10) was described by Rachford and Rice (1952), who considered Equations (7-6) and (7-7) in the form

$$\sum_{i=1}^{m} y_i - \sum_{i=1}^{m} x_i = 0. \tag{7-11}$$

113

Then substitution of Equations (7-8) and (7-9) gives

$$\sum_{i=1}^{m} \frac{(K_i - 1)w_i}{(K_i - 1)\alpha + 1} = 0 \qquad (7-12)$$

which, for given K_i, is easily solved for α by iterative techniques such as the Newton-Raphson method.

The equilibrium ratios are not fixed in a separation calculation and, even for an isothermal system, they are functions of the phase compositions. Further, the enthalpy balance, Equation (7-3), must be simultaneously satisfied and, unless specified, the flash temperature simultaneously determined. Thus, the equilibrium separation problem is described by[*]

$$G_1(T,\underline{x},\underline{y},\alpha) = \sum_{i=1}^{m} \frac{(K_i - 1)w_i}{(K_i - 1)\alpha + 1} = 0 \qquad (7-13)$$

$$G_2(T,\underline{x},\underline{y},\alpha,\tfrac{Q}{F}) = 1 + \frac{Q}{Fh^F} - \frac{h^V}{h^F} - (1-\alpha)\frac{h^L}{h^F} = 0 \qquad (7-14)$$

together with Equations (7-8) and (7-9) to give x_i and y_i, and appropriate thermodynamic functions to give equilibrium ratios and enthalpies.

For the special case of a bubble-point calculation (incipient vaporization), α is 0 (also $Q = 0$) and Equation (7-13) becomes

$$\sum_{i=1}^{m} K_i w_i = 1. \qquad (7-15)$$

Similarly, for a dew-point calculation (incipient condensation) α is 1 (again $Q = 0$) and Equation (7-13) leads to

$$\sum_{i=1}^{m} \frac{w_i}{K_i} = 1 \qquad (7-16)$$

In both of these cases, Equation (7-14) becomes trivial as h^V or h^L is identical to h^F.

For an isothermal vapor-liquid flash, T is known and Equation (7-14) is used only to find Q after the other equations have

---

[*] $\underline{x}$ and $\underline{y}$ are vectors containing the liquid mole fractions and vapor mole fractions respectively.

been satisfied. For an adiabatic flash, Q is 0 and Equation (7-14) must be satisfied simultaneously with the other equations of the system.

The same fundamental development as presented here for vapor-liquid flash calculations can be applied to liquid-liquid equilibrium separations. In this case, the feed splits into an extract at rate E and a raffinate at rate R, which are in equilibrium with each other. The compositions of these phases are indicated by mole fractions x_i^E and x_i^R. Since heat effects in liquid-liquid separations are small, only the isothermal separation at given T is considered, and the enthalpy balance is discarded. The equation system representing this separation is

$$G(\underline{x}^R, \underline{x}^E, \alpha) = \sum_{i=1}^{m} \frac{(K_i - 1)w_i}{(K_i - 1)\alpha + 1} = 0 \qquad (7\text{-}17)$$

where α is E/F and x_i^E is given by Equation (7-9) and x_i^R by Equation (7-8). However, for liquid-liquid equilibria, the equilibrium ratios K_i are strong functions of both phase compositions. The system is thus far more difficult to solve than the superficially similar system of equations for the isothermal vapor-liquid flash. In fact, some of the arguments leading to the selection of the Rachford-Rice form for Equation (7-17) do not apply strictly in the case of two liquid phases. Nevertheless, this form does avoid spurious roots at $\alpha = 0$ or 1 and has been shown, by extensive experience, to be markedly superior to alternatives.

The equation systems representing equilibrium separation calculations can be considered multidimensional, nonlinear objective functions

$$\underline{G}(\underline{X}) = \underline{0} \qquad (7\text{-}18)$$

Iterative Calculation Procedures

where the vector function $\underline{G}$ consists of Equations (7-13) and (7-14) plus Equations (7-8) and (7-9) for all components. The vector $\underline{X}$ of variables includes T, α, and the composition vectors $\underline{x}$ and $\underline{y}$.*

---

*For liquid-liquid separations x_i^R and x_i^E replace x_i and y_i.

115

These systems are solved by a step-limited Newton-Raphson iteration, which, because of its second-order convergence characteristic, avoids the problem of "creeping" often encountered with first-order methods (Law and Bailey, 1967):

$$\underline{X}^{(r+1)} = \underline{X}^{(r)} - t \ \underline{\underline{J}}^{-1}\underline{G}(\underline{X}^{(r)}).$$

(7-19)

Here the superscript (r) represents the iteration number and $\underline{\underline{J}}$ is the Jacobian derivative matrix whose elements are

$$J_{ij} \equiv \left[\frac{\partial G_i}{\partial X_j}\right]_{X_{k \neq j}}$$

(7-20)

The scalar t is in the range 0 to 1 and provides step-limiting, or damping, when required to obtain a convergent iteration. Such step-limiting is often helpful because the direction of correction provided by the Newton-Raphson procedure, that is, the relative magnitudes of the elements of the vector $\underline{\underline{J}}^{-1}\underline{G}$, is very frequently more reliable than the magnitude of the correction (Naphtali, 1964). In application, t is initially set to 1, and remains at this value as long as the Newton-Raphson corrections serve to decrease the norm (magnitude) of $\underline{G}$, that is, for

$$\| \underline{G}(\underline{X}^{(r+1)}) \| \ < \ \| \underline{G}(\underline{X}^{(r)}) \|$$

If any iteration yields a $\underline{X}^{(r+1)}$ such that

$$\| \underline{G}(\underline{X}^{(r+1)}) \| \ \geq \ \| \underline{G}(\underline{X}^{(r)}) \|$$

then the value of t is decreased and $\underline{X}^{(r+1)}$ is recalculated (note: $\underline{\underline{J}}$ need not be re-evaluated). Here the reduction used involves multiplication of t by 0.7; other choices are possible. This reduction in t is repeated until $\|\underline{G}\|$ decreases, in which case the iteration continues, or until t becomes too small (here < 0.2), in which case the iteration is abandoned. Whenever $\|\underline{G}\|$ decreases, the value of t is reset to 1.*

---

*There is justification for allowing t to increase beyond 1, and in many particular applications this may be desirable. Here a more conservative approach is used to reduce the chance of unstable iterations.

Strict application of the Newton-Raphson procedure involves use of the full vector $\underline{X}$ of variables, including $\underline{x}$ and $\underline{y}$, with the augmentation of $\underline{G}$ by Equations (7-8) and (7-9) in the form

$$G_{2+i}(T,\underline{x},\underline{y},\alpha) \equiv x_i - \frac{w_i}{(K_i - 1)\alpha + 1} = 0 \qquad (7\text{-}21)$$

$$G_{2+m+i}(T,\underline{x},\underline{y},\alpha) \equiv y_i - \frac{K_i w_i}{(K_i - 1)\alpha + 1} = 0 \qquad (7\text{-}22)$$

The procedure would then require calculation of $(2m+2)^2$ partial derivatives per iteration, requiring $2m+2$ evaluations of the thermodynamic functions K_i per iteration.* Since the computation effort is essentially proportional to the number of K_i evaluations, this form of iteration is excessively expensive, even if it converges rapidly. Fortunately, simpler forms exist that are almost always much more efficient in application.

If the objective function is considered two-dimensional, consisting of Equations (7-13) and (7-14) and the vector $\underline{X}$ includes only T and α, then the only change in the iteration is that the derivatives of K_i with respect to composition are ignored in establishing the Newton-Raphson corrections to T and α. The new compositions can then be determined from Equations (7-8) and (7-9). Such a simplified procedure sacrifices little in convergence rate for vapor-liquid systems, where the contributions of composition-derivatives to changes in T and α are almost always small. This approach requires only two evaluations of K_i per iteration and still avoids creeping since it is essentially second-order in the limit as convergence is approached.

For liquid-liquid systems, the separations are isothermal and the objective function is one-dimensional, consisting of Equation (7-17). However, the composition dependence of the

---

*With derivatives with respect to temperature and composition of necessity found by finite difference approximation,

$$\left[\frac{\partial G_i}{\partial X_j}\right] \approx \frac{G_i(X_j + \Delta X_j, \underline{X}_{i \neq j}) - G_i(X_j, \underline{X}_{i \neq j})}{\Delta X_j}$$

the evaluations must be at the base vector $\underline{X}^{(r)}$ and at each of $2m+1$ separate perturbations of this vector. Derivatives with respect to α can be found analytically.

equilibrium ratios often dominates the behavior of G. The simplified (here one-dimensional) Newton-Raphson iteration then does not have second-order convergence behavior with respect to composition corrections; convergence can be slow, especially for systems near the plait point. Although the full Newton-Raphson method, including composition derivatives, could be used in this case, it has been found more effective to use a simple linear acceleration method (Wegstein, 1955) applied to compositions, coupled with a Newton-Raphson iteration for α. This procedure is described later in this chapter.

It is important to stress that unnecessary thermodynamic function evaluations must be avoided in equilibrium separation calculations. Thus, for example, in an adiabatic vapor-liquid flash, no attempt should be made iteratively to correct compositions (and K_i's) at current estimates of T and α before proceeding with the Newton-Raphson iteration. Similarly, in liquid-liquid separations, iterations on phase compositions at the current estimate of phase ratio (α), or at some estimate of the conjugate phase composition, are almost always counterproductive. Each thermodynamic function evaluation (set of K_i) should be used to improve estimates of all variables in the system.

Bubble and Dew-Point Computation

For bubble and dew-point calculations we have, respectively, the objective functions

$$\sum_{i=1}^{m} K_i w_i = 1 \qquad (7\text{-}15)$$

and

$$\sum_{i=1}^{m} \frac{w_i}{K_i} = 1 \qquad (7\text{-}16)$$

with the single unknown variable either the temperature (at given pressure) or the pressure (at given temperature).

The Newton-Raphson approach, being essentially a point-slope method, converges most rapidly for near linear objective functions. Thus it is helpful to note that K_i tends to vary as $1/P$ and as $\exp(1/T)$. For bubble-point-temperature calculation, we can define an objective function

$$G\left(\frac{1}{T}\right) \equiv \ln\left\{\sum_{i=1}^{m} K_i w_i\right\} = 0 \qquad\qquad (7\text{-}23)$$

that has more nearly linear behavior than does Equation (7-15) as a function of T. Similarly, for dew-point temperature

$$G\left(\frac{1}{T}\right) \equiv \ln\left\{\sum_{i=1}^{m} \frac{w_i}{K_i}\right\} = 0. \qquad\qquad (7\text{-}24)$$

For bubble-point and dew-point pressure calculations, the appropriate forms are, respectively*

$$G\left(\frac{1}{P}\right) \equiv \sum_{i=1}^{m} K_i w_i - 1 = 0 \qquad\qquad (7\text{-}25)$$

and

$$G(P) \equiv \sum_{i=1}^{m} \frac{w_i}{K_i} - 1 = 0. \qquad\qquad (7\text{-}26)$$

In application of the Newton-Raphson iteration to these objective functions [Equations (7-23) through (7-26)], the near linear nature of the functions makes the use of step-limiting unnecessary.

The bubble and dew-point temperature calculations have been implemented by the FORTRAN IV subroutine BUDET and the pressure calculations by subroutine BUDEP, which are described and listed in Appendix F. These subroutines calculate the unknown temperature or pressure, given feed composition and the fixed pressure or temperature. They provide for input of initial estimates of the temperature or pressure sought, but converge quickly from any estimates within the range of validity of the thermodynamic framework. Standard initial estimates are provided by the subroutines.

---

*At low or moderate pressures, a Newton-Raphson iteration is not required, and the bubble and dew-point pressure iteration can be, respectively,

$$P^{(r+1)} = P^{(r)} \sum_{i=1}^{m} K_i w_i$$

and

$$P^{(r+1)} = P^{(r)} \Big/ \sum_{i=1}^{m} \frac{w_i}{K_i}.$$

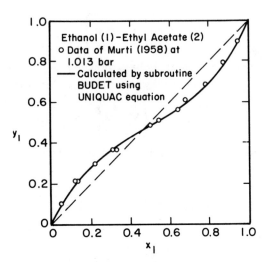

Figure 7-1. Incipient equilibrium vapor-phase compositions calculated with subroutine BUDET.

Convergence is usually accomplished in 2 to 4 iterations. For example, an average of 2.6 iterations was required for 9 bubble-point-temperature calculations over the complete composition range for the azeotropic system ehtanol-ethyl acetate. Standard initial estimates were used. Figure 1 shows results for the incipient vapor-phase compositions together with the experimental data of Murti and van Winkle (1958). For this case, calculated bubble-point temperatures were never more than 0.4 K from observed values.

Vapor-Liquid Equilibrium Separation Computation

The vapor-liquid equilibrium separation calculations considered here are for two cases, isothermal and adiabatic, both at fixed pressure.

For the isothermal flash, the step-limited Newton-Raphson iteration is applied to the single objective function

$$G(\underline{x},\underline{y},\alpha) = \sum_{i=1}^{m} \frac{(K_i - 1)w_i}{(K_i - 1)\alpha + 1} = 0 \qquad (7\text{-}13)$$

to determine new estimates of the vaporization α. For this purpose, an analytic derivative of G with respect to α is used.

$$\left(\frac{\partial G}{\partial \alpha}\right)_{K_i} = -\sum_{i=1}^{m} \frac{1}{w_i} \left[\frac{(K_i - 1)w_i}{(K_i - 1)\alpha + 1}\right]^2 . \tag{7-27}$$

Equations (7-8) and (7-9) are then used to calculate the compositions, which are normalized and used in the thermodynamic subroutines to find new equilibrium ratios, K_i.* These K_i values are then used in the next Newton-Raphson iteration. The iterative process continues until the magnitude of the objective function $|G|$ is less than a convergence criterion, ε. If initial estimates of $\underline{x}$, $\underline{y}$, and α are not provided externally (for instance from previous calculations of the same separation under slightly different conditions), they are taken to be

$$x_i = y_i = w_i \tag{7-28}$$

and

$$\alpha = \frac{T - T_B}{T_D - T_B} \tag{7-29}$$

where T_B and T_D are the bubble and dew-point temperatures of the feed.

In the case of the adiabatic flash, application of a two-dimensional Newton-Raphson iteration to the objective functions represented by Equations (7-13) and (7-14), with $Q/F = 0$, is used to provide new estimates of α and T simultaneously. The derivatives with respect to α in the Jacobian matrix are found analytically while those with respect to T are found by finite-difference approximation

$$\left[\frac{\partial G_1}{\partial T}\right]_{\alpha,\underline{x},\underline{y}} \simeq \frac{G_1(T + \Delta T, \underline{x}, \underline{y}, \alpha) - G_1(T, \underline{x}, \underline{y}, \alpha)}{\Delta T} . \tag{7-30}$$

Again, Equations (7-8) and (7-9) are then used to calculate new compositions. These compositions, normalized, and the new value for T are utilized in thermodynamic subroutine calls to find equilibrium ratios and enthalpies for use in the next iteration.

---

*Normalization simply replaces x_i by $x_i \Big/ \sum_{i=1}^{m} x_i$.

121

Convergence of the iteration requires the norm of the objective vector $|\underline{G}|$ to be less than the convergence criterion, ε.* The initial estimates used, if not provided externally, are, in addition to Equation (7-28)

$$\alpha = \frac{h^F - h^L(\underline{w}, T_B)}{h^V(\underline{w}, T_D) - h^L(\underline{w}, T_B)} \qquad (7\text{-}31)$$

and

$$T = T_B + \alpha(T_D - T_B) \qquad (7\text{-}32)$$

where all enthalpies are evaluated at feed composition.

Both vapor-liquid flash calculations are implemented by the FORTRAN IV subroutine FLASH, which is described and listed in Appendix F. This subroutine can accept vapor and liquid feed streams simultaneously. It provides for input of estimates of vaporization, vapor and liquid compositions, and, for the adiabatic calculation, temperature, but makes its own initial estimates as specified above in the absence (0 values) of the external estimates. No cases have been encountered in which convergence is not achieved from internal initial estimates.

The convergence rate depends somewhat on the problem and on the initial estimates used. For mixtures that are not extremely wide-boiling, convergence is usually accomplished in three or four iterations,† even in the presence of relatively strong liquid-phase nonidealities. For example, cases 1 through 4 in Table 1 are typical of relatively close-boiling mixtures; the latter three exhibit significant liquid-phase nonidealities. Cases 3 and 4 show strong vapor-phase nonidealities as well.

---

*A value of ε of 10^{-3} gives solutions that do not change significantly upon further reduction of ε.

†Each iteration of an isothermal calculation requires only one call of the thermodynamic subroutine VALIK. Each iteration of an adiabatic calculation requires two calls of VALIK plus two calls of the enthalpy subroutine ENTH for each phase. Additional iterations give no improvement in accuracy.

Table 7-1

Examples of Vapor-Liquid Separation Calculations
Conducted with Subroutine FLASH

Case Number	Flash Type	Components	Feed Mole Fraction	Feed Pressure (bar)	Feed Temperature (K)	Flash Pressure (bar)	Flash Temperature (K)	V/F	Mole Fractions Liquid	Mole Fractions Vapor	No. of Iterations
1	Adia-batic	Benzene	0.250	15.0	420.0	3.0	387.0	0.252	0.251	0.246	3
		Cyclohexane	0.250						0.262	0.216	
		Methylcyclo-hexane	0.250						0.246	0.261	
		n-Hexane	0.250						0.241	0.277	
2	Adia-batic	Ethanol	0.250	30.0	450.0	10.0	425.3	0.160	0.207	0.474	4
		n-Propanol	0.150						0.148	0.160	
		Cyclohexane	0.200						0.203	0.184	
		Toluene	0.400						0.442	0.182	
3	Iso-thermal	Formic acid	0.350	--	--	1.0	381.0	0.782	0.346	0.351	3
		Acetic acid	0.350						0.383	0.341	
		Water	0.300						0.271	0.308	
4	Adia-batic	Acetic acid	0.020	10.0	425.0	1.0	356.0	0.167	0.023	0.004	5
		Ethanol	0.100						0.074	0.240	
		Ethyl acetate	0.080						0.042	0.272	
		Water	0.800						0.863	0.484	
5*	Adia-batic	n-Butane	0.250	15.0	470.0	7.0	442.9	0.409	0.112	0.450	6
		i-Butane	0.150						0.065	0.273	
		n-Pentane	0.100						0.067	0.148	
		n-Decane	0.500						0.756	0.129	
6	Isothermal or (Adia-batic)**	Hydrogen	0.100	(30.4)	(298.0)	30.4	298.0 (297.7)	0.285	0.002	0.345	3 (5)
		Carbon monoxide	0.100						0.006	0.337	
		Methane**	0.100						0.017	0.308	
	Fed as vapor & liquid streams	Acetone	0.350						0.487	0.007	
		Benzene	0.350						0.488	0.003	

*UNIQUAC interaction parameters were not determined, but were assumed to be zero; a reasonable first approximation for this system. **Quantities in parentheses refer to adiabatic flash.

Calculations for wide-boiling mixtures are a little more difficult to converge, especially for mixtures having very light or noncondensable components together with relatively nonvolatile components and lacking components of intermediate volatility. Flash calculations for these mixtures usually require four to eight iterations. Cases 5 and 6 in Table 1 have feeds of this type, including noncondensable components in Case 6. Within the limits of the thermodynamic framework used here, no case has been encountered where FLASH has required more than 12 iterations for satisfactory convergence.

Liquid-Liquid Separation Computation

Liquid-liquid equilibrium separation calculations are superficially similar to isothermal vapor-liquid flash calculations. They also use the objective function, Equation (7-13), in a step-limited Newton-Raphson iteration for α, which is here E/F. However, because of the very strong dependence of equilibrium ratios on phase compositions, a computation as described for isothermal flash processes can converge very slowly, especially near the plait point. (Sometimes 50 or more iterations are required.)

We have repeatedly observed that the slowly converging variables in liquid-liquid calculations following the isothermal flash procedure are the mole fractions of the two solvent components* in the conjugate liquid phases. In addition, we have found that the mole fractions of these components, as well as those of the other components, follow roughly linear relationships with certain measures of deviation from equilibrium, such as the differences in component activities (or fugacities) in the extract and the raffinate.

---

*The solvent components usually have a low mutual solubility and are present in reasonably large mole fractions in the system. If solvents are not so designated, we take as the "solvent components" those two components, present in significant mole fraction in the system, that have the lowest binary solubilities.

For liquid-liquid separations, the basic Newton-Raphson iteration for α is converged for equilibrium ratios (K_i) determined at the previous composition estimate. (It helps, and costs very little, to converge this iteration quite tightly.) Then, using new compositions from this converged inner iteration loop, new values for equilibrium ratios are obtained. This procedure is applied directly for the first three iterations of composition. If convergence has not occurred after three iterations, the mole fractions of all components in both phases are accelerated linearly with the deviation function

$$\sum_{i=1}^{m} |\gamma_i^E x_i^E - \gamma_i^R x_i^R| \qquad (7-33)$$

using the method described by Wegstein (1955). This acceleration is applied at alternate iterations, and takes place during calculation of the new compositions. These compositions are used in thermodynamic evaluations to obtain new values for γ_i^E, γ_i^R and $K_i = \gamma_i^R/\gamma_i^E$. With this modification, convergence is usually achieved in less than ten iterations for cases with compositions of conjugate phases reasonably far from a plait-point region.

This computation procedure has been implemented by FORTRAN IV subroutine ELIPS, which is described and listed in Appendix G. This subroutine provides for designation of "solvent" components; if not designated, they are determined internally.

The rate of convergence is strongly affected by the distance of the feed composition from any plait-point region. For calculations yielding conjugate phases far removed from the plait point, or where no plait point exists (type-II systems), convergence is usually attained in six to eight iterations.* Cases 1 and 5 in Table 2 for type-I ternary systems, and Cases 3 and 4 for a type-II ternary system, represent calculations of this kind. Figure 2 shows the results of separation calculations for the ternary system water-acrylonitrile-acetonitrile compared with the data of Volpicelli (1968). Quarternary (and higher-order) systems, as in Cases 7 through 9, show similar convergence behavior.

*Each iteration requires only one call of the thermodynamic liquid-liquid subroutine LILIK. The inner iteration loop requires no thermodynamic subroutine calls; thus is uses extremely little computation effort.

125

Table 7-2

Examples of Liquid-Liquid Equilibrium Saturation
Calculations Conducted with Subroutine ELIPS

Case Number	Components	Feed Mole Fraction	Feed Temperature (K)	Separation E/R	Separation Mole Fractions R Phase	Separation Mole Fractions E Phase	No. of iterations
1	Water	0.600	333	0.477	0.952	0.214	7
	Acrylonitrile	0.350			0.034	0.696	
	Acetonitrile	0.050			0.014	0.090	
2*	Water	0.600	333	0.644	0.837	0.469	14
	Acrylonitrile	0.150			0.043	0.209	
	Acetonitrile	0.250			0.120	0.322	
3	Furfural	0.400	298	0.600	0.914	0.058	6
	2,2,4-trimethyl-						
	pentane	0.300			0.028	0.481	
	Cyclohexane	0.300			0.058	0.461	
4	Furfural	0.100	298	0.949	0.872	0.059	6
	2,2,4-Trimethyl-						
	pentane	0.100			0.008	0.105	
	Cyclohexane	0.800			0.120	0.836	
5	Furfural	0.400	298	0.541	0.739	0.112	7
	2,2,4-Trimethyl-						
	pentane	0.400			0.080	0.672	
	Benzene	0.200			0.181	0.216	
6*	Furfural	0.400	298	0.402	0.537	0.197	12
	2,2,4-Trimethyl-						
	pentane	0.270			0.144	0.457	
	Benzene	0.330			0.319	0.346	
7	Furfural	0.400	298	0.569	0.814	0.086	7
	2,2,4-Trimethyl-						
	pentane	0.300			0.044	0.494	
	Benzene	0.100			0.088	0.109	
	Cyclohexane	0.200			0.054	0.311	
8	Furfural	0.400	298	0.557	0.788	0.091	7
	2,2,4-Trimethyl-						
	pentane	0.200			0.033	0.333	
	Benzene	0.100			0.090	0.108	
	Cyclohexane	0.300			0.089	0.468	
9	Furfural	0.300	298	0.686	0.604	0.161	10
	2,2,4-Trimethyl-						
	pentane	0.100			0.032	0.131	
	Benzene	0.200			0.187	0.206	
	Cyclohexane	0.400			0.177	0.502	

*Feed conditions in the region of the plait point of type I system.

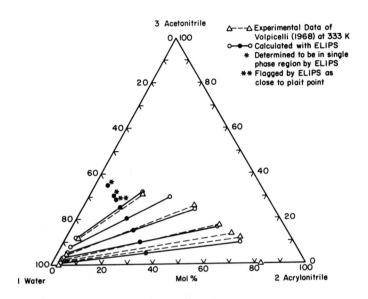

Figure 7-2. Conjugate liquid phase compositions for water-acrylonitrile-acetonitrile system calculated with subroutine ELIPS for feeds shown by ● .

As the feed composition approaches a plait point, the rate of convergence of the calculation procedure is markedly reduced. Typically, 10 to 20 iterations are required, as shown in Cases 2 and 6 for ternary type-I systems. Very near a plait point, convergence can be extremely slow, requiring 50 iterations or more. ELIPS checks for these situations, terminates without a solution, and returns an error flag (ERR=7) to avoid unwarranted computational effort.** This is not a significant disadvantage since liquid-liquid separations are not intentionally conducted near plait points.

Outside the two-phase region, ELIPS yields a value of 0 for E/F on the R-phase side and 1 for E/F on the E-phase side. Convergence to these values again requires about eight or fewer iterations, except near the plait-point region where convergence is somewhat slower.

**The criterion used for "too near the plait point" is that ratio of K's for the two "solvent" components is less than seven with the feed composition in the two-phase region.

The calculational procedure employed in ELIPS, when used with the particular initial phase-composition estimated included in the subroutine, has converged satisfactorily for all systems we have encountered (except very near plait points as noted). The subroutine is well suited to the typical problems of liquid-liquid separation calculations wehre good estimates of equilibrium phase compositions are not available.* However, if very good initial estimates of conjugate-phase compositions are available à priori, more effective procedures, with second-order convergence, can probably be developed for special applications such as tracing the entire boundary of a two-phase region.

---

*In the highly nonlinear equilibrium situations characteristic of liquid separations, the use of à priori initial estimates of phase compositions that are not very close to the true compositions of these phases can lead to divergence of iterative computations or to spurious convergence upon feed composition. The special estimates used in ELIPS, which are essentially pure phases of the "solvent components" (98%, with 2% of the other solvent) are chosen to avoid these problems in this iterative procedure.

References

Friday, J. R., Smith, P. D., A.I.Ch.E. J., 10, 698 (1964).

Goldstein, R. P., Stanfield, R. B., Ind. Eng. Chem. Process Des. Dev., 9, 78 (1970).

Hanson, D. H., Duffin, J. H., Somerville, G. F., "Computation of Multistage Separation Processes," Reinhold, New York (1962).

Holland, C. D., "Fundamentals and Modeling of Separation Processes," Prentice-Hall, Englewood Cliffs, N.J. (1975).

Law, V. J., Bailey, R. V., Chem. Eng. Sci., 18, 189 (1967).

Murti, P. S., Van Winkle, M., J. Chem. Eng. Data, 3, 72 (1958).

Naphtali, L. M., Sandholm, D. P., A.I.Ch.E. J., 17, 148 (1971).

Naphtali, L. M., Chem. Eng. Progr., 60, 70 (1964).

Newman, J. S., Hydrocarbon Process. Petrol. Ref., 42, 141 (1963).

Null, H. R., "Phase Equilibria in Process Design," Wiley-Interscience (1970).

Rachford, H. H., Jr., Rice, J. D., J. Petrol. Technol., $\underline{4}$, sec. 1, p. 19, sec. 2, p. 3 (1952).

Renon, H., Asselineav, L., Cohen, G., Raimbault, C. "Cacul sur Ordinateur des Equilibres Liquide-Vapeur et Liquide-Liquide," Editions Technip, Paris (1971).

Ricker, N. L., Grens, E. A. II, A.I.Ch.E. J., $\underline{20}$. 238 (1974).

Seppala, R. E., Luus, R., J. Franklin Inst., $\underline{293}$, 325 (1972).

Smith, B. D., "Design of Equilibrium Stage Processes," McGraw-Hill, New York (1963).

Tomich, J. F., A.I.Ch.E. J., $\underline{16}$, 229 (1970).

Volpicelli, G., J. Chem. Eng. Data, $\underline{13}$, 150 (1968).

Wegstein, J. H., Commun. A.C.M., $\underline{1}$, 9 (1955).

Appendix A

CALCULATION OF VAPOR-PHASE NONIDEALITIES

In the generalized method of Hayden and O'Connell (1975),[†] the pure-component and cross second virial coefficients B_{ij} are given by the sum of two contributions

$$B_{ij} = B_{ij}^F + B_{ij}^D \qquad (A-1)$$

where

$$B_{ij}^F = (B_{nonpolar}^F)_{ij} + (B_{polar}^F)_{ij} \qquad (A-2)$$

$$B_{ij}^D = (B_{metastable})_{ij} + (B_{bound})_{ij} + (B_{chemical})_{ij} \qquad (A-3)$$

Here, superscript F denotes relatively "free" molecules (weak physical forces), and D denotes relatively "bound" or "dimerized" molecules ("chemical" forces).

Individual contributions to the second virial coefficient are calculated from temperature-dependent correlations:

$$(B_{nonpolar}^F)_{ij} = b_{0ij}(0.94 - \frac{1.47}{T_{ij}^{*'}} - \frac{0.85}{T_{ij}^{*'2}} - \frac{1.015}{T_{ij}^{*'3}}) \qquad (A-4)$$

$$(B_{polar}^F)_{ij} = -b_{0ij}\mu_{ij}^{*'}(0.74 - \frac{3.0}{T_{ij}^{*'}} + \frac{2.1}{T_{ij}^{*'2}} + \frac{2.1}{T_{ij}^{*'3}}) \qquad (A-5)$$

$$(B_{metastable})_{ij} + (B_{bound})_{ij} = b_{0ij}A_{ij}\ exp(\frac{\Delta h_{ij}}{T_{ij}^*}) \qquad (A-6)$$

$$(B_{chemical})_{ij} = b_{0ij}E_{ij}[1 - exp(\frac{1500\ \eta_{ij}}{T})] \qquad (A-7)$$

$$1/T_{ij}^{*'} = 1/T_{ij}^* - 1.6\ \omega_{ij} \qquad (A-8)$$

$$T_{ij}^* = T/(\varepsilon_{ij}/k) \qquad (A-9)$$

[†]See References, Chapter 3

Temperature-independent parameters used in Equations (4)-(9) are

$$b_{0ij} = 1.26184 \; \sigma_{ij}^3 \quad (cm^3/g\text{-}mole) \tag{A-10}$$

$$
\begin{aligned}
\mu_{ij}^{*\,\prime} &= \mu_{ij}^{*} & \mu_{ij}^{*} &< 0.04 \\
&= 0 & 0.04 &\leq \mu_{ij}^{*} < 0.25 \\
&= \mu_{ij}^{*} - 0.25 & \mu_{ij}^{*} &\geq 0.25
\end{aligned}
\tag{A-11}
$$

$$A_{ij} = -0.3 - 0.05 \; \mu_{ij}^{*} \tag{A-12}$$

$$\Delta h_{ij} = 1.99 + 0.2 \; \mu_{ij}^{*2} \tag{A-13}$$

$$\mu_{ij}^{*} = \frac{7243.8 \; \mu_i \mu_j}{(\varepsilon_{ij}/k)\sigma_{ij}^3} \tag{A-14}$$

$$E_{ij} = \exp\{\eta_{ij}(\frac{650}{(\varepsilon_{ij}/k)+300} - 4.27)\} \quad \text{for } \eta_{ij} < 4.5 \tag{A-15a}$$

or

$$E_{ij} = \exp\{\eta_{ij}(\frac{42800}{(\varepsilon_{ij}/k)+22400.} - 4.27)\} \quad \text{for } \eta_{ij} \geq 4.5 \tag{A-15b}$$

where

T = temperature, K
(ε_{ij}/k) = characteristic energy for the i-j interaction, K
σ_{ij} = molecular size, Å
μ_i = dipole moment of component i, Debye
η_{ij} = association parameter (i=j); solvation parameter (i≠j)
ω_{ij} = nonpolar acentric factor

For i=j, parameters (ε_{ii}/k), σ_{ii}, and ω_{ii} are predicted from pure-component properties

$$\omega_{ii} = 0.006026 \; R_{D_i} + 0.02096 \; R_{D_i}^2 - 0.001366 \; R_{D_i}^3 \tag{A-16}$$

$$(\varepsilon_{ii}/k) = (\varepsilon_{ii}/k)'\{1 - \xi c_1 [1 - \frac{\xi(1+c_1)}{2}]\} \tag{A-17}$$

$$\sigma_{ii} = \sigma_{ii}'(1 + \xi c_2)^{1/3} \tag{A-18}$$

131

where

$$(\varepsilon_{ii}/k)' = T_{c_i}\left[0.748 + 0.91\,\omega_{ii} - \frac{0.4\eta_{ii}}{2 + 20\omega_{ii}}\right] \qquad \text{(A-19)}$$

$$\sigma'_{ii} = (2.44 - \omega_{ii})(1.0133 T_{c_i}/P_{c_i})^{1/3} \qquad \text{(A-20)}$$

$$\xi = 0 \quad \text{for } \mu_i < 1.45 \text{ or} \qquad \text{(A-21a)}$$

$$\xi = \frac{1.7941 \times 10^7 \mu_i^4}{[(2.882 - \frac{1.882\omega_{ii}}{0.03+\omega_{ii}})T_{c_i}\sigma'^6_{ii}(\varepsilon_{ii}/k)']} \quad \text{for } \mu_i \geq 1.45 \quad \text{(A-21b)}$$

$$c_1 = \frac{16 + 400\,\omega_{ii}}{10 + 400\,\omega_{ii}} \qquad \text{(A-22)}$$

$$c_2 = \frac{3}{10 + 400\,\omega_{ii}} \qquad \text{(A-23)}$$

Pure-component parameters required in Equations (16) through (23) are

T_{c_i} = critical temperature of component i, K

P_{c_i} = critical pressure of component i, bars

R_{D_i} = mean radius of gyration of component i, Å

Cross parameters (ε_{ij}/k), σ_{ij}, and ω_{ij} $(i{\neq}j)$ are calculated using suitable mixing rules and pure-component parameters given by Equations (16) through (23).

$$\omega_{ij} = \frac{1}{2}(\omega_{ii} + \omega_{jj}) \qquad \text{(A-24)}$$

$$(\varepsilon_{ij}/k) = (\varepsilon_{ij}/k)'(1 + \xi'c_1') \qquad \text{(A-25)}$$

$$\sigma_{ij} = \sigma'_{ij}(1 - \xi'c_2') \qquad \text{(A-26)}$$

where

$$(\varepsilon_{ij}/k)' = 0.7[(\varepsilon_{ii}/i)(\varepsilon_{jj}/k)]^{1/2} + \frac{0.6}{[1/(\varepsilon_{ii}/k)+(1/(\varepsilon_{jj}/k)]}$$

$$\text{(A-27)}$$

$$\sigma'_{ij} = (\sigma_{ii}\sigma_{jj})^{1/2} \tag{A-28}$$

$$\xi' = \frac{\mu_i^2(\epsilon_{jj}/k)^{2/3}\sigma_{jj}^4}{(\epsilon_{ij}/k)'\sigma_{ij}^6} \quad \text{for } \mu_i \geq 2 \text{ and } \mu_j = 0 \tag{A-29a}$$

or

$$\xi' = \frac{\mu_j^2(\epsilon_{kk}/k)^{2/3}\sigma_{ii}^4}{(\epsilon_{ij}/k)'\sigma_{ij}^6} \quad \text{for } \mu_j \geq 2 \text{ and } \mu_i = 0 \tag{A-29b}$$

or

$$\xi' = 0.0 \text{ for all other values of } \mu_i \text{ and } \mu_j \tag{A-29c}$$

$$c_1' = \frac{16 + 400\ \omega_{ij}}{10 + 400\ \omega_{ij}} \tag{A-30}$$

$$c_2' = \frac{3}{10 + 400\ \omega_{ij}} \tag{A-31}$$

Chemical Theory

As discussed in Chapter 3, the virial equation is suitable for describing vapor-phase nonidealities of nonassociating (or weakly associating) fluids at moderate densities. Equation (1) gives the second virial coefficient which is used directly in Equation (3-10b) to calculate the fugacity coefficients.

However, when carboxylic acids are present in a mixture, fugacity coefficients must be calculated using the chemical theory. Chemical theory leads to a fugacity coefficient dependent on true equilibrium concentrations, as shown by Equation (3-13).

Equilibrium constants, K_{ij}, for all possible dimerization reactions are calculated from the metastable, bound, and chemical contributions to the second virial coefficients, B_{ij}^D, as given by Equations (6) and (7). The equilibrium constants, K_{ij}, are calculated using Equation (3-15).

The total free contribution to the second virial coefficient, B_{ij}^F, given by Equations (2), (4), and (5), is used to calculate fugacity coefficients, $\phi_i^\ddagger$ and $\phi_{ij}^\ddagger$, for the monomers

133

and dimers in the equilibrium mixture. These fugacity coefficients take into account physical nonidealities due to interactions of the true species. Two assumptions are made: (1)
The Lewis fugacity rule is used for calculating the fugacity
coefficients of the true species, and (2) the second virial coefficients for monomer i and dimer ii are given by B_{ii}^F, while
the second virial coefficients for dimer ij are given by B_{ij}^F.

With these assumptions, we combine Equations (3-12), (3-14),
and (3-15) to obtain

$$K_{ij} \equiv \frac{1}{P} \frac{z_{ij}}{z_i z_j} \frac{\exp(B_{ij}^F P/RT)}{\exp(B_{ii}^F P/RT) \exp(B_{jj}^F P/RT)} = \frac{-B_{ij}^D (2 - \delta_{ij})}{RT} \quad (A-32)$$

For a given temperature and pressure, let

$$C_{ij} = -(2 - \delta_{ij}) B_{ij}^D (P/RT) \exp\{\frac{P}{RT}(B_{ii}^F + B_{jj}^F - B_{ij}^F)\} \quad (A-33)$$

Equation (32) then becomes

$$C_{ij} = \frac{z_{ij}}{z_i z_j} \quad (A-34)$$

The total number of stoichiometric (nonreacted) components
is designated by m; there are $m(m+1)/2$ equations of the form
given by Equations (33) and (34).

Using material balances for all species and the relations

$$\sum_{i=1}^{m} y_i = 1 \qquad \text{and} \qquad \sum_{i=1}^{m} z_i + \sum_{i=1}^{m} \sum_{j=1}^{i} z_{ij} = 1,$$

we obtain an equation for the apparent (stoichiometric) mole
fractions in terms of the true mole fractions.

$$y_i = \frac{z_i(1 + \sum_{j=1}^{i} z_{ij}/z_i + \sum_{j=i}^{m} z_{ji}/z_i)}{[1 + \frac{1}{2} \sum_{k=1}^{m} (\sum_{j=1}^{k} z_{kj} + \sum_{j=k}^{m} z_{jk})]} \quad (A-35)$$

We define

$$\lambda_i \equiv \sum_{j=1}^{i} z_{ij} + \sum_{j=i}^{m} z_{ji} = z_i \{ \sum_{j=1}^{i} C_{ij} z_j + \sum_{j=i}^{m} C_{ji} z_j \} \tag{A-36}$$

From Equation (35), an iteration function can be developed in the form

$$z_i^{(r+1)} = \frac{y_i \{ 1 + \frac{1}{2} \sum_{k=1}^{m} \lambda_k^{(r)} \}}{\{ 1 + \sum_{j=1}^{i} C_{ij} z_j^{(r)} + \sum_{j=i}^{m} C_{ji} z_j^{(r)} \}} \tag{A-37}$$

for each component. Superscript (r) indicates the rth iteration. A slight (20%) damping of this iterative procedure has been found to increase the rate of convergence.

Convergence of this iteration is influenced by initial estimates for the true mole fractions, z_i^0. The following rules have been found to lead to rapid convergence in all cases.

$$z_i^0 = y_i \text{ if } C_{ii} \text{ and } C_{ij} \leq 0.5 \text{ for all } j \tag{A-38}$$

This implies that species i does not solvate or associate significantly. Otherwise, we set z_i^0 equal to the minimum of one of the following for all j.

Either
$$z_i^0 = \frac{\sqrt{1+8C_{ii}y_i} - 1}{4C_{ii}} \text{ for } C_{ii} > 0.5 \tag{A-39}$$

or
$$z_i^0 = y_i (1+C_{ij}y_j)^{-1} \tag{A-40a}$$
$$\text{for } C_{ij} > 0.5 \text{ and } y_j > y_i$$
$$z_j^0 = y_j (1+C_{ij}z_i^0)^{-1} \tag{A-40b}$$

or
$$z_j^0 = y_j (1+C_{ij}y_i)^{-1} \tag{A-41a}$$
$$\text{for } C_{ij} > 0.5 \text{ and } y_i > y_j$$
$$z_i^0 = y_i (1+C_{ij}z_j^0)^{-1} \tag{A-41b}$$

These initial estimates are used in the iteration function, Equation (37), to obtain values of the z_i's that do not change significantly from one iteration to the next. These true mole fractions, with Equation (3-13), yield the desired fugacity

135

coefficients.

Vapor Enthalpy Corrections for
Associating Compounds

There are two enthalpy corrections for strongly associating vapors. The dominant term is due to the combined enthalpies of reaction of the stoichiometric species, Δh^D, to form the true species. The second correction, Δh^F, accounts for the physical interactions of these true species.

The enthalpy changes due to dimerization are determined from the van't Hoff relation. For a dimerization reaction between species i and j

$$\frac{\Delta h^D_{ij}}{R} = - \frac{d \ln K_{ij}}{d(1/T)} \qquad (A-42)$$

where Δh^D_{ij} is the enthalpy of reaction per mole of dimer i-j formed. When the expression for K_{ij} given by Equation (32) is substituted into Equation (42), we obtain

$$\Delta h^D_{ij} = -RT(1 - \frac{T}{B^D_{ij}} \frac{d B^D_{ij}}{dT}) \qquad (A-43)$$

The total enthalpy correction due to chemical reactions is the sum of all the enthalpies of dimerization for each i-j pair multiplied by the mole fraction of dimer i-j. Since this gives the enthalpy correction for one mole of true species, we multiply this quantity by the ratio of the true number of moles to the stoichiometric number of moles. This gives

$$\Delta h^D = -(\frac{n_t}{n_a}) RT \ [\sum_{i=1}^{m} \sum_{j=1}^{i} z_{ij} (1 - \frac{T}{B^D_{ij}} \frac{dB^D_{ij}}{dT})] \qquad (A-44)$$

where the quantity

$$(\frac{n_t}{n_a}) = \frac{1}{1 + \frac{1}{2} \sum_{i=1}^{m} (\sum_{j=1}^{i} z_{ij} + \sum_{j=i}^{m} z_{ji})} \qquad (A-45)$$

is determined from material balances over the true and stoichiometric species.

The enthalpy correction due to the physical interactions of

136

the true species, Δh^F, is calculated from Equation (5-7) and the virial equation of state. However, the virial coefficients for the true species are given only by the total free contribution, B^F_{ij}. The equation of state for the associated mixture is written

$$\frac{PV}{RT} = n_t + n_t \frac{B^f_{mix} P}{RT} \qquad (A-46)$$

where, consistent with previous assumptions

$$B^F_{mix} = \sum_{i=1}^{m} z_i B^F_{ii} + \sum_{i=1}^{m} \sum_{j=1}^{i} z_{ij} B^F_{ij} \qquad (A-47)$$

Equation (5-7) applied to Equation (46) yields

$$\Delta h^F = P[B^F_{mix} - T \frac{dB^F_{mix}}{dT}] \qquad (A-48)$$

which can be written

$$\Delta h^F = J(\frac{n_t}{n_t^a}) P[\sum_{i=1}^{m} z_i (B^F_i - T \frac{dB^F_{ii}}{dT}) + \sum_{i=1}^{m} \sum_{j=1}^{i} z_{ij} (B^F_{ij} - T\frac{dB^F_{ij}}{dT})] \qquad (A-49)$$

where J is the appropriate conversion factor, equal to 0.1000 joules/cm^3-bar. Equation (49) also includes the correction term (n_t/n_t^a); this yields a molar enthalpy correction which is based on one stoichiometric mole.

Equations (44) and (49) are the same as Equations (5-11a) and (5-11b).

Appendix B

STANDARD-STATE FUGACITIES AT ZERO PRESSURE

Correlations for standard-state fugacities at zero pressure, for the temperature range 200° to 600°K, were generated for pure fluids using the best available vapor-pressure data. The subsequent representations are probably reliable within the range of data used (always less broad than 200° to 600°K), but they are only approximations outside that range. The functions are, however, always monotonic in temperature, to provide appropriate corrections when iterative programs choose temperature excursions outside the range of data.

The correlations were generated by first choosing from the literature the best sets of vapor-pressure data for each fluid. These were converted from vapor pressure P^S to fugacity using the vapor-phase corrections (for pure components), discussed in Chapter 3; then the Poynting correction was applied to adjust to zero pressure:

$$f^{(P0)} = P^S \phi^S \exp \frac{1}{RT} \int_{P^S}^{0} v^L dP \qquad (B-1)$$

where $f^{(P0)}$ is the fugacity of pure liquid at zero pressure and ϕ^S is the fugacity coefficient at saturation, all at temperature T. The liquid molar volume v^L is assumed constant over the range of the integral. This volume was calculated using Rackett's equation as modified by Spencer and Danner (1972). The modified Rackett equation for the saturated-liquid molar volume as used here is

$$v = \frac{RT_c}{P_c} z_a^\tau \qquad (B-2)$$

where

$$\tau = 1 + (1 - T/T_c)^{2/7} \quad \text{for } T/T_c \leq 0.75 \qquad (B-3a)$$

$$\tau = 1.60 + 0.00693026 \ (T/T_c - 0.655)^{-1} \quad \text{for } T/T_c \geq 0.75 \quad (B-3b)$$

The saturated molar liquid volume and its temperature derivative are continuous at T/T_c at 0.75.

138

For a large number of liquids, the quantity Z_a is tabulated by Spencer and Danner. We list Z_a for 92 pure fluids in Appendix C. These values are sufficiently accurate for present purposes but are not reliable for high accuracy in molar volumes.

Extrapolated "data" were then generated from 200°K to the lowest real data point, and from the highest real data point (limited to $T_R \leq 0.85$) to 600°K. Real experimental data above $0.85T_c$ were eliminated; beyond that point, the vapor-phase correction, as calculated here, is inadequate and the liquid molar volume is no longer constant with pressure.

The low-pressure extrapolated data points were generated by linear extrapolation of the lowest 4-6 points on a plot of $\log f^{(P0)}$ versus $\frac{1}{T}$. This extrapolation is not highly precise, but guarantees a monotonic function.

At pressures above the highest real data point, the extrapolated data were generated by the correlation of Lyckman et al. (1965), modified slightly to eliminate any discontinuity between the real and generated data. This modification is small, only a few percent, well within the uncertainties of the Lyckman method. The Lyckman correlation was always used within its recommended limits of validity--that is, at reduced temperatures no greater than 1.5 to 2.0.

The data, both real and generated, were then fit to a function of the form

$$\ln f_i^{(P0)} = C_{1i} + C_{2i}/T + C_{3i}T + C_{4i}\ln T + C_{5i}T^2 \qquad (B-4)$$

where the C's are constants. For some liquids not all constants C were used. Equation (2) excludes any form of the Antoine equation which, although popular for fitting data over a narrow range, extrapolates very poorly over a wide span of temperature.

We use Equation (2) primarily with five parameters, or with four parameters, excluding C_5. When data were sparse or of poor precision, a linear two-parameter fit ($C_3 = C_4 = C_5 = 0$) was used. Fitting of the data was achieved by minimization of the objective function χ:

$$\chi = \sum_{\text{all points}} \frac{[f_{exp}^{(P0)}]^2 [\ln f_{exp}^{(P0)} - \ln f_{calc}^{(P0)}]^2}{\sigma^2} \qquad \text{(B-5)}$$

where σ is an estimated uncertainty for each datum. The estimated uncertainties which seemed to give the best fit of the equation to the experimental and the high-temperature generated data were:

Extrapolated Data at Low Temperature

$f^{(P0)} < 1.0$ mm Hg $\qquad\qquad\qquad \sigma = f^{(P0)}$

1.0 mm Hg $< f^{(P0)} < 10.0$ mm Hg $\quad \sigma = 0.1\ f^{(P0)}$

$f^{(P0)} > 10.0$ mm Hg $\qquad\qquad\qquad \sigma = 0.01\ f^{(P0)}$

Experimental Data

$P^S < 10$ mm Hg $\qquad\qquad\qquad\qquad \sigma = 0.2$ mm Hg

10 mm Hg $< P^S < 1$ atm $\qquad\qquad \sigma = 0.1$ mm Hg

$P^S > 1$ atm $\qquad\qquad\qquad\qquad\quad \sigma = 0.002\ f^{(P0)}$

Extrapolated Data at High Temperature

All points $\qquad\qquad\qquad\qquad\qquad \sigma = 0.05\ f^{(P0)}$

Large values of the uncertainty are assigned to the generated points, because the primary purpose of the generated points is not to have them fit accurately, but rather to maintain a reasonable slope of the function in the range outside the experimental points.

Large errors in the low-pressure points often have little effect on phase-equilibrium calculations; e.g., when the pressure is a few millitorr, it usually does not matter if we are off by 100 or even 1000%. By contrast, the high-pressure end should be reliable; large errors should be avoided when the data are extrapolated beyond the critical temperature.

Appendix C presents the best set of constants for Equation (2). Also shown are the temperature limits of the real experimental data. Users must exercise caution when using the correlation outside the range of real data; such use should, in general, be avoided.

Judgment had to be exercised in data selection. For each fluid, all available data were first fit simultaneously and second, in groups of authors. Data that were obviously very old, data that were obviously in error, and data that were inconsistent with the rest of the data, were removed.

Using what appeared to be the most reliable data available, we investigated the propriety of fitting with 2, 4, and 5-parameter forms of Equation (2) as well as the effect of different uncertainty factors σ. The two-parameter form is useful only when the data are sparse and scattered. The differences between the four-parameter form and five-parameter form are not large but, in general, the five-parameter form appears to give superior fits and superior values of the derivative, thus giving better agreement with experimental enthalpies for those cases where good data were available.

However, because the differences are not large, there are some cases where a four-parameter fit was used instead of a five-parameter fit, to avoid maxima or minima with respect to temperature.

Our choice of σ comes from the following: it is likely that, because the experiments are simple, experimental measurements between 10 millimeters Hg and one atmosphere have high reliability; we presume these to have an uncertainty no greater than 0.1 torr. Experimental measurements below 10 torr may be of lower accuracy, due to the difficulties of making measurements at low pressure; for these we have assigned $\sigma = 0.2$ torr. For experimental measurements above one atmosphere, it appears likely that the measurements were made using a precise Bourden tube gauge, such as a Heise gauge or a deadweight tester. These measurements are somewhat less accurate than those one can achieve with a mercury column; for these higher-pressure data we have assigned an uncertainty of 0.2%, which corresponds to about the precision obtainable in the middle range of a Heise gauge.

Below the temperature of the lowest experimental datum, standard-state fugacities were obtained by simple extrapolation. Uncertainties assigned to these fugacities are largest when the fugacities are smallest, for two reasons: (1) the extrapolation

141

is farther, and (2) a large uncertainty in a very low fugacity range often has little effect on vapor-liquid equilibrium calculations. The uncertainties used are 100% of the standard-state fugacities below 1 millimeter, and 1% of standard-state fugacities above 10 millimeters.

At temperatures above those corresponding to the highest experimental pressures, data were generated using the Lyckman correlation; all of these were assigned an uncertainty of 5% of the standard-state fugacity at zero pressure. Frequently, this uncertainty amounts to one half or more atmosphere for the lowest point, and to 1 to 5 atmospheres for the highest point.

Because the precision assigned to the upper and lower extrapolated points is relatively poor, it is possible to obtain a maximum or minimum in the curve, even when fitting all real and extrapolated data from 200° to 600°C. Extrema can occur anywhere, but generally they occur very close to either the lower or the upper end. A check of the sign of the slope at 200°C and 600°C easily indicated the presence of an extremum. When an extremum occurred, a new fit was established to avoid it.

References

Lyckman, E. W., Eckert, C. A., Prausnitz, J. M., Chem. Eng. Sci. 20, 685 (1965).

Spencer, C. F., Danner, R. P., J. Chem. Eng. Data, 17, 236 (1972).

Appendix C

TABLES OF PARAMETERS

Appendix C presents properties and parameters for 92 pure fluids and characteristic binary-mixture parameters for 150 binary pairs.

Appendix C-1 gives corresponding-state parameters and UNI-QUAC surface and volume parameters.

CODE = six-letter abbreviated name
MOL WT = molecular weight
TC = critical temperature, K
PC = critical pressure, bars
VC = critical volume, cm^3/g-vol
VSTR = O'Connell characteristic volume parameter, cm^3/g-mol[†]
ZRA = Rackett equation parameter
RD = mean radius of gyration, Å
DM = dipole moment, D
R = UNIQUAC r
Q = UNIQUAC q
QP = UNIQUAC q'

Appendix C-2 gives constants for the zero-pressure, pure-liquid, standard-state fugacity equation for condensable components and constants for the hypothetical liquid standard-state fugacity equation for noncondensable components

$$\ln(f^{(PO)}) \text{ or } \ln(f^{OL}) = C1 + C2/T + C3*T + C4*\ln(T) + C5*T^2$$

with fugacity in bars and T in kelvins.

Appendix C-3 gives constants for the ideal-gas, heat-capacity equation

---

[†]VSTR is useful for estimating partial molar volumes at infinite dilution but is not used here because of Equation (4-17).

$$C_p = D1 + D2/T + D3*T + D4*\ln(t)$$

with C_p in J/mol-K and T in kelvin.

Appendix C-4 gives association and solvation parameters for 92 fluids. Many of these parameters are estimated.

Appendix C-5 lists selected UNIQUAC binary parameters and characteristic binary parameters for noncondensable-condensable interactions for 150 binary pairs. For any binary pair, the parameters shown are believed to be the best now available. Parameters listed here were chosen from the more extensive lists in Appendix C-6 and C-7. A12 and A21 correspond to the UNIQUAC parameters, a_{12} and a_{21}, for systems not flagged with asterisk. The asterisk indicates a noncondensable component, and the parameters for these systems are those used in Equation (4-21); A12 = $\delta_{12}^{(0)}$ and A21 = $\delta_{12}^{(1)}$.

Appendix C-6 gives parameters for all the condensable binary systems we have here investigated; literature references are also given for experimental data. Parameters given are for each set of data analyzed; they often reflect in temperature (or pressure) range, number of data points, and experimental accuracy. Best calculated results are usually obtained when the parameters are obtained from experimental data at conditions of temperature, pressure, and composition close to those where the calculations are performed. However, sometimes, if the experimental data at these conditions are of low quality, better calculated results may be obtained with parameters obtained from good experimental data measured at other conditions.

   A12 = UNIQUAC binary interaction parameter a_{12}, K
   A21 = UNIQUAC binary interaction parameter a_{21}, K
   VL = vapor-liquid equilibrium data
   MS = mutual solubility data
   AZ = azeotropic data
VARIANCE OF FIT = (sum of squared, weighted residuals)/(number of
         degrees of freedom)

Appendix C-7 gives interaction parameters for noncondensable components with condensable components. (These are also included in Appendix C-5). Binary data sources are given.

Appendix C-1

Corresponding-State Parameters and
UNIQUAC Surface and Volume Parameters

* * PURE COMPONENT DATA * *

NO	NAME	CODE	FORMULA	MOL WT	TC K	PC BARS	VC CC/MOL	VSTR CC/MOL	ZRA	RD A	DM D	UNIQUAC PARAMETERS R	UNIQUAC PARAMETERS Q
1	HYDROGEN	HYDRGN	H2	4.003	43.60	20.80	51.5	51.5	0.0000	0.371	0.00	0.00	0.00
2	NITROGEN	NTRGN	N2	28.016	126.26	33.99	90.1	90.1	0.0000	0.547	0.00	0.00	0.00
3	OXYGEN	OXYGEN	O2	32.000	154.76	50.82	74.4	73.4	0.0000	0.604	0.00	0.00	0.00
4	CARBON MONOXIDE	CRBMOX	CO	28.010	132.93	34.99	93.1	93.1	0.0000	0.558	0.13	0.00	0.00
5	ARGON	ARGON	AR	39.944	150.71	48.65	75.2	74.6	0.0000	0.000	0.00	0.00	0.00
6	METHANE	METHAN	CH4	16.040	190.58	41.91	99.5	99.5	0.0000	1.123	0.00	0.00	0.00
7	CARBON TETRACHLORIDE	CRBTCL	CCL4	153.840	556.30	45.57	276.0	276.0	0.2721	3.458	0.00	3.33	2.82
8	CHLOROFORM	CLRFRM	CHCL3	119.390	536.54	54.74	240.0	219.6	0.2748	3.178	1.02	2.70	2.34
9	DICHLOROMETHANE	DCLMTN	CH2CL2	84.930	510.37	60.67	193.0	168.8	0.2620	2.432	1.54	2.14	1.92
10	FORMIC ACID	FRMACD	CH2O2	46.030	574.00	61.00	130.0	120.0	0.1880	1.480	1.52	1.54	1.48
11	NITRO-METHANE	NTRMTN	CH3NO2	61.040	588.00	63.10	173.0	150.0	0.2305	2.306	3.44	2.01	1.87
12	METHANOL	MTHNOL	CH4O	32.040	512.58	80.94	118.0	101.5	0.2318	1.536	1.71	1.43	1.43
*13	CARBON DIOXIDE	CRBDOX	CO2	44.010	241.00	53.80	105.0	80.0	0.2000	0.992	0.18	1.32	1.28
14	CARBON DISULFIDE	CRBDSF	CS2	76.130	552.00	79.00	170.0	165.0	0.2800	1.424	0.00	2.08	1.81
15	TRICHLOROETHYLENE	TCLETN	C2HCL3	131.390	571.00	46.20	249.0	258.9	0.2600	3.759	1.10	3.31	2.86
*16	ACETYLENE	ACTYLN	C2H2	26.040	244.30	44.70	130.0	130.0	0.1900	1.095	0.00	1.52	1.39
17	ACETONITRILE	ACTNTR	C2H3N	41.050	548.00	48.30	173.0	150.0	0.2000	1.821	3.94	1.87	1.72
18	ETHYLENE	ETHLYN	C2H4	28.050	282.36	50.32	124.0	127.3	0.2810	1.538	0.00	1.57	1.49
19	1.2-DICHLOROETHANE	DCLETN	C2H4CL2	98.970	560.93	53.71	225.0	226.0	0.2660	2.851	0.00	2.88	2.52
20	ACETALDEHYDE	ACTALD	C2H4O	60.050	461.00	55.40	169.0	150.0	0.2380	2.021	2.70	1.90	1.80

* * PURE COMPONENT DATA * *

NO	NAME	CODE	FORMULA	MOL WT	TC K	PC BARS	VC CC/MOL	VSTR CC/MOL	ZRA	RD A	DM D	UNIQUAC PARAMETERS R	Q
21	ACETIC ACID	ACTACD	C2H4O2	60.050	594.80	57.85	171.0	158.0	0.2240	2.595	1.74	2.23	2.04
22	ETHYL IODIDE	ETHIOD	C2H5I	155.970	550.00	56.00	235.0	225.0	0.2775	2.930	1.90	2.84	2.38
23	NITRO-ETHANE	NTRETH	C2H5NO2	75.070	600.00	49.70	228.0	210.0	0.2350	2.760	3.62	2.68	2.41
24	ETHANE	ETHANE	C2H6	30.070	305.42	48.80	148.0	158.0	0.2789	1.831	0.00	1.80	1.70
25	ETHANOL	ETHNOL	C2H6O	46.070	516.26	63.80	167.0	154.0	0.2520	2.250	1.69	2.11	1.97
26	DIMETHYL AMINE	DMTAMN	C2H7N	45.080	437.80	53.10	187.0	180.0	0.2645	2.264	1.03	2.33	2.09
27	ETHYLENE GLYCOL	ETHGLY	C2H6O2	62.070	646.00	75.50	186.0	168.4	0.2500	2.470	2.20	2.41	2.25
28	PROPYLENE	PRPYLN	C3H6	56.100	364.76	48.19	181.0	181.0	0.2785	2.228	0.37	2.25	2.02
29	ACETONE	ACETON	C3H6O	58.800	509.10	47.60	211.0	200.0	0.2470	2.740	2.86	2.57	2.34
30	METHYL ACETATE	MTHACT	C3H6O2	74.080	506.90	46.90	228.0	220.0	0.2560	2.862	1.72	2.80	2.58
31	PROPIONIC ACID	PRPACD	C3H6O2	74.080	612.70	53.70	230.0	220.0	0.2500	3.050	1.75	2.90	2.58
32	1-NITROPROPANE	NTPRP1	C3H7NO2	89.090	610.00	42.10	283.0	270.0	0.2400	3.150	3.75	3.36	2.95
33	2-NITROPROPANE	NTPRP2	C3H7NO2	89.090	605.00	43.00	279.0	270.0	0.2470	3.170	3.76	3.36	2.95
34	PROPANE	PROPAN	C3H8	44.090	369.82	42.49	200.0	200.0	0.2763	2.426	0.00	2.48	2.24
35	N-PROPANOL	NPRPNL	C3H8O	60.090	536.71	51.70	218.2	210.0	0.2485	2.736	1.68	2.78	2.51
36	I-PROPANOL	IPRPNL	C3H8O	60.090	508.32	47.64	220.4	216.0	0.2540	2.726	1.66	2.78	2.51
37	TRIMETHYLAMINE	TRMTAM	C3H9N	59.110	433.30	40.70	254.0	250.0	0.2690	2.736	0.61	2.99	2.64
38	BUTENE-1	BUTENE	C4H8	56.100	419.32	40.20	240.0	240.0	0.2736	2.746	0.34	2.92	2.56
39	2-BUTANONE	BTNON2	C4H8O	72.100	535.60	41.50	257.0	250.0	0.2510	3.139	2.70	3.25	2.88
40	TETRAHYDROFURAN	TTRHFR	C4H8O	72.100	541.00	51.90	224.0	237.0	0.2675	2.600	1.63	2.94	2.40

* * PURE COMPONENT DATA * *

NO	NAME	CODE	FORMULA	MOL WT	TC K	PC BARS	VC CC/MOL	VSTR CC/MOL	ZRA	RD A	DM D	UNIQUAC PARAMETERS R	Q
41	DIOXANE	DIOANE	C4H8O2	88.100	586.00	51.40	238.0	238.0	0.2660	3.110	0.00	3.07	2.28
42	ETHYL ACETATE	ETHACT	C4H8O2	88.100	523.30	38.30	286.0	280.0	0.2538	3.348	1.78	3.48	3.12
43	N-BUTANE	NBTANE	C4H10	58.120	425.18	37.97	255.0	255.0	0.2728	2.889	0.00	3.15	2.78
44	I-BUTANE	IBTANE	C4H10	58.120	408.14	36.48	263.0	263.0	0.2750	2.896	0.13	3.15	2.77
45	N-BUTANOL	NBTNOL	C4H10O	74.120	562.93	44.13	274.6	265.0	0.2590	3.225	1.66	3.45	3.05
46	I-BUTANOL	IBTNOL	C4H10O	74.120	547.73	42.95	272.2	263.0	0.2590	3.140	1.64	3.45	3.05
47	SEC-BUTANOL	SBTNOL	C4H10O	74.120	535.95	41.94	269.0	263.0	0.2565	3.182	1.66	3.45	3.05
48	TERT-BUTANOL	TBTNOL	C4H10O	74.120	506.15	39.72	274.5	269.0	0.2570	3.019	1.62	3.45	3.05
49	DI-ETHYL ETHER	DETETH	C4H10O	74.120	465.80	36.10	274.0	281.2	0.2650	3.140	1.17	3.39	3.02
50	ETHYL CELLOSOLVE	ETHCEL	C4H10O2	90.120	570.00	41.80	296.0	280.0	0.2570	3.310	3.00	3.70	3.29
51	DIETHYL AMINE	DETAMN	C4H11N	73.140	496.70	37.10	301.0	290.0	0.2630	3.250	0.92	3.68	3.17
52	FURFURAL	FRFRAL	C5H4O2	96.090	657.00	48.70	268.0	250.0	0.2430	3.170	2.30	3.17	2.48
53	PYRIDINE	PYRIDN	C5H5N	79.100	620.00	56.40	254.0	240.0	0.2660	3.050	2.20	3.00	2.16
54	ISOPRENE	ISOPRN	C5H8	68.120	484.00	37.40	266.0	266.0	0.2610	3.300	0.30	3.36	3.01
55	CYCLOPENTANE	CYCPTN	C5H10	70.130	511.76	45.08	260.0	260.0	0.2687	3.120	0.00	3.30	2.47
56	ISOPENTANE	ISOPTN	C5H12	72.150	460.43	33.81	308.0	308.0	0.2716	3.313	0.13	3.82	3.31
57	N-PENTANE	NORPTN	C5H12	72.150	469.65	33.55	311.0	309.0	0.2685	3.385	0.00	3.82	3.31
58	CHLOROBENZENE	CHLBZN	C6H5CL	112.560	632.40	45.20	308.0	308.0	0.2640	3.568	1.75	3.79	2.84
59	NITRO-BENZENE	NTRBZN	C6H5NO2	123.110	725.00	42.20	340.0	321.0	0.2430	3.640	4.28	4.13	3.14
60	BENZENE	BENZEN	C6H6	78.110	562.16	48.98	260.0	255.0	0.2696	3.004	0.00	3.19	2.40

* * PURE COMPONENT DATA * *

NO	NAME	CODE	FORMULA	MOL WT	TC K	PC BARS	VC CC/MOL	VSTR CC/MOL	ZRA	RD A	DM D	UNIQUAC PARAMETERS R	Q
61	PHENOL	PHENOL	C6H6O	94.110	694.25	61.34	254.0	250.0	0.2780	3.550	1.41	3.55	2.70
62	ANILINE	ANILIN	C6H7N	93.120	698.80	53.00	274.0	285.0	0.2620	3.393	1.49	3.72	2.82
63	CYCLOHEXANONE	CYHXON	C6H10O	98.150	629.00	38.50	320.0	318.0	0.2465	3.410	3.00	4.07	3.11
64	CYCLOHEXANE	CYHXAN	C6H12	84.160	553.54	40.75	308.0	311.0	0.2729	3.261	0.00	3.97	3.01
65	HEXENE-1	HEXENE	C6H12	84.160	504.03	31.72	368.0	368.0	0.2660	3.647	0.50	4.27	3.64
66	METHYLCYCLOPENTANE	MTCYPN	C6H12	84.160	532.79	37.85	319.0	319.0	0.2700	3.167	0.00	3.97	3.01
67	CYCLOHEXANOL	CYHXNL	C6H12O	100.160	625.00	37.50	330.0	326.0	0.2450	3.434	1.40	4.27	3.51
68	METHYLISOBUTYLKETONE	MISBKT	C6H12O	100.160	571.50	32.70	371.0	360.0	0.2600	3.740	2.60	4.60	4.03
69	N-BUTYL ACETATE	NBTACT	C6H12O2	116.160	579.10	31.10	395.0	380.0	0.2570	4.170	1.90	4.83	4.20
70	N-HEXANE	NHEXAN	C6H14	86.170	507.43	30.12	363.0	369.0	0.2635	3.812	0.00	4.50	3.86
71	2,3-DIMETHYL BUTANE	DMBT23	C6H14	86.170	499.90	31.30	358.0	358.0	0.2700	3.521	0.15	4.50	3.86
72	TRIETHYLAMINE	TRETAM	C6H15N	101.190	535.40	30.40	394.0	390.0	0.2700	3.930	0.66	5.01	4.26
73	TOLUENE	TOLUEN	C7H8	92.130	591.79	41.09	316.0	312.0	0.2646	3.443	0.36	3.92	2.97
74	METHYL CYCLOHEXANE	MTCYHX	C7H14	98.180	572.19	34.72	344.0	344.0	0.2699	3.747	0.00	4.64	3.55
75	N-HEPTANE	NHEPTN	C7H16	100.200	540.26	27.36	426.0	425.0	0.2611	4.267	0.00	5.17	4.40
76	STYRENE	STYREN	C8H8	104.150	647.59	39.99	373.0	373.0	0.2630	3.800	0.00	4.37	3.30
77	ETHYL BENZENE	ETHBZN	C8H10	106.160	617.17	36.09	374.0	374.0	0.2626	3.821	0.58	4.60	3.51
78	M-XYLENE	MXYLEN	C8H10	106.160	617.05	35.41	376.0	362.5	0.2593	3.897	0.40	4.66	3.54
79	O-XYLENE	OXYLEN	C8H10	106.160	630.37	37.33	369.0	363.5	0.2633	3.789	0.62	4.66	3.54
80	P-XYLENE	PXYLEN	C8H10	106.160	616.26	35.11	378.0	375.0	0.2589	3.796	0.00	4.66	3.54

* * PURE COMPONENT DATA * *

****

NO	NAME	CODE	FORMULA	MOL WT	TC K	PC BARS	VC CC/MOL	VSTR CC/MOL	ZRA	RD A	DM D	UNIQUAC PARAMETERS R	Q
81	N-OCTANE	NOCTAN	C8H18	114.220	568.83	24.92	486.0	489.0	0.2567	4.680	0.00	5.85	4.94
82	224-TRIMETHYLPENTANE	TMP224	C8H18	114.220	543.96	25.68	482.0	482.0	0.2672	4.171	0.00	5.85	4.94
83	N-DECANE	NDECAN	C10H22	142.280	617.65	20.96	602.0	602.0	0.2503	5.539	0.00	7.20	6.02
84	N-HEXADECANE	NHXDCN	C16H34	226.440	717.00	14.20	950.0	970.0	0.2380	8.318	0.00	11.24	9.26
85	HYDROGEN CHLORIDE	HYDCHL	HCL	36.490	324.54	82.60	87.6	83.3	0.2625	0.299	1.07	1.00	1.00
86	WATER	WATER	H2O	18.020	647.37	221.20	56.0	46.4	0.2380	0.615	1.83	0.92	1.40
87	HYDROGEN SULFIDE	HYDSUL	H2S	34.080	373.54	90.05	97.7	90.0	0.2851	0.604	0.90	1.00	1.00
88	AMMONIA	AMMNIA	NH3	17.030	405.54	112.80	72.5	65.2	0.2465	0.853	1.47	1.00	1.00
89	SULFUR DIOXIDE	SULDIO	SO2	64.060	430.65	78.94	122.0	115.0	0.2660	1.674	1.61	1.55	1.45
90	ANISOLE	ANISOL	C7H8O	108.140	641.00	40.66	336.0	0.	0.2574	3.692	1.20	4.17	3.21
91	ACRYLONITRILE	ACRNTR	C3H3N	53.064	536.00	44.41	210.0	0.	0.2275	2.443	3.50	2.31	2.05
92	VINYL ACETATE	VNLACT	C4H6O2	86.091	525.00	42.44	265.0	0.	0.2573	3.089	1.70	3.25	2.90

****

*NOTE - MODIFIED PARAMETERS, INCLUDING CRITICAL TEMPERATURE, PRESSURE, VOLUME, DIPOLE MOMENT,
ETC., ARE USED FOR CARBON DIOXIDE AND ACETYLENE IN ORDER TO IMPROVE THE HAYDEN-AND-O-CONNELL
SECOND VIRIAL COEFFICIENT CORRELATION FOR THESE COMPONENTS.

SIZE PARAMETER Q-PRIME (QP) FOR WATER AND ALCOHOLS

WATER	1.00	C4-ALCOHOLS	0.88
CH3OH	0.96	C5-ALCOHOLS	1.15
C2H5OH	0.92	C6-ALCOHOLS	1.78
C3-ALCOHOLS	0.89	C7-ALCOHOLS	2.71

Q-PRIME EQUALS Q (QP=Q) FOR ALL OTHER COMPONENTS.

Appendix C-2

Constants for Standard-State Fugacity Equation

** CONSTANTS FOR ZERO PRESSURE REFERENCE STATE FUGACITY EQUATION **

$$LN(FO) = C1 + C2/T + C3*T + C4*LN(T) + C5*T**2 \quad /// \quad (T\ IN\ DEG\ K,\ FO\ IN\ BARS)$$

NO	NAME	T-RNG	C1	C2	C3	C4	C5
1	HYDROGEN	0- 0	2.0067E+01	-3.2848E+02	0.	-2.5980E+00	0.
2	NITROGEN	0- 0	2.3320E+01	-9.5124E+02	0.	-2.5980E+00	0.
3	OXYGEN	0- 0	2.4251E+01	-1.1660E+03	0.	-2.5980E+00	0.
4	CARBON MONOXIDE	0- 0	2.3483E+01	-1.0015E+03	0.	-2.5980E+00	0.
5	ARGON	0- 0	2.4139E+01	-1.1354E+03	0.	-2.5980E+00	0.
6	METHANE	0- 0	2.4599E+01	-1.4358E+03	0.	-2.5980E+00	0.
7	CARBON TETRACHLORIDE	251-463	3.8895E+01	-5.3628E+03	-5.7784E-03	-3.7490E+00	3.1663E-06
8	CHLOROFORM	273-373	1.7424E+02	-8.1400E+03	6.5975E-02	-2.901E+01	-3.0001E-05
9	DICHLOROMETHANE	233-313	-3.0509E+02	3.1744E+03	-1.5950E-01	5.8953E+01	6.2252E-05
10	FORMIC ACID	281-383	-2.0871E+02	4.9951E+02	-6.1595E-02	3.8691E+01	0.
11	NITRO-METHANE	288-409	-3.5043E+02	3.5330E+03	-1.7689E-01	6.7044E+01	7.0865E-05
12	METHANOL	285-323	3.3387E+02	-1.2679E+04	1.3761E-01	-5.7722E+01	-5.9496E-05
13	CARBON DIOXIDE	216-253	6.3208E+00	-2.6423E+03	-3.9322E-02	2.7347E+00	2.6718E-05
14	CARBON DISULFIDE	195-448	1.8618E+02	-7.6592E+03	7.9048E-02	-3.1888E+01	-3.5119E-05
15	TRICHLORDETHYLENE	229-359	2.7094E+02	-9.8879E+03	1.3848E-01	-4.8280E+01	-7.0755E-05
16	ACETYLENE	214-258	-6.5283E+01	-1.1762E+03	-7.8870E-02	1.6178E+01	4.5152E-05
17	ACETONITRILE	288-377	4.1947E+02	-1.4100E+04	1.8197E-01	-7.4006E+01	-7.8134E-05
18	ETHYLENE	191-236	2.7951E+01	-2.2833E+03	-9.2408E-03	-2.5685E+00	8.4359E-06
19	1,2-DICHLORDETHANE	249-456	-2.4603E+02	2.9426E+03	-7.1143E-02	4.4781E+01	0.
20	ACETALDEHYDE	272-293	5.3137E+02	-1.5846E+04	2.3046E-01	-9.4477E+01	-9.5666E-05

** CONSTANTS FOR ZERO PRESSURE REFERENCE STATE FUGACITY EQUATION **

$LN(FO) = C1 + C2/T + C3*T + C4*LN(T) + C5*T**2$ / / / (T IN DEG K, FO IN BARS)

NO	NAME	T-RNG	C1	C2	C3	C4	C5
21	ACETIC ACID	273-503	3.8698E+02	-1.5091E+04	1.6774E-01	-6.7642E+01	-7.2738E-05
22	ETHYL IODIDE	253-333	1.4879E+01	-3.7711E+03	1.1605E-02	-1.1008E+00	-1.3086E-05
23	NITRO-ETHANE	252-387	2.7700E+02	-1.1339E+04	1.2358E-01	-4.8126E+01	-5.8807E-05
24	ETHANE	200-250	4.4641E+00	-2.0177E+03	-2.2423E-02	1.9367E+00	1.4062E-05
25	ETHANOL	271-376	-9.0910E+01	-3.4659E+03	-6.2301E-02	2.0486E+01	2.0664E-05
26	DIMETHYL AMINE	201-280	3.3491E+01	-5.0011E+03	-2.5441E-02	-1.7380E+00	1.6134E-05
27	ETHYLENE GLYCOL	271-376	1.6142E+01	-8.6798E+03	0.	0.	0.
28	PROPYLENE	195-309	1.0078E+02	-4.3747E+03	3.9936E-02	-1.6503E+01	-1.9271E-05
29	ACETONE	273-366	-2.3066E+02	6.8603E+02	-1.4358E-01	4.6384E+01	6.3961E-05
30	METHYL ACETATE	253-423	5.8575E+00	-4.9733E+03	-3.2652E-02	3.1157E+00	1.7275E-05
31	PROPIONIC ACID	318-501	1.3201E+03	-4.8232E+04	2.8472E-01	-2.1938E+02	0.
32	1-NITROPROPANE	263-404	6.6219E+02	-2.0913E+04	2.9480E-01	-1.1812E+02	-1.2259E-04
33	2-NITROPROPANE	254-393	1.0235E+03	-2.8440E+04	4.8386E-01	-1.8557E+02	-2.1231E-04
34	PROPANE	216-300	1.0491E+01	-2.8237E+03	-1.9487E-02	1.0303E+00	1.1171E-05
35	N-PROPANOL	273-453	-1.0789E+03	1.8583E+04	-5.3858E-01	2.0250E+02	2.2251E-04
36	I-PROPANOL	273-431	-9.1961E+02	1.5250E+04	-4.6193E-01	1.7315E+02	1.8922E-04
37	TRIMETHYLAMINE	192-313	1.0233E+02	-5.3622E+03	3.5547E-02	-1.6253E+01	-1.8193E-05
38	BUTENE-1	191-348	1.5052E+00	-3.2001E+03	-2.4887E-02	2.8950E+00	1.3017E-05
39	2-BUTANONE	224-352	-1.8254E+02	-1.8187E+02	-1.0790E-01	3.6725E+01	4.5490E-05
40	TETRAHYDROFURAN	293-455	4.0939E+02	-1.2980E+04	1.8626E-01	-7.2961E+01	-8.0006E-05

** CONSTANTS FOR ZERO PRESSURE REFERENCE STATE FUGACITY EQUATION **

LN(FO) = C1 + C2/T + C3*T + C4*LN(T) + C5*T**2 // (T IN DEG K, FO IN BARS)

NO	NAME	T-RNG	C1	C2	C3	C4	C5
41	DIOXANE	285-374	-1.4321E+02	3.6222E+02	-3.7330E-02	2.6361E+01	0.
42	ETHYL ACETATE	253-443	-1.2913E+02	-2.2599E+03	-9.6853E-02	2.8020E+01	4.3325E-05
43	N-BUTANE	212-330	1.6627E+02	-6.5056E+03	6.9340E-02	-2.7986E+01	-3.2221E-05
44	I-BUTANE	201-344	6.7026E+01	-4.4025E+03	1.1954E-02	-9.5337E+00	-4.0595E-06
45	N-BUTANOL	288-472	2.1605E+01	-8.0399E+03	-2.8619E-02	1.4677E+00	8.9569E-06
46	I-BUTANOL	287-464	-9.0037E+02	1.4109E+04	-4.5526E-01	1.6990E+02	1.8670E-04
47	SEC-BUTANOL	273-453	-2.8814E+02	-9.5651E+02	-1.9165E-01	5.9205E+01	8.3282E-05
48	TERT-BUTANOL	293-426	-3.0511E+02	-4.6564E+02	-2.0836E-01	6.2765E+01	9.3294E-05
49	DI-ETHYL ETHER	212-353	-1.1108E+02	-1.6936E+03	-8.7474E-02	2.4377E+01	4.0310E-05
50	ETHYL CELLOSOLVE	298-408	-1.0291E+01	-6.0540E+03	-2.0846E-02	5.5880E+00	0.
51	DIETHYL AMINE	240-418	-2.9626E+01	-3.2272E+03	-2.1227E-02	8.0047E+00	0.
52	FURFURAL	313-443	-1.3302E+03	2.4308E+04	-6.6073E-01	2.4831E+02	2.8068E-04
53	PYRIDINE	316-388	-7.3720E+01	-4.2467E+03	-7.1711E-02	1.8047E+01	3.2360E-05
54	ISOPRENE	193-305	-3.6272E+01	-2.4266E+03	-2.1155E-02	8.8491E+00	0.
55	CYCLOPENTANE	225-323	4.7418E+01	-4.9853E+03	3.5435E-03	-5.6813E+00	-2.9840E-06
56	ISOPENTANE	217-383	-7.1803E+01	-2.2593E+03	-6.7898E-02	1.6942E+01	3.3389E-05
57	N-PENTANE	208-398	6.9020E+01	-5.3625E+03	9.9221E-03	-9.4897E+00	-3.8363E-06
58	CHLOROBENZENE	335-452	5.1200E+01	-6.7499E+03	-2.3118E-03	-5.6437E+00	1.6116E-06
59	NITRO-BENZENE	298-483	-4.5060E+02	4.1182E+03	-2.1635E-01	8.5297E+01	8.3112E-05
60	BENZENE	284-473	9.7209E+01	-6.9761E+03	1.9082E-02	-1.4212E+01	-6.7182E-06

* * CONSTANTS FOR ZERO PRESSURE REFERENCE STATE FUGACITY EQUATION * *

****

$$LN(FO) = C1 + C2/T + C3*T + C4*LN(T) + C5*T**2 \quad /// \quad (T \text{ IN DEG K, FO IN BARS})$$

NO	NAME	T-RNG	C1	C2	C3	C4	C5
61	PHENOL	335-556	4.7203E+02	-2.3326E+04	8.7395E-02	-7.4933E+01	-9.3166E-06
62	ANILINE	350-457	-1.2252E+03	2.5717E+04	-5.0703E-01	2.2237E+02	1.8493E-04
63	CYCLOHEXANONE	274-428	6.0058E+02	-2.0267E+04	2.4430E-01	-1.0566E+02	-9.6292E-05
64	CYCLOHEXANE	293-354	-1.2138E+02	-1.8845E+03	-8.5395E-02	2.5948E+01	3.6812E-05
65	HEXENE-1'	289-337	-1.1929E+02	-1.6667E+03	-8.4539E-02	2.5525E+01	3.6525E-05
66	METHYLCYCLOPENTANE	219-345	5.3797E+01	-5.6900E+03	9.4455E-04	-6.4352E+00	-4.5651E-07
67	CYCLOHEXANOL	370-433	1.1353E+02	-1.2187E+04	-4.6087E-03	-1.3746E+01	0.
68	METHYLISOBUTYLKETONE	294-390	-1.0095E+03	1.6772E+04	-5.3720E-01	1.9111E+02	2.3544E-04
69	N-BUTYL ACETATE	315-399	1.7741E+02	-1.0025E+04	5.7383E-02	-2.8568E+01	-2.5981E-05
70	N-HEXANE	286-342	-3.5052E+02	3.3183E+03	-2.0259E-01	6.8498E+01	8.8924E-05
71	2,3-DIMETHYL BUTANE	263-423	8.1261E+01	-5.9351E+03	1.6510E-02	-1.1729E+01	-7.1816E-06
72	TRIETHYLAMINE	273-367	-1.7289E+03	3.4922E+04	-8.5675E-01	3.2163E+02	3.6387E-04
73	TOLUENE	273-500	2.0899E+01	-5.7902E+03	-2.0741E-02	7.1440E-02	1.1510E-05
74	METHYL CYCLOHEXANE	273-343	1.4687E+02	-8.3627E+03	4.1169E-02	-2.3246E+01	-1.6077E-05
75	N-HEPTANE	295-402	-1.7613E+01	-4.6698E+03	-3.5093E-02	6.9580E+00	1.4503E-05
76	STYRENE	298-418	1.2110E+03	-3.5413E+04	5.1556E-01	-2.1627E+02	-2.0887E-04
77	ETHYL BENZENE	329-410	9.6541E+01	-7.7957E+03	2.3659E-02	-1.4175E+01	-1.1680E-05
78	M-XYLENE	266-413	2.0481E+02	-1.0496E+04	7.1922E-02	-3.3837E+01	-3.1063E-05
79	O-XYLENE	285-418	2.0411E+02	-1.0627E+04	7.0104E-02	-3.3601E+01	-2.9854E-05
80	P-XYLENE	298-412	1.6273E+02	-9.4273E+03	5.3274E-02	-2.6213E+01	-2.3502E-05

****

* * CONSTANTS FOR ZERO PRESSURE REFERENCE STATE FUGACITY EQUATION * *

****

LN(FO) = C1 + C2/T + C3*T + C4*LN(T) + C5*T**2 /// (T IN DEG K, FO IN BARS)

NO	NAME	T-RNG	C1	C2	C3	C4	C5
81	N-OCTANE	326-399	-1.3781E+01	-5.8350E+03	-5.1518E-02	7.3940E+00	2.9243E-05
82	224-TRIMETHYLPENTANE	297-453	5.6300E+01	-6.1924E+03	2.9238E-04	-6.7366E+00	3.8963E-07
83	N-DECANE	290-446	6.0687E+01	-6.0039E+03	4.6404E-02	-1.0084E+01	-3.2489E-05
84	N-HEXADECANE	410-559	-4.3622E+02	3.0918E+03	-1.7886E-01	8.1029E+01	5.7550E-05
85	HYDROGEN CHLORIDE	201-264	3.7383E+01	-2.8016E+03	-1.6625E-03	-4.2252E+00	3.1210E-06
86	WATER	263-548	5.7042E+01	-7.0048E+03	3.5888E-03	-6.6589E+00	-8.5054E-07
87	HYDROGEN SULFIDE	190-298	2.1163E+02	-6.7998E+03	9.0769E-02	-3.6797E+01	-3.8744E-05
88	AMMONIA	195-340	1.6411E+01	-3.5319E+03	-1.8522E-02	3.9773E-01	9.9178E-06
89	SULFUR DIOXIDE	203-356	-4.4074E+02	4.9426E+03	-2.9311E-01	8.7745E+01	1.4571E-04
90	ANISOLE	383-437	-9.2661E+01	1.8837E+04	-3.8166E-01	1.6850E+02	1.3631E-04
91	ACRYLONITRILE	293-343	-5.9041E+01	-1.8052E+03	-1.9300E-02	1.2096E+01	0.
92	VINYL ACETATE	295-345	-3.1153E+02	2.4210E+03	-1.7358E-01	6.0878E+01	7.2286E-05

****

NOTE - T-RNG GIVES THE TEMPERATURE RANGE (K) OF THE EXPERIMENTAL DATA USED TO FIT THE
CONSTANTS. CONSTANTS FOR NONCONDENSABLES (COMPONENTS 1-6) WERE DETERMINED
FROM A GENERALIZED CORRELATION FOR THE HYPOTHETICAL REFERENCE FUGACITY.

Appendix C-3

Constants for the Ideal-Gas Heat-Capacity Equation

$$CPO = D1 + D2/T + D3*T + D4*LN(T) \quad /// \quad (T\ IN\ DEG\ K,\ CPO\ IN\ J/DEG-MOL)$$

NO	NAME	D1	D2	D3	D4
1	HYDROGEN	9.4505E+01	-2.7389E+03	1.0707E-02	-1.0478E+01
2	NITROGEN	6.9748E+01	-8.7072E+02	1.5637E-02	-7.4377E+00
3	OXYGEN	-1.6820E+01	1.7113E+03	2.2906E-03	6.9771E+00
4	CARBON MONOXIDE	6.0174E+01	-5.4642E+02	1.4930E-02	-5.9096E+00
5	ARGON	2.0790E+01	0.	0.	0.
6	METHANE	-1.8604E+02	8.1426E+03	2.4559E-02	3.2841E+01
7	CARBON TETRACHLORIDE	3.1389E+01	-5.9494E+03	-1.2207E-02	1.3338E+01
8	CHLOROFORM	-2.8425E+02	5.2118E+03	-4.5733E-02	6.0703E+01
9	DICHLOROMETHANE	-3.8435E+02	9.9458E+03	-4.1803E-02	7.2728E+01
10	FORMIC ACID	-3.7801E+02	1.0374E+04	-3.0997E-02	6.9802E+01
11	NITRO-METHANE	-7.0535E+02	2.0541E+04	-6.0718E-02	1.2494E+02
12	METHANOL	-3.7655E+02	1.3407E+04	-2.0846E-03	6.6014E+01
13	CARBON DIOXIDE	-1.0919E+02	2.4663E+03	-9.7531E-03	2.4740E+01
14	CARBON DISULFIDE	-5.5870E+01	2.3521E+02	-1.1946E-02	1.8278E+01
15	TRICHLOROETHYLENE	-2.4707E+02	2.5222E+03	-3.8890E-02	5.8004E+01
16	ACETYLENE	7.2685E+01	-6.1536E+03	1.5577E-02	-2.2409E+00
17	ACETONITRILE	-4.2507E+02	1.3269E+04	-2.2581E-02	7.7156E+01
1	rLENE	-5.3036E+02	1.5454E+04	-3.7645E-02	9.3597E+01
7	1,2-DICHLOROETHANE	-4.2218E+02	1.0950E+04	-2.1340E-02	8.2713E+01
20	ACETALDEHYDE	-6.5025E+02	2.0890E+04	-4.3099E-02	1.1368E+02

** CONSTANTS FOR IDEAL GAS HEAT CAPACITY EQUATION **

CP0 = D1 + D2/T + D3*T + D4*LN(T) /// (T IN DEG K, CP0 IN J/DEG-MOL)

NO	NAME	D1	D2	D3	D4
21	ACETIC ACID	-6.7620E+02	1.7691E+04	-5.7811E-02	1.2297E+02
22	ETHYL IODIDE	-7.0995E+02	2.0556E+04	-5.0692E-02	1.2652E+02
23	NITRO-ETHANE	-9.2341E+02	2.7114E+04	-8.6238E-02	1.7680E+02
24	ETHANE	-5.5227E+02	1.6821E+04	-9.2425E-03	9.6761E+01
25	ETHANOL	-7.4460E+02	2.1117E+04	-4.6915E-02	1.3208E+02
26	DIMETHYL AMINE	-1.0662E+03	3.2074E+04	-8.3408E-02	1.8474E+02
27	ETHYLENE GLYCOL	-9.3692E+02	2.5413E+04	-8.4588E-02	1.6740E+02
28	PROPYLENE	-8.4423E+02	2.5372E+04	-5.4926E-02	1.4733E+02
29	ACETONE	-9.8587E+02	3.1012E+04	-6.4734E-02	1.7130E+02
30	METHYL ACETATE	-2.2064E+02	-3.4684E+03	4.0393E-02	5.4158E+01
31	PROPIONIC ACID	-9.8717E+02	2.6352E+04	-8.2603E-02	1.7777E+02
32	1-NITROPROPANE	-1.3050E+03	3.5782E+04	-1.1123E-01	2.3171E+02
33	2-NITROPROPANE	-1.3237E+03	3.4957E+04	-1.2057E-01	2.3604E+02
34	PROPANE	-9.3975E+02	2.6187E+04	-5.2652E-02	1.6521E+02
35	N-PROPANOL	-1.1784E+03	3.5836E+04	-8.5452E-02	2.0549E+02
36	I-PROPANOL	-1.2304E+03	3.3989E+04	-1.1101E-01	2.1734E+02
37	TRIMETHYLAMINE	-1.4836E+03	4.3141E+04	-1.2873E-01	2.5783E+02
38	BUTENE-1	-1.0754E+03	2.9447E+04	-7.2903E-02	1.9026E+02
39	2-BUTANONE	-1.2918E+03	4.0853E+04	-9.3316E-02	2.2561E+02
40	TETRAHYDROFURAN	-1.6691E+03	4.6609E+04	-1.3550E-01	2.8605E+02

```
* * CONSTANTS FOR IDEAL GAS HEAT CAPACITY EQUATION * *
*********************************************************************************

         CPO = D1 + D2/T + D3*T + D4*LN(T)  / / /  (T IN DEG K, CPO IN J/DEG-MOL)
NO   NAME                 D1            D2            D3            D4
*********************************************************************************
```

NO	NAME	D1	D2	D3	D4
41	DIOXANE	-3.2035E+03	8.8618E+04	-4.9545E-01	5.5157E+02
42	ETHYL ACETATE	-4.4329E+02	-6.1859E+02	2.0987E-02	9.6210E+01
43	N-BUTANE	-1.3954E+03	4.0917E+04	-9.8673E-02	2.4308E+02
44	I-BUTANE	-1.2348E+03	3.3104E+04	-7.9316E-02	2.1839E+02
45	N-BUTANOL	6.9227E+03	-2.5547E+05	1.5016E+00	-1.1260E+03
46	I-BUTANOL	-1.9568E+03	5.1652E+04	-2.6073E-01	3.4575E+02
47	SEC-BUTANOL	-6.6078E+02	9.6986E+03	2.6127E-03	1.2973E+02
48	TERT-BUTANOL	-1.3766E+03	3.6499E+04	-1.1237E-01	2.4590E+02
49	DI-ETHYL ETHER	-4.6484E+02	1.6148E+04	1.2219E-01	8.6598E+01
50	ETHYL CELLOSOLVE	8.4268E+03	-2.7267E+05	2.0069E+00	-1.4004E+03
51	DIETHYL AMINE	-1.3600E+03	3.2979E+04	-9.5371E-02	2.4382E+02
52	FURFURAL	-1.8242E+03	6.2060E+04	-2.9946E-01	4.0289E+02
53	PYRIDINE	-1.5158E+03	3.9327E+04	-1.5780E-01	2.6488E+02
54	ISOPRENE	-8.0084E+02	1.3616E+04	-5.7187E-02	1.5534E+02
55	CYCLOPENTANE	-2.0888E+03	6.0001E+04	-1.7483E-01	3.5500E+02
56	ISOPENTANE	-1.5865E+03	4.3537E+04	-1.0497E-01	2.7913E+02
57	N-PENTANE	-1.7162E+03	4.9945E+04	-1.2489E-01	2.9946E+02
58	CHLOROBENZENE	-1.4370E+03	3.4524E+04	-1.5757E-01	2.5735E+02
59	NITRO-BENZENE	-2.0044E+03	4.9332E+04	-2.3049E-01	5.8054E+02
60	BENZENE	-1.4851E+03	3.6934E+04	-1.4521E-01	2.6084E+02

```
*********************************************************************************
```

157

* * CONSTANTS FOR IDEAL GAS HEAT CAPACITY EQUATION * *

$CPO = D1 + D2/T + D3*T + D4*LNT)$ / / / (T IN DEG K, CPO IN J/DEG-MOL)

NO	NAME	D1	D2	D3	D4
61	PHENOL	-1.4257E+03	3.2418E+04	-1.5504E-01	2.5745E+02
62	ANILINE	-1.4535E+03	3.1901E+04	-1.5159E-01	2.6327E+02
63	CYCLOHEXANONE	-2.9404E+03	9.1966E+04	-2.6256E-01	4.9453E+02
64	CYCLOHEXANE	-2.8942E+03	8.7140E+04	-2.5048E-01	4.8844E+02
65	HEXENE-1	-1.7378E+03	4.9073E+04	-1.2593E-01	3.0594E+02
66	METHYLCYCLOPENTANE	-2.3196E+03	6.4792E+04	-1.9966E-01	3.9871E+02
67	CYCLOHEXANOL	-3.0866E+03	9.1437E+04	-2.8815E-01	5.2376E+02
68	METHYLISOBUTYLKETONE	-1.9861E+03	5.8446E+04	-1.5182E-01	3.4712E+02
69	N-BUTYL ACETATE	-1.0652E+03	1.6704E+04	-2.8596E-02	2.0580E+02
70	N-HEXANE	-2.0153E+03	5.8049E+04	-1.4836E-01	3.5242E+02
71	2,3-DIMETHYL BUTANE	-2.0157E+03	5.5804E+04	-1.5195E-01	3.5354E+02
72	TRIETHYLAMINE	-2.3495E+03	6.2894E+04	-2.0566E-01	4.1371E+02
73	TOLUENE	-1.9064E+03	5.0672E+04	-1.7556E-01	3.3215E+02
74	METHYL CYCLOHEXANE	-2.9947E+03	8.6254E+04	-2.5828E-01	5.1206E+02
75	N-HEPTANE	-2.3035E+03	6.5832E+04	-1.6950E-01	4.0354E+02
76	STYRENE	-1.7580E+03	4.2911E+04	-1.6605E-01	3.1350E+02
77	ETHYL BENZENE	-2.0858E+03	5.3949E+04	-1.9146E-01	3.6687E+02
78	M-XYLENE	-2.1318E+03	5.9084E+04	-1.8318E-01	3.7135E+02
79	O-XYLENE	-1.9902E+03	5.5515E+04	-1.6271E-01	3.4953E+02
80	P-XYLENE	-2.1823E+03	6.2080E+04	-1.8543E-01	3.7844E+02

* * CONSTANTS FOR IDEAL GAS HEAT CAPACITY EQUATION * *

****

$CPD = D1 + D2/T + D3*T + D4*LN(T)$ / / / (T IN DEG K, CPD IN J/DEG-MOL)

NO	NAME	D1	D2	D3	D4
81	N-OCTANE	-2.6084E+03	7.4238E+04	-1.9335E-01	4.5735E+02
82	224-TRIMETHYLPENTANE	-2.6084E+03	7.4238E+04	-1.9335E-01	4.5735E+02
83	N-DECANE	-3.2404E+03	9.1928E+04	-2.4462E-01	5.6857E+02
84	N-HEXADECANE	-5.0941E+03	1.4341E+05	-3.9161E-01	8.9540E+02
85	HYDROGEN CHLORIDE	5.7139E+01	-6.6165E+02	9.7100E-03	-5.0306E+00
86	WATER	2.5573E+01	8.9786E+02	1.3705E-02	1.5825E-01
87	HYDROGEN SULFIDE	2.2946E+01	7.7244E+02	1.8856E-02	5.2914E-01
88	AMMONIA	-8.7547E+01	4.0985E+03	1.1720E-02	1.8583E+01
89	SULFUR DIOXIDE	-2.1296E+02	6.6886E+03	-2.9518E-02	4.1984E+01
90	ANISOLE	0.	0.	0.	0.
91	ACRYLONITRILE	0.	0.	0.	0.
92	VINYL ACETATE	0.	0.	0.	0.

****

Association and Solvation Parameters for 92 Fluids

```
*********************************************************************************

HYDRGN 0.00

NITRGN 0.00
       HYDRGN 0.00

OXYGEN 0.00
       HYDRGN 0.00   NITRGN 0.00

CRBMOX 0.00
       HYDRGN 0.00   NITRGN 0.00   OXYGEN 0.00

ARGON  0.00
       HYDRGN 0.00   NITRGN 0.00   OXYGEN 0.00   CRBMOX 0.00

METHAN 0.00
       HYDRGN 0.00   NITRGN 0.00   OXYGEN 0.00   CRBMOX 0.00   ARGON  0.00

CRBTCL 0.00
       HYDRGN 0.00   NITRGN 0.00   OXYGEN 0.00   CRBMOX 0.00   ARGON  0.00

CLRFRM 0.00
       HYDRGN 0.00   NITRGN 0.00   OXYGEN 0.00   CRBMOX 0.00   ARGON  0.00   METHAN 0.00   CRBTCL 0.00

DCLMTN 0.00
       HYDRGN 0.00   NITRGN 0.00   OXYGEN 0.00   CRBMOX 0.00   ARGON  0.00   METHAN 0.00   CRBTCL 0.00
       CLRFRM 0.00

FRMACD 4.50
       HYDRGN 0.00   NITRGN 0.00   OXYGEN 0.00   CRBMOX 0.00   ARGON  0.00   METHAN 0.00   CRBTCL 0.15
       CLRFRM 1.50   DCLMTN 0.50

NTRMTN 1.66
       HYDRGN 0.00   NITRGN 0.00   OXYGEN 0.00   CRBMOX 0.00   ARGON  0.00   METHAN 0.00   CRBTCL 0.30
       CLRFRM 0.80   DCLMTN 0.40   FRMACD 2.20

MTHNOL 1.63
       HYDRGN 0.00   NITRGN 0.00   OXYGEN 0.00   CRBMOX 0.00   ARGON  0.00   METHAN 0.00   CRBTCL 0.00
       CLRFRM 0.10   DCLMTN 0.00   FRMACD 2.50   NTRMTN 1.20

CRBDOX 0.16
       HYDRGN 0.00   NITRGN 0.00   OXYGEN 0.00   CRBMOX 0.00   ARGON  0.00   METHAN 0.00   CRBTCL 0.00
       CLRFRM 0.10   DCLMTN 0.00   FRMACD 0.50   NTRMTN 0.40   MTHNOL 0.30

*********************************************************************************
```

```
* * ASSOCIATION AND SOLVATION PARAMETERS * *
*********************************************************************************

CRBDSF 0.34
   HYDRGN 0.00    NITRGN 0.00    OXYGEN 0.00    CRBMOX 0.00    ARGON 0.00    METHAN 0.00    CRBTCL 0.00
   CLRFRM 0.00    DCLMTN 0.00    FRMACD 0.25    NTRMTN 0.20    MTHNOL 0.15    CRBDOX 0.20

TCLETN 0.00
   HYDRGN 0.00    NITRGN 0.00    OXYGEN 0.00    CRBMOX 0.00    ARGON 0.00    METHAN 0.00    CRBTCL 0.00
   CLRFRM 0.00    DCLMTN 0.00    FRMACD 0.30    NTRMTN 0.20    MTHNOL 0.00    CRBDOX 0.00    CRBDSF 0.00

ACTYLN 0.16
   HYDRGN 0.00    NITRGN 0.00    OXYGEN 0.00    CRBMOX 0.00    ARGON 0.00    METHAN 0.00    CRBTCL 0.00
   CLRFRM 0.00    DCLMTN 0.00    FRMACD 0.60    NTRMTN 0.50    MTHNOL 0.30    CRBDOX 0.16    CRBDSF 0.20
   TCLETN 0.00

ACTNTR 1.65
   HYDRGN 0.00    NITRGN 0.00    OXYGEN 0.00    CRBMOX 0.00    ARGON 0.00    METHAN 0.00    CRBTCL 0.30
   CLRFRM 1.80    DCLMTN 0.60    FRMACD 0.20    NTRMTN 2.00    MTHNOL 1.50    CRBDOX 0.00    CRBDSF 0.00
   TCLETN 0.20    ACTYLN 0.50

ETHLYN 0.00
   HYDRGN 0.00    NITRGN 0.00    OXYGEN 0.00    CRBMOX 0.00    ARGON 0.00    METHAN 0.00    CRBTCL 0.00
   CLRFRM 0.00    DCLMTN 0.00    FRMACD 0.30    NTRMTN 0.20    MTHNOL 0.00    CRBDOX 0.00    CRBDSF 0.00
   TCLETN 0.00    ACTYLN 0.00    ACTNTR 0.20

DCLETN 0.00
   HYDRGN 0.00    NITRGN 0.00    OXYGEN 0.00    CRBMOX 0.00    ARGON 0.00    METHAN 0.00    CRBTCL 0.00
   CLRFRM 0.00    DCLMTN 0.00    FRMACD 0.00    NTRMTN 0.00    MTHNOL 0.00    CRBDOX 0.00    CRBDSF 0.00
   TCLETN 0.00    ACTYLN 0.00    ACTNTR 0.00    ETHLYN 0.00

ACTALD 0.58
   HYDRGN 0.00    NITRGN 0.00    OXYGEN 0.00    CRBMOX 0.00    ARGON 0.00    METHAN 0.00    CRBTCL 0.15
   CLRFRM 1.10    DCLMTN 0.00    FRMACD 1.60    NTRMTN 1.40    MTHNOL 0.80    CRBDOX 0.20    CRBDSF 0.00
   TCLETN 0.00    ACTYLN 0.00    ACTNTR 2.38    ETHLYN 0.00    DCLETN 0.00

ACTACD 4.50
   HYDRGN 0.00    NITRGN 0.00    OXYGEN 0.00    CRBMOX 0.00    ARGON 0.00    METHAN 0.00    CRBTCL 0.15
   CLRFRM 1.50    DCLMTN 0.50    FRMACD 4.50    NTRMTN 2.20    MTHNOL 2.50    CRBDOX 0.50    CRBDSF 0.25
   TCLETN 0.30    ACTYLN 0.60    ACTNTR 2.50    ETHLYN 0.30    DCLETN 0.00    ACTALD 1.60

ETHIOD 0.00
   HYDRGN 0.00    NITRGN 0.00    OXYGEN 0.00    CRBMOX 0.00    ARGON 0.00    METHAN 0.00    CRBTCL 0.00
   CLRFRM 0.00    DCLMTN 0.00    FRMACD 0.00    NTRMTN 0.00    MTHNOL 0.00    CRBDOX 0.00    CRBDSF 0.00
   TCLETN 0.00    ACTYLN 0.00    ACTNTR 0.00    ETHLYN 0.00    DCLETN 0.00    ACTALD 0.00    ACTACD 0.00

*********************************************************************************
```

161

```
* * ASSOCIATION AND SOLVATION PARAMETERS * *
***************************************************************************************************

NTRETH 1.66
  HYDRGN 0.00    NITRGN 0.00    OXYGEN 0.00    CRBMOX 0.00    ARGON  0.00    METHAN 0.00    CRBTCL 0.30
  CLRFRM 0.80    DCLMTN 0.40    FRMACD 2.20    NTRMTN 1.66    MTHNOL 1.20    CRBDOX 0.50    CRBDSF 0.20
  TCLETN 0.20    ACTYLN 0.50    ACTNTR 2.00    ETHLYN 0.20    DCLETN 0.00    ACTALD 1.40    ACTACD 2.20
  ETHIOD 0.00

ETHANE 0.00
  HYDRGN 0.00    NITRGN 0.00    OXYGEN 0.00    CRBMOX 0.00    ARGON  0.00    METHAN 0.00    CRBTCL 0.00
  CLRFRM 0.00    DCLMTN 0.00    FRMACD 0.00    NTRMTN 0.00    MTHNOL 0.00    CRBDOX 0.00    CRBDSF 0.00
  TCLETN 0.00    ACTYLN 0.00    ACTNTR 0.00    ETHLYN 0.00    DCLETN 0.00    ACTALD 0.00    ACTACD 0.00
  ETHIOD 0.00    NTRETH 0.00

ETHNOL 1.40
  HYDRGN 0.00    NITRGN 0.00    OXYGEN 0.00    CRBMOX 0.00    ARGON  0.00    METHAN 0.00    CRBTCL 0.00
  CLRFRM 0.10    DCLMTN 0.00    FRMACD 2.50    NTRMTN 1.20    MTHNOL 1.55    CRBDOX 0.30    CRBDSF 0.00
  TCLETN 0.00    ACTYLN 0.30    ACTNTR 1.50    ETHLYN 0.00    DCLETN 0.00    ACTALD 0.80    ACTACD 2.50
  ETHIOD 0.00    NTRETH 1.20

DMTAMN 0.17
  HYDRGN 0.00    NITRGN 0.00    OXYGEN 0.00    CRBMOX 0.00    ARGON  0.00    METHAN 0.00    CRBTCL 0.00
  CLRFRM 0.30    DCLMTN 0.00    FRMACD 2.20    NTRMTN 1.40    MTHNOL 0.20    CRBDOX 0.20    CRBDSF 0.10
  TCLETN 0.20    ACTYLN 0.50    ACTNTR 1.40    ETHLYN 0.20    DCLETN 0.00    ACTALD 0.85    ACTACD 2.20
  ETHIOD 0.00    NTRETH 1.40    ETHANE 0.00    ETHNOL 0.20

ETHGLY 1.55
  HYDRGN 0.00    NITRGN 0.00    OXYGEN 0.00    CRBMOX 0.00    ARGON  0.00    METHAN 0.00    CRBTCL 0.00
  CLRFRM 0.10    DCLMTN 0.00    FRMACD 2.50    NTRMTN 1.20    MTHNOL 1.55    CRBDOX 0.30    CRBDSF 0.15
  TCLETN 0.00    ACTYLN 0.30    ACTNTR 0.00    ETHLYN 0.00    DCLETN 0.00    ACTALD 0.00    ACTACD 0.00
  ETHIOD 0.00    NTRETH 0.00    ETHANE 0.00    ETHNOL 0.00    DMTAMN 0.00

PRPYLN 0.00
  HYDRGN 0.00    NITRGN 0.00    OXYGEN 0.00    CRBMOX 0.00    ARGON  0.00    METHAN 0.00    CRBTCL 0.00
  CLRFRM 0.00    DCLMTN 0.00    FRMACD 0.30    NTPMTN 0.20    MTHNOL 0.20    CRBDOX 0.00    CRBDSF 0.00
  TCLETN 0.00    ACTYLN 0.00    ACTNTR 0.20    ETHLYN 0.00    DCLETN 0.00    ACTALD 0.00    ACTACD 0.30
  ETHIOD 0.00    NTRETH 0.20    ETHANE 0.00    ETHNOL 0.00    DMTAMN 0.20    ETHGLY 0.00

ACETON 0.90
  HYDRGN 0.00    NITRGN 0.00    OXYGEN 0.00    CRBMOX 0.00    ARGON  0.00    METHAN 0.00    CRBTCL 0.20
  CLRFRM 1.26    DCLMTN 0.40    FRMACD 1.80    NTRMTN 1.63    MTHNOL 1.00    CRBDOX 0.00    CRBDSF 0.00
  TCLETN 0.00    ACTYLN 0.00    ACTNTR 1.30    ETHLYN 0.00    DCLETN 0.00    ACTALD 1.00    ACTACD 1.80
  ETHIOD 0.00    NTRETH 1.63    ETHANE 0.00    ETHNOL 0.00    DMTAMN 1.00    ETHGLY 1.00    PRPYLN 0.00

MTHACT 0.85
  HYDRGN 0.00    NITRGN 0.00    OXYGEN 0.00    CRBMOX 0.00    ARGON  0.00    METHAN 0.00    CRBTCL 0.25
***************************************************************************************************
```

```
                        * * ASSOCIATION AND SOLVATION PARAMETERS * *
********************************************************************************************

MTHACT 0.85
  CLRFRM 1.55     DCLMTN 0.50     FRMACD 2.00     NTRMTN 1.80     MTHNOL 1.30     CRBDOX 0.00     CRBDSF 0.00
  TCLETN 0.00     ACTYLN 0.00     ACTNTR 1.40     ETHLYN 0.00     DCLETN 0.00     ACTALD 0.75     ACTACD 2.00
  ETHIOD 0.00     NTRETH 1.80     ETHANE 0.00     ETHNOL 1.30     DMTAMN 1.10     ETHGLY 1.30     PRPYLN 0.00
  ACETON 1.10

PRPACD 4.50
  HYDRGN 0.00     NITRGN 0.00     OXYGEN 0.00     CRBMOX 0.00     ARGON  0.00     METHAN 0.00     CRBTCL 0.15
  CLRFRM 1.50     DCLMTN 0.50     FRMACD 4.50     NTRMTN 2.20     MTHNOL 2.50     CRBDOX 0.50     CRBDSF 0.25
  TCLETN 0.30     ACTYLN 0.60     ACTNTR 2.50     ETHLYN 0.30     DCLETN 0.00     ACTALD 1.60     ACTACD 4.50
  ETHIOD 0.00     NTRETH 2.20     ETHANE 0.00     ETHNOL 2.50     DMTAMN 2.20     ETHGLY 2.50     PRPYLN 0.30
  ACETON 1.80     MTHACT 2.00

NTPRP1 1.66
  HYDRGN 0.00     NITRGN 0.00     OXYGEN 0.00     CRBMOX 0.00     ARGON  0.00     METHAN 0.00     CRBTCL 0.30
  CLRFRM 0.80     DCLMTN 0.40     FRMACD 2.20     NTRMTN 1.66     MTHNOL 1.20     CRBDOX 0.40     CRBDSF 0.20
  TCLETN 0.20     ACTYLN 0.50     ACTNTR 2.00     ETHLYN 0.30     DCLETN 0.00     ACTALD 1.40     ACTACD 2.20
  ETHIOD 0.00     NTRETH 1.66     ETHANE 0.00     ETHNOL 1.20     DMTAMN 1.40     ETHGLY 1.20     PRPYLN 0.20
  ACETON 1.63     MTHACT 1.80

NTPRP2 1.66
  HYDRGN 0.00     NITRGN 0.00     OXYGEN 0.00     CRBMOX 0.00     ARGON  0.00     METHAN 0.00     CRBTCL 0.30
  CLRFRM 0.80     DCLMTN 0.40     FRMACD 2.20     NTRMTN 1.66     MTHNOL 1.20     CRBDOX 0.40     CRBDSF 0.20
  TCLETN 0.20     ACTYLN 0.56     ACTNTR 2.00     ETHLYN 1.20     DCLETN 0.00     ACTALD 1.40     ACTACD 2.20
  ETHIOD 0.00     NTRETH 1.66     ETHANE 0.00     ETHNOL 1.66     DMTAMN 1.40     ETHGLY 1.20     PRPYLN 0.20
  ACETON 1.63

PROPAN 0.00
  HYDRGN 0.00     NITRGN 0.00     OXYGEN 0.00     CRBMOX 0.00     ARGON  0.00     METHAN 0.00     CRBTCL 0.00
  CLRFRM 0.00     DCLMTN 0.00     FRMACD 0.00     NTRMTN 0.00     MTHNOL 0.00     CRBDOX 0.00     CRBDSF 0.00
  TCLETN 0.00     ACTYLN 0.00     ACTNTR 0.00     ETHLYN 0.00     DCLETN 0.00     ACTALD 0.00     ACTACD 0.00
  ETHIOD 0.00     NTRETH 0.00     ETHANE 0.00     ETHNOL 0.00     DMTAMN 0.00     ETHGLY 0.00     PRPYLN 0.00
  ACETON 0.00     MTHACT 0.00     PRPACD 0.00     NTPRP1 0.00     NTPRP2 0.00

NPRPNL 1.40
  HYDRGN 0.00     NITRGN 0.00     OXYGEN 0.00     CRBMOX 0.00     ARGON  0.00     METHAN 0.00     CRBTCL 0.00
  CLRFRM 0.10     DCLMTN 0.00     FRMACD 2.50     NTRMTN 1.20     MTHNOL 1.55     CRBDOX 0.30     CRBDSF 0.15
  TCLETN 0.00     ACTYLN 0.30     ACTNTR 1.50     ETHLYN 0.00     DCLETN 0.00     ACTALD 0.80     ACTACD 2.50
  ETHIOD 0.00     NTRETH 1.20     ETHANE 0.00     ETHNOL 1.55     DMTAMN 0.20     ETHGLY 1.55     PRPYLN 0.00
  ACETON 1.00     MTHACT 1.30     PRPACD 2.50     NTPRP1 1.20     NTPRP2 1.20     PROPAN 0.00

IPRPNL 1.32
  HYDRGN 0.00     NITRGN 0.00     OXYGEN 0.00     CRBMOX 0.00     ARGON  0.00     METHAN 0.00     CRBTCL 0.00
  CLRFRM 0.10     DCLMTN 0.00     FRMACD 2.50     NTRMTN 1.20     MTHNOL 1.55     CRBDOX 0.30     CRBDSF 0.15

********************************************************************************************
```

```
*******************************************************************************************

IPRPNL 1.32
   TCLETN 0.00   ACTYLN 0.30   ACTNTR 1.50   ETHLYN 0.00   DCLETN 0.00   ACTALD 0.80   ACTACD 2.50
   ETHIOD 0.00   NTRETH 1.20   ETHANE 0.00   ETHNOL 1.55   DMTAMN 0.20   ETHGLY 1.55   PRPYLN 0.00
   ACETON 1.00   MTHACT 1.30   PRPACD 2.50   NTPRP1 1.20   NTPRP2 1.20   PROPAN 0.00   NPRPNL 1.55

TRMTAM 0.06
   HYDRGN 0.00   NITRGN 0.00   OXYGEN 0.00   CRBMOX 0.00   ARGON  0.00   METHAN 0.00   CRBTCL 0.00
   CLRFRM 0.30   DCLMTN 0.00   FRMACD 2.20   NTRMTN 1.40   MTHNOL 0.20   CRBDOX 0.20   CRBDSF 0.10
   TCLETN 0.00   ACTYLN 0.50   ACTNTR 1.40   ETHLYN 0.20   DCLETN 0.00   ACTALD 0.85   ACTACD 2.20
   ETHIOD 0.00   NTRETH 1.40   ETHANE 0.00   ETHNOL 0.20   DMTAMN 0.00   ETHGLY 0.20   PRPYLN 0.20
   ACETON 1.00   MTHACT 1.10   PRPACD 2.20   NTPRP1 1.40   NTPRP2 1.40   PROPAN 0.00   NPRPNL 0.20
   IPRPNL 0.20

BUTENE 0.00
   HYDRGN 0.00   NITRGN 0.00   OXYGEN 0.00   CRBMOX 0.00   ARGON  0.00   METHAN 0.00   CRBTCL 0.00
   CLRFRM 0.00   DCLMTN 0.00   FRMACD 0.30   NTRMTN 0.20   MTHNOL 0.00   CRBDOX 0.00   CRBDSF 0.00
   TCLETN 0.00   ACTYLN 0.00   ACTNTR 0.20   ETHLYN 0.00   DCLETN 0.00   ACTALD 0.00   ACTACD 0.30
   ETHIOD 0.00   NTRETH 0.20   ETHANE 0.00   ETHNOL 0.00   DMTAMN 0.00   ETHGLY 0.00   PRPYLN 0.00
   ACETON 0.00   MTHACT 0.00   PRPACD 0.30   NTPRP1 0.20   NTPRP2 0.20   PROPAN 0.00   NPRPNL 0.00
   IPRPNL 0.00   TRMTAM 0.20

BTNON2 0.90
   HYDRGN 0.00   NITRGN 0.00   OXYGEN 0.00   CRBMOX 0.00   ARGON  0.00   METHAN 0.00   CRBTCL 0.20
   CLRFRM 1.26   DCLMTN 0.40   FRMACD 1.80   NTRMTN 1.80   MTHNOL 1.00   CRBDOX 0.00   CRBDSF 0.00
   TCLETN 0.00   ACTYLN 0.00   ACTNTR 1.30   ETHLYN 1.30   DCLETN 0.00   ACTALD 1.00   ACTACD 1.80
   ETHIOD 0.00   NTRETH 1.63   ETHANE 0.00   ETHNOL 0.00   DMTAMN 1.00   ETHGLY 1.00   PRPYLN 0.00
   ACETON 0.90   MTHACT 0.00   PRPACD 1.80   NTPRP1 1.63   NTPRP2 1.63   PROPAN 0.00   NPRPNL 1.00
   IPRPNL 1.00   TRMTAM 1.00   BUTENE 0.00

TTRHFR 0.00
   HYDRGN 0.00   NITRGN 0.00   OXYGEN 0.00   CRBMOX 0.00   ARGON  0.00   METHAN 0.00   CRBTCL 0.10
   CLRFRM 0.95   DCLMTN 0.35   FRMACD 1.20   NTRMTN 1.25   MTHNOL 0.50   CRBDOX 0.00   CRBDSF 0.00
   TCLETN 0.00   ACTYLN 0.00   ACTNTR 1.00   ETHLYN 1.00   DCLETN 0.00   ACTALD 0.40   ACTACD 1.20
   ETHIOD 0.00   NTRETH 1.25   ETHANE 0.00   ETHNOL 0.00   DMTAMN 0.70   ETHGLY 0.50   PRPYLN 0.00
   ACETON 0.00   MTHACT 0.00   PRPACD 1.20   NTPRP1 1.25   NTPRP2 1.25   PROPAN 0.00   NPRPNL 0.50
   IPRPNL 0.50   TRMTAM 0.70   BUTENE 0.00   BTNON2 0.00

DIOANE 0.00
   HYDRGN 0.00   NITRGN 0.00   OXYGEN 0.00   CRBMOX 0.00   ARGON  0.00   METHAN 0.00   CRBTCL 0.10
   CLRFRM 0.95   DCLMTN 0.95   FRMACD 0.00   NTRMTN 1.20   MTHNOL 1.25   CRBDOX 0.50   CRBDSF 0.00
   TCLETN 0.00   ACTYLN 0.00   ACTNTR 1.00   ETHLYN 1.00   DCLETN 0.00   ACTALD 0.40   ACTACD 1.20
   ETHIOD 0.00   NTRETH 1.25   ETHANE 0.00   ETHNOL 0.50   DMTAMN 0.20   ETHGLY 0.50   PRPYLN 0.00
   ACETON 0.00   MTHACT 0.00   PRPACD 1.20   NTPRP1 1.25   NTPRP2 1.25   PROPAN 0.00   NPRPNL 0.50
   IPRPNL 0.50   TRMTAM 0.70   BUTENE 0.00   BTNON2 0.00   TTRHFR 0.00

*******************************************************************************************
```

* * ASSOCIATION AND SOLVATION PARAMETERS * *

DETAMN 0.21

DETAMN 0.21	TRMTAM 0.20	BUTENE 0.20	BTNON2 1.00	TTRHFR 0.00	DIOANE 0.00	ETHACT 1.10
IPRPNL 0.20	IBTANE 0.00	NBTNOL 0.20	IBTNOL 0.20	SBTNOL 0.20	TBTNOL 0.20	DETETH 0.70
NBTANE 0.00						
ETHCEL 0.20						

FRFRAL 0.58

FRFRAL 0.58		OXYGEN 0.00	CRBMOX 0.00	ARGON 0.00	METHAN 0.00	CRBTCL 0.15
HYDRGN 0.00	NITRGN 0.00	FRMACD 1.60	NTRMTN 1.40	MTHNOL 0.80	CRBDOX 0.20	CRBDSF 0.10
CLRFRM 1.10	DCLMTN 0.40	ACTNTR 2.38	ETHLYN 0.00	DCLETN 0.00	ACTALD 0.55	ACTACD 1.60
TCLETN 0.00	ACTYLN 0.00	ETHANE 0.00	ETHNOL 0.80	DMTAMN 0.85	ETHGLY 0.80	PRPYLN 0.80
ETHIOD 0.00	NTRETH 1.40	PRPACD 1.60	NTPRP1 1.40	NTPRP2 1.40	PROPAN 0.00	NPRPNL 0.80
ACETON 1.00	MTHACT 0.75	BUTENE 0.00	BTNON2 1.00	TTRHFR 0.40	DIOANE 0.40	ETHACT 0.75
IPRPNL 0.80	TRMTAM 0.85	NBTNOL 0.80	IBTNOL 0.80	SBTNOL 0.80	TBTNOL 0.80	DETETH 0.40
NBTANE 0.00	IBTANE 0.00					
ETHCEL 0.80	DETAMN 0.85					

PYRDIN 0.00

PYRDIN 0.00		OXYGEN 0.00	CRBMOX 0.00	ARGON 0.00	METHAN 0.00	CRBTCL 0.15
HYDRGN 0.00	NITRGN 0.00	FRMACD 1.00	NTRMTN 1.30	MTHNOL 0.40	CRBDOX 0.00	CRBDSF 0.00
CLRFRM 0.30	DCLMTN 0.00	ACTNTR 1.30	ETHLYN 0.15	DCLETN 0.00	ACTALD 0.40	ACTACD 1.00
TCLETN 0.05	ACTYLN 0.00	ETHANE 0.00	ETHNOL 0.40	DMTAMN 0.20	ETHGLY 0.40	PRPYLN 0.05
ETHIOD 0.00	NTRETH 1.30	PRPACD 1.00	NTPRP1 1.30	NTPRP2 1.30	PROPAN 0.00	NPRPNL 0.40
ACETON 0.00	MTHACT 0.30	BUTENE 0.05	BTNON2 0.00	TTRHFR 0.00	DIOANE 0.00	ETHACT 0.30
IPRPNL 0.40	TRMTAM 0.20	NBTNOL 0.40	IBTNOL 0.40	SBTNOL 0.40	TBTNOL 0.40	DETETH 0.00
NBTANE 0.00	IBTANE 0.00	FRFRAL 0.40				
ETHCEL 0.40	DETAMN 0.20					

ISOPRN 0.00

ISOPRN 0.00		OXYGEN 0.00	CRBMOX 0.00	ARGON 0.00	METHAN 0.00	CRBTCL 0.00
HYDRGN 0.00	NITRGN 0.00	FRMACD 0.30	NTRMTN 0.30	MTHNOL 0.00	CRBDOX 0.00	CRBDSF 0.00
CLRFRM 0.00	DCLMTN 0.00	ACTNTR 0.2	ETHLYN 0.00	DCLETN 0.00	ACTALD 0.00	ACTACD 0.30
TCLETN 0.00	ACTYLN 0.00	ETHANE 0.00	ETHNOL 0.00	DMTAMN 0.20	ETHGLY 0.00	PRPYLN 0.00
ETHIOD 0.00	NTRETH 0.20	PRPACD 0.30	NTPRP1 0.20	NTPRP2 0.20	PROPAN 0.00	NPRPNL 0.00
ACETON 0.00	MTHACT 0.00	BUTENE 0.00	BTNON2 0.00	TTRHFR 0.00	DIOANE 0.00	ETHACT 0.00
IPRPNL 0.00	TRMTAM 0.20	NBTNOL 0.00	IBTNOL 0.00	SBTNOL 0.00	TBTNOL 0.00	DETETH 0.00
NBTANE 0.00	IBTANE 0.00	FRFRAL 0.00	PYRDIN 0.05			
ETHCEL 0.00	DETAMN 0.20					

CYCPTN 0.00

CYCPTN 0.00		OXYGEN 0.00	CRBMOX 0.00	ARGON 0.00	METHAN 0.00	CRBTCL 0.00
HYDRGN 0.00	NITRGN 0.00	FRMACD 0.00	NTRMTN 0.00	MTHNOL 0.00	CRBDOX 0.00	CRBDSF 0.00
CLRFRM 0.00	DCLMTN 0.00	ACTNTR 0.00	ETHLYN 0.00	DCLETN 0.00	ACTALD 0.00	ACTACD 0.00
TCLETN 0.00	ACTYLN 0.00	ETHANE 0.00	ETHNOL 0.00	DMTAMN 0.00	ETHGLY 0.00	PRPYLN 0.00
ETHIOD 0.00	NTRETH 0.00	PRPACD 0.00	NTPRP1 0.00	NTPRP2 0.00	PROPAN 0.00	NPRPNL 0.00
ACETON 0.00	MTHACT 0.00	BUTENE 0.00	BTNON2 0.00	TTRHFR 0.00	DIOANE 0.00	ETHACT 0.00
IPRPNL 0.00	TRMTAM 0.00	NBTNOL 0.00	IBTNOL 0.00	SBTNOL 0.00	TBTNOL 0.00	DETETH 0.00
NBTANE 0.00	IBTANE 0.00	FRFRAL 0.00	PYRDIN 0.00	ISOPRN 0.00		
ETHCEL 0.00	DETAMN 0.00					

167

```
* * ASSOCIATION AND SOLVATION PARAMETERS * *
*****************************************************************************************

ISOPTN 0.00
   HYDRGN 0.00    NITRGN 0.00    OXYGEN 0.00    CRBMOX 0.00    ARGON  0.00    METHAN 0.00    CRBTCL 0.00
   CLRFRM 0.00    DCLMTN 0.00    FRMACD 0.00    NTRMTN 0.00    MTHNOL 0.00    CRBDOX 0.00    CRBDSF 0.00
   TCLETN 0.00    ACTYLN 0.00    ACTNTR 0.00    ETHLYN 0.00    DCLETN 0.00    ACTALD 0.00    ACTACD 0.00
   ETHIOD 0.00    NTRETH 0.00    ETHANE 0.00    ETHNOL 0.00    DMTAMN 0.00    ETHGLY 0.00    PRPYLN 0.00
   ACETON 0.00    MTHACT 0.00    PRPACD 0.00    NTPRP1 0.00    NTPRP2 0.00    PROPAN 0.00    NPRPNL 0.00
   IPRPNL 0.00    TRMTAM 0.00    BUTENE 0.00    BTNON2 0.00    TTRHFR 0.00    DIOANE 0.00    ETHACT 0.00
   NBTANE 0.00    IBTANE 0.00    NBTNOL 0.00    IBTNOL 0.00    SBTNOL 0.00    TBTNOL 0.00    DETETH 0.00
   ETHCEL 0.00    DETAMN 0.00    FRFRAL 0.00    PYRDIN 0.00    ISOPRN 0.00    CYCPTN 0.00    DETETH 0.00

NORPTN 0.00
   HYDRGN 0.00    NITRGN 0.00    OXYGEN 0.00    CRBMOX 0.00    ARGON  0.00    METHAN 0.00    CRBTCL 0.00
   CLRFRM 0.00    DCLMTN 0.00    FRMACD 0.00    NTRMTN 0.00    MTHNOL 0.00    CRBDOX 0.00    CRBDSF 0.00
   TCLETN 0.00    ACTYLN 0.00    ACTNTR 0.00    ETHLYN 0.00    DCLETN 0.00    ACTALD 0.00    ACTACD 0.00
   ETHIOD 0.00    NTRETH 0.00    ETHANE 0.00    ETHNOL 0.00    DMTAMN 0.00    ETHGLY 0.00    PRPYLN 0.00
   ACETON 0.00    MTHACT 0.00    PRPACD 0.00    NTPRP1 0.00    NTPRP2 0.00    PROPAN 0.00    NPRPNL 0.00
   IPRPNL 0.00    TRMTAM 0.00    BUTENE 0.00    BTNON2 0.00    TTRHFR 0.00    DIOANE 0.00    ETHACT 0.00
   NBTANE 0.00    IBTANE 0.00    NBTNOL 0.00    IBTNOL 0.00    SBTNOL 0.00    TBTNOL 0.00    DETETH 0.00
   ETHCEL 0.00    DETAMN 0.00    FRFRAL 0.00    PYRDIN 0.00    ISOPRN 0.00    CYCPTN 0.00    ISOPTN 0.00

CHLBZN 0.00
   HYDRGN 0.00    NITRGN 0.00    OXYGEN 0.00    CRBMOX 0.00    ARGON  0.00    METHAN 0.00    CRBTCL 0.25
   CLRFRM 0.10    DCLMTN 0.00    FRMACD 0.40    NTRMTN 0.50    MTHNOL 0.00    CRBDOX 0.00    CRBDSF 0.00
   TCLETN 0.00    ACTYLN 0.30    ACTNTR 0.70    ETHLYN 0.10    DCLETN 0.00    ACTALD 0.40    ACTACD 0.40
   ETHIOD 0.00    NTRETH 0.50    ETHANE 0.00    ETHNOL 0.00    DMTAMN 0.00    ETHGLY 0.00    PRPYLN 0.10
   ACETON 0.50    MTHACT 0.60    PRPACD 0.40    NTPRP1 0.50    NTPRP2 0.50    PROPAN 0.00    NPRPNL 0.00
   IPRPNL 0.00    TRMTAM 0.00    BUTENE 0.10    BTNON2 0.50    TTRHFR 0.30    DIOANE 0.30    ETHACT 0.60
   NBTANE 0.00    IBTANE 0.00    NBTNOL 0.00    IBTNOL 0.00    SBTNOL 0.00    TBTNOL 0.00    DETETH 0.30
   ETHCEL 0.00    DETAMN 0.00    FRFRAL 0.40    PYRDIN 0.00    ISOPRN 0.10    CYCPTN 0.00    ISOPTN 0.00
   NORPTN 0.00

NTRBZN 1.66
   HYDRGN 0.00    NITRGN 0.00    OXYGEN 0.00    CRBMOX 0.00    ARGON  0.00    METHAN 0.00    CRBTCL 0.30
   CLRFRM 0.80    DCLMTN 0.40    FRMACD 2.20    NTRMTN 1.66    MTHNOL 1.20    CRBDOX 0.40    CRBDSF 0.20
   TCLETN 0.20    ACTYLN 0.50    ACTNTR 2.00    ETHLYN 0.30    DCLETN 0.30    ACTALD 1.40    ACTACD 2.20
   ETHIOD 0.00    NTRETH 1.66    ETHANE 0.00    ETHNOL 1.20    DMTAMN 1.40    ETHGLY 1.20    PRPYLN 0.20
   ACETON 1.63    MTHACT 1.80    PRPACD 2.20    NTPRP1 1.66    NTPRP2 1.66    PROPAN 0.00    NPRPNL 1.20
   IPRPNL 1.20    TRMTAM 0.00    BUTENE 0.20    BTNON2 1.63    TTRHFR 1.25    DIOANE 1.25    ETHACT 1.80
   NBTANE 0.00    IBTANE 0.00    NBTNOL 1.20    IBTNOL 1.20    SBTNOL 1.20    TBTNOL 1.20    DETETH 1.25
   ETHCEL 1.20    DETAMN 1.40    FRFRAL 1.40    PYRDIN 1.30    ISOPRN 0.20    CYCPTN 0.00    ISOPTN 0.00
   NORPTN 0.00    CHLBZN 0.50

BENZEN 0.00
   HYDRGN 0.00    NITRGN 0.00    OXYGEN 0.00    CRBMOX 0.00    ARGON  0.00    METHAN 0.00    CRBTCL 0.25
*****************************************************************************************
```

BENZEN 0.00

CLRFRM 0.10	DCLMTN 0.00	FRMACD 0.40	NTRMTN 0.50	MTHNOL 0.00	CRBDOX 0.00	CRBDSF 0.00
TCLETN 0.10	ACTYLN 0.30	ACTNTR 0.70	ETHLYN 0.10	DCLETN 0.00	ACTALD 0.40	ACTACD 0.40
ETHIOD 0.00	NTRETH 0.50	ETHANE 0.00	ETHNOL 0.00	DMTAMN 0.00	ETHGLY 0.00	PRPYLN 0.10
ACETON 0.50	MTHACT 0.60	PRPACD 0.40	NTPRP1 0.50	NTPRP2 0.50	PROPAN 0.00	NPRPNL 0.00
IPRPNL 0.00	TRMTAM 0.00	BUTENE 0.10	BTNON2 0.50	TTRHFR 0.30	DIOANE 0.30	ETHACT 0.60
NBTANE 0.00	IBTANE 0.00	NBTNOL 0.00	IBTNOL 0.00	SBTNOL 0.00	TBTNOL 0.00	DETETH 0.30
ETHCEL 0.00	DETAMN 0.00	FRFRAL 0.40	PYRDIN 0.00	ISOPRN 0.10	CYCPTN 0.00	ISOPTN 0.00
NORPTN 0.00	CHLBZN 0.00	NTRBZN 0.50				

PHENOL 0.32

HYDRGN 0.00	NITRGN 0.00	OXYGEN 0.00	CRBMOX 1.20	ARGON 0.00	METHAN 0.00	CRBTCL 0.25
CLRFRM 0.10	DCLMTN 0.00	FRMACD 2.50	NTRMTN 1.20	MTHNOL 1.55	CRBDOX 0.30	CRBDSF 0.00
TCLETN 0.00	ACTYLN 0.30	ACTNTR 1.50	ETHLYN 0.10	DCLETN 0.00	ACTALD 0.80	ACTACD 2.50
ETHIOD 0.00	NTRETH 1.20	ETHANE 0.00	ETHNOL 1.55	DMTAMN 0.20	ETHGLY 1.55	PRPYLN 1.55
ACETON 1.00	MTHACT 1.30	PRPACD 2.50	NTPRP1 1.00	NTPRP2 1.20	PROPAN 0.00	NPRPNL 1.55
IPRPNL 1.55	TRMTAM 0.20	BUTENE 0.10	BTNON2 1.00	TTRHFR 0.50	DIOANE 0.50	ETHACT 1.30
NBTANE 0.20	IBTANE 0.00	NBTNOL 1.55	IBTNOL 1.55	SBTNOL 1.55	TBTNOL 1.55	DETETH 0.50
ETHCEL 1.55	DETAMN 0.20	FRFRAL 0.80	PYRDIN 0.40	ISOPRN 0.10	CYCPTN 0.00	ISOPTN 0.00
NORPTN 0.00	CHLBZN 0.00	NTRBZN 0.00	BENZEN 0.00			

ANILIN 0.20

HYDRGN 0.00	NITRGN 0.00	OXYGEN 0.00	CRBMOX 0.00	ARGON 0.00	METHAN 0.00	CRBTCL 0.25
CLRFRM 0.30	DCLMTN 0.00	FRMACD 2.20	NTRMTN 1.40	MTHNOL 0.20	CRBDOX 0.20	CRBDSF 0.10
TCLETN 0.20	ACTYLN 0.20	ACTNTR 1.40	ETHLYN 0.20	DCLETN 0.20	ACTALD 0.85	ACTACD 2.20
ETHIOD 0.00	NTRETH 1.40	ETHANE 0.00	ETHNOL 0.20	DMTAMN 0.20	ETHGLY 0.20	PRPYLN 0.20
ACETON 1.00	MTHACT 1.10	PRPACD 2.20	NTPRP1 1.00	NTPRP2 1.40	PROPAN 0.00	NPRPNL 0.20
IPRPNL 0.20	TRMTAM 0.20	BUTENE 0.20	BTNON2 1.00	TTRHFR 0.70	DIOANE 0.70	ETHACT 1.10
NBTANE 0.00	IBTANE 0.00	NBTNOL 0.20	IBTNOL 0.20	SBTNOL 0.20	TBTNOL 0.20	DETETH 0.70
ETHCEL 0.20	DETAMN 0.20	FRFRAL 0.85	PYRDIN 0.20	ISOPRN 0.20	CYCPTN 0.00	ISOPTN 0.00
NORPTN 0.00	CHLBZN 0.00	NTRBZN 1.40	BENZEN 0.00	PHENOL 0.20		

CYHXON 0.90

HYDRGN 0.00	NITRGN 0.00	OXYGEN 0.00	CRBMOX 0.00	ARGON 0.00	METHAN 0.00	CRBTCL 0.20
CLRFRM 1.26	DCLMTN 0.40	FRMACD 1.80	NTRMTN 1.63	MTHNOL 1.00	CRBDOX 0.00	CRBDSF 0.00
TCLETN 0.00	ACTYLN 0.00	ACTNTR 1.30	ETHLYN 1.00	DCLETN 0.00	ACTALD 1.00	ACTACD 1.80
ETHIOD 0.00	NTRETH 1.63	ETHANE 0.00	ETHNOL 1.00	DMTAMN 1.00	ETHGLY 1.00	PRPYLN 1.00
ACETON 0.90	MTHACT 1.10	PRPACD 1.80	NTPRP1 1.63	NTPRP2 1.63	PROPAN 0.00	NPRPNL 1.00
IPRPNL 1.00	TRMTAM 1.00	BUTENE 0.00	BTNON2 0.90	TTRHFR 0.00	DIOANE 0.00	ETHACT 1.10
NBTANE 0.00	IBTANE 0.00	NBTNOL 1.00	IBTNOL 1.00	SBTNOL 1.00	TBTNOL 1.00	DETETH 0.00
ETHCEL 1.00	DETAMN 1.00	FRFRAL 1.00	PYRDIN 0.00	ISOPRN 0.00	CYCPTN 0.00	ISOPTN 0.00
NORPTN 0.00	CHLBZN 0.50	NTRBZN 1.63	BENZEN 0.50	PHENOL 1.00	ANILIN 1.00	

CYHXAN 0.00

HYDRGN 0.00	NITRGN 0.00	OXYGEN 0.00	CRBMOX 0.00	ARGON 0.00	METHAN 0.00	CRBTCL 0.00

CYHXAN 0.00

CLRFRM 0.00	DCLMTN 0.00	FRMACD 0.00	NTRMTN 0.00	MTHNOL 0.00	CRBDOX 0.00	CRBDSF 0.00
TCLETN 0.00	ACTYLN 0.00	ACTNTR 0.00	ETHLYN 0.00	DCLETN 0.00	ACTALD 0.00	ACTACD 0.00
ETHIOD 0.00	NTRETH 0.00	ETHANE 0.00	ETHNOL 0.00	DMTAMN 0.00	ETHGLY 0.00	PRPYLN 0.00
ACETON 0.00	MTHACT 0.00	PRPACD 0.00	NTPRP1 0.00	NTPRP2 0.00	DIOANE 0.00	NPRPNL 0.00
IPRPNL 0.00	TRMTAM 0.00	BUTENE 0.00	BTNON2 0.00	TTRHFR 0.00	TBTNOL 0.00	ETHACT 0.00
NBTANE 0.00	IBTANE 0.00	NBTNOL 0.00	IBTNOL 0.00	SBTNOL 0.00	CYCPTN 0.00	DETETH 0.00
ETHCEL 0.00	DETAMN 0.00	FRFRAL 0.00	PYRDIN 0.00	ISOPRN 0.00	ANILIN 0.00	ISOPTN 0.00
NORPTN 0.00	CHLBZN 0.00	NTRBZN 0.00	BENZEN 0.00	PHENOL 0.00		CYHXON 0.00

HEXENE 0.00

HYDRGN 0.00	NITRGN 0.00	OXYGEN 0.00	CRBMOX 0.00	ARGON 0.00	METHAN 0.00	CRBTCL 0.00
CLRFRM 0.00	DCLMTN 0.00	FRMACD 0.30	NTRMTN 0.20	MTHNOL 0.00	CRBDOX 0.00	CRBDSF 0.00
TCLETN 0.00	ACTYLN 0.00	ACTNTR 0.20	ETHLYN 0.00	DCLETN 0.00	ACTALD 0.00	ACTACD 0.30
ETHIOD 0.00	NTRETH 0.20	ETHANE 0.00	ETHNOL 0.00	DMTAMN 0.20	ETHGLY 0.00	PRPYLN 0.00
ACETON 0.00	MTHACT 0.00	PRPACD 0.30	NTPRP1 0.20	NTPRP2 0.20	PROPAN 0.00	NPRPNL 0.00
IPRPNL 0.00	TRMTAM 0.20	BUTENE 0.00	BTNON2 0.00	TTRHFR 0.00	DIOANE 0.00	ETHACT 0.00
NBTANE 0.00	IBTANE 0.00	NBTNOL 0.00	IBTNOL 0.00	SBTNOL 0.00	TBTNOL 0.00	DETETH 0.00
ETHCEL 0.00	DETAMN 0.00	FRFRAL 0.00	PYRDIN 0.05	ISOPRN 0.00	CYCPTN 0.00	ISOPTN 0.00
NORPTN 0.00	CHLBZN 0.00	NTRBZN 0.10	BENZEN 0.10	PHENOL 0.10	ANILIN 0.20	CYHXON 0.00
CYHXAN 0.00						

MTCYPN 0.00

HYDRGN 0.00	NITRGN 0.00	OXYGEN 0.00	CRBMOX 0.00	ARGON 0.00	METHAN 0.00	CRBTCL 0.00
CLRFRM 0.00	DCLMTN 0.00	FRMACD 0.00	NTRMTN 0.00	MTHNOL 0.00	CRBDOX 0.00	CRBDSF 0.00
TCLETN 0.00	ACTYLN 0.00	ACTNTR 0.00	ETHLYN 0.00	DCLETN 0.00	ACTALD 0.00	ACTACD 0.00
ETHIOD 0.00	NTRETH 0.00	ETHANE 0.00	ETHNOL 0.00	DMTAMN 0.00	ETHGLY 0.00	PRPYLN 0.00
ACETON 0.00	MTHACT 0.00	PRPACD 0.00	NTPRP1 0.00	NTPRP2 0.00	PROPAN 0.00	NPRPNL 0.00
IPRPNL 0.00	TRMTAM 0.00	BUTENE 0.00	BTNON2 0.00	TTRHFR 0.00	DIOANE 0.00	ETHACT 0.00
NBTANE 0.00	IBTANE 0.00	NBTNOL 0.00	IBTNOL 0.00	SBTNOL 0.00	TBTNOL 0.00	DETETH 0.00
ETHCEL 0.00	DETAMN 0.00	FRFRAL 0.00	PYRDIN 0.00	ISOPRN 0.00	CYCPTN 0.00	ISOPTN 0.00
NORPTN 0.00	CHLBZN 0.00	NTRBZN 0.00	BENZEN 0.00	PHENOL 0.00	ANILIN 0.00	CYHXON 0.00
CYHXAN 0.00	HEXENE 0.00					

CYHXNL 1.55

HYDRGN 0.00	NITRGN 0.00	OXYGEN 0.00	CRBMOX 0.00	ARGON 0.00	METHAN 0.00	CRBTCL 0.00
CLRFRM 0.10	DCLMTN 0.00	FRMACD 2.50	NTRMTN 1.20	MTHNOL 1.55	CRBDOX 0.30	CRBDSF 0.15
TCLETN 0.00	ACTYLN 0.00	ACTNTR 1.50	ETHLYN 0.00	DCLETN 0.00	ACTALD 0.80	ACTACD 2.50
ETHIOD 0.00	NTRETH 1.20	ETHANE 1.20	ETHNOL 1.55	DMTAMN 1.20	ETHGLY 1.55	PRPYLN 0.00
ACETON 1.00	MTHACT 1.30	PRPACD 2.50	NTPRP1 1.20	NTPRP2 1.20	PROPAN 0.50	NPRPNL 1.55
IPRPNL 1.55	TRMTAM 0.20	BUTENE 0.00	BTNON2 1.00	TTRHFR 0.50	DIOANE 0.50	ETHACT 1.30
NBTANE 0.00	IBTANE 0.00	NBTNOL 1.55	IBTNOL 1.55	SBTNOL 1.55	TBTNOL 1.55	DETETH 0.50
ETHCEL 1.55	DETAMN 0.20	FRFRAL 0.80	PYRDIN 0.40	ISOPRN 1.00	CYCPTN 0.00	ISOPTN 0.00
NORPTN 0.00	CHLBZN 0.00	NTRBZN 0.00	BENZEN 0.00	PHENOL 1.55	ANILIN 0.20	CYHXON 1.00

CYHXNL 1.55   CYHXAN 0.00   HEXENE 0.00   MTCYPN 0.00

MISBKT 0.90

HYDRGN 0.00	NITRGN 0.00	OXYGEN 0.00	CRBMOX 0.00	ARGON 0.00	METHAN 0.00	CRBTCL 0.20
CLRFRM 1.26	DCLMTN 0.00	FRMACD 1.80	NTRMTN 1.63	MTHNOL 1.00	CRBDOX 1.00	CRBDSF 0.00
TCLETN 0.00	ACTYLN 0.00	ACTNTR 1.30	ETHLYN 0.00	DCLETN 0.00	ACTALD 1.00	ACTACD 1.80
ETHIOD 0.00	NTRETH 1.63	ETHANE 0.00	ETHNOL 0.00	DMTAMN 0.00	ETHGLY 1.00	PRPYLN 0.00
ACETON 0.90	MTHACT 1.10	PRPACD 1.80	NTPRP1 1.63	NTPRP2 1.63	PROPAN 0.00	NPRPNL 1.00
IPRPNL 1.00	TRMTAM 1.00	BUTENE 0.00	BTNON2 0.90	TTRHFR 0.00	DIOANE 0.00	ETHACT 1.10
NBTANE 0.00	IBTANE 0.00	NBTNOL 1.00	IBTNOL 1.00	SBTNOL 1.00	TBTNOL 1.00	DETETH 0.00
ETHCEL 1.00	DETAMN 1.00	FRFRAL 1.00	PYRDIN 1.00	ISOPRN 0.00	CYCPTN 0.00	ISOPTN 0.00
NORPTN 0.00	CHLBZN 0.50	NTRBZN 1.63	BENZEN 0.50	PHENOL 1.00	ANILIN 1.00	CYHXON 0.00
CYHXAN 0.00	HEXENE 0.00	MTCYPN 0.00	CYHXNL 1.00			CYHXON 0.90

NBTACT 0.53

HYDRGN 0.00	NITRGN 0.00	OXYGEN 0.00	CRBMOX 0.00	ARGON 0.00	METHAN 0.00	CRBTCL 0.25
CLRFRM 1.55	DCLMTN 0.50	FRMACD 2.00	NTRMTN 1.80	MTHNOL 1.30	CRBDOX 0.00	CRBDSF 0.00
TCLETN 0.00	ACTYLN 0.00	ACTNTR 1.40	ETHLYN 0.00	DCLETN 0.00	ACTALD 0.75	ACTACD 2.00
ETHIOD 0.00	NTRETH 1.80	ETHANE 0.00	ETHNOL 1.30	DMTAMN 1.10	ETHGLY 1.30	PRPYLN 0.00
ACETON 1.10	MTHACT 0.53	PRPACD 2.00	NTPRP1 1.80	NTPRP2 1.80	PROPAN 0.00	NPRPNL 1.30
IPRPNL 1.30	TRMTAM 1.00	BUTENE 0.00	BTNON2 1.10	TTRHFR 0.00	DIOANE 0.00	ETHACT 0.53
NBTANE 0.00	IBTANE 0.00	NBTNOL 1.30	IBTNOL 0.30	SBTNOL 1.30	TBTNOL 1.30	DETETH 0.00
ETHCEL 1.30	DETAMN 1.10	FRFRAL 0.75	PYRDIN 0.30	ISOPRN 0.00	CYCPTN 0.00	ISOPTN 0.00
NORPTN 0.00	CHLBZN 0.60	NTRBZN 1.80	BENZEN 0.60	PHENOL 1.30	ANILIN 1.10	CYHXON 1.10
CYHXAN 0.00	HEXENE 0.00	MTCYPN 0.00	CYHXNL 1.30	MISBKT 1.10		

NHEXAN 0.00

HYDRGN 0.00	NITRGN 0.00	OXYGEN 0.00	CRBMOX 0.00	ARGON 0.00	METHAN 0.00	CRBTCL 0.00
CLRFRM 0.00	DCLMTN 0.00	FRMACD 0.00	NTRMTN 0.00	MTHNOL 0.00	CRBDOX 0.00	CRBDSF 0.00
TCLETN 0.00	ACTYLN 0.00	ACTNTR 0.00	ETHLYN 0.00	DCLETN 0.00	ACTALD 0.00	ACTACD 0.00
ETHIOD 0.00	NTRETH 0.00	ETHANE 0.00	ETHNOL 0.00	DMTAMN 0.00	ETHGLY 0.00	PRPYLN 0.00
ACETON 0.00	MTHACT 0.00	PRPACD 0.00	NTPRP1 0.00	NTPRP2 0.00	PROPAN 0.00	NPRPNL 0.00
IPRPNL 0.00	TRMTAM 0.00	BUTENE 0.00	BTNON2 0.00	TTRHFR 0.00	DIOANE 0.00	ETHACT 0.00
NBTANE 0.00	IBTANE 0.00	NBTNOL 0.00	IBTNOL 0.00	SBTNOL 0.00	TBTNOL 0.00	DETETH 0.00
ETHCEL 0.00	DETAMN 0.00	FRFRAL 0.00	PYRDIN 0.00	ISOPRN 0.00	CYCPTN 0.00	ISOPTN 0.00
NORPTN 0.00	CHLBZN 0.00	NTRBZN 0.00	BENZEN 0.00	PHENOL 0.00	ANILIN 0.00	CYHXON 0.00
CYHXAN 0.00	HEXENE 0.00	MTCYPN 0.00	CYHXNL 0.00	MISBKT 0.00	NBTACT 0.00	

DMBT23 0.00

HYDRGN 0.00	NITRGN 0.00	OXYGEN 0.00	CRBMOX 0.00	ARGON 0.00	METHAN 0.00	CRBTCL 0.00
CLRFRM 0.00	DCLMTN 0.00	FRMACD 0.00	NTRMTN 0.00	MTHNOL 0.00	CRBDOX 0.00	CRBDSF 0.00
TCLETN 0.00	ACTYLN 0.00	ACTNTR 0.00	ETHLYN 0.00	DCLETN 0.00	ACTALD 0.00	ACTACD 0.00
ETHIOD 0.00	NTRETH 0.00	ETHANE 0.00	ETHNOL 0.00	DMTAMN 0.00	ETHGLY 0.00	PRPYLN 0.00

171

```
* * ASSOCIATION AND SOLVATION PARAMETERS * *
```

DMBT23 0.00

ACETON 0.00	MTHACT 0.00	PRPACD 0.00	NTPRP1 0.00	NTPRP2 0.00	PROPAN 0.00	NPRPNL 0.00
IPRPNL 0.00	TRMTAM 0.00	BUTENE 0.00	BTNON2 0.00	TTRHFR 0.00	DIOANE 0.00	ETHACT 0.00
NBTANE 0.00	IBTANE 0.00	NBTNOL 0.00	IBTNOL 0.00	SBTNOL 0.00	TBTNOL 0.00	DETETH 0.00
ETHCEL 0.00	DETAMN 0.00	FRFRAL 0.00	PYRDIN 0.00	ISOPRN 0.00	CYCPTN 0.00	ISOPTN 0.00
NORPTN 0.00	CHLBZN 0.00	NTRBZN 0.00	BENZEN 0.00	PHENOL 0.00	ANILIN 0.00	CYHXON 0.00
CYHXAN 0.00	HEXENE 0.00	MTCYPN 0.00	CYHXNL 0.00	MISBKT 0.00	NBTACT 0.00	NHEXAN 0.00

TRETAM 0.06

HYDRGN 0.00	NITRGN 0.00	OXYGEN 0.00	CRBMOX 0.00	ARGON 0.00	METHAN 0.00	CRBTCL 0.00
CLRFRM 0.30	DCLMTN 0.00	FRMACD 2.20	NTRMTN 1.40	MTHNOL 0.20	CRBDOX 0.20	CRBDSF 0.10
TCLETN 0.20	ACTYLN 0.50	ACTNTR 1.40	ETHLYN 0.20	DCLETN 0.00	ACTALD 0.85	ACTACD 2.20
ETHIOD 0.00	NTRETH 1.40	ETHANE 0.00	ETHNOL 0.20	DMTAMN 0.20	ETHGLY 0.20	PRPYLN 0.20
ACETON 1.00	MTHACT 1.10	PRPACD 1.40	NTPRP1 1.40	NTPRP2 1.40	PROPAN 0.00	NPRPNL 0.20
IPRPNL 0.20	TRMTAM 0.20	BUTENE 0.20	BTNON2 1.00	TTRHFR 0.70	DIOANE 0.70	ETHACT 1.10
NBTANE 0.00	IBTANE 0.00	NBTNOL 0.20	IBTNOL 0.20	SBTNOL 0.20	TBTNOL 0.20	DETETH 0.70
ETHCEL 0.70	DETAMN 0.20	FRFRAL 0.85	PYRDIN 0.20	ISOPRN 0.20	CYCPTN 0.00	ISOPTN 0.00
NORPTN 0.00	CHLBZN 0.00	NTRBZN 0.50	BENZEN 0.20	PHENOL 0.20	ANILIN 0.20	CYHXON 1.00
CYHXAN 0.00	HEXENE 0.20	MTCYPN 0.00	CYHXNL 0.20	MISBKT 1.00	NBTACT 1.10	NHEXAN 0.00
DMBT23 0.00						

TOLUEN 0.00

HYDRGN 0.00	NITRGN 0.00	OXYGEN 0.00	CRBMOX 0.00	ARGON 0.00	METHAN 0.00	CRBTCL 0.25
CLRFRM 0.10	DCLMTN 0.00	FRMACD 0.40	NTRMTN 0.50	MTHNOL 0.50	CRBDOX 0.00	CRBDSF 0.00
TCLETN 0.00	ACTYLN 0.30	ACTNTR 0.70	ETHLYN 0.10	DCLETN 0.10	ACTALD 0.40	ACTACD 0.40
ETHIOD 0.00	NTRETH 0.50	ETHANE 0.00	ETHNOL 0.00	DMTAMN 0.00	ETHGLY 0.00	PRPYLN 0.10
ACETON 0.50	MTHACT 0.00	PRPACD 0.40	NTPRP1 0.50	NTPRP2 0.50	PROPAN 0.00	NPRPNL 0.00
IPRPNL 0.00	TRMTAM 0.00	BUTENE 0.10	BTNON2 0.50	TTRHFR 0.30	DIOANE 0.30	ETHACT 0.60
NBTANE 0.00	IBTANE 0.00	NBTNOL 0.00	IBTNOL 0.00	SBTNOL 0.00	TBTNOL 0.00	DETETH 0.30
ETHCEL 0.30	DETAMN 0.00	FRFRAL 0.40	PYRDIN 0.00	ISOPRN 0.10	CYCPTN 0.00	ISOPTN 0.00
NORPTN 0.00	CHLBZN 0.00	NTRBZN 0.50	BENZEN 0.00	PHENOL 0.00	ANILIN 0.00	CYHXON 0.50
CYHXAN 0.00	HEXENE 0.10	MTCYPN 0.00	CYHXNL 0.00	MISBKT 0.50	NBTACT 0.60	NHEXAN 0.00
DMBT23 0.00	TRETAM 0.00					

MTCYHX 0.00

HYDRGN 0.00	NITRGN 0.00	OXYGEN 0.00	CRBMOX 0.00	ARGON 0.00	METHAN 0.00	CRBTCL 0.00
CLRFRM 0.00	DCLMTN 0.00	FRMACD 0.00	NTRMTN 0.00	MTHNOL 0.00	CRBDOX 0.00	CRBDSF 0.00
TCLETN 0.00	ACTYLN 0.00	ACTNTR 0.00	ETHLYN 0.00	DCLETN 0.00	ACTALD 0.00	ACTACD 0.00
ETHIOD 0.00	NTRETH 0.00	ETHANE 0.00	ETHNOL 0.00	DMTAMN 0.00	ETHGLY 0.00	PRPYLN 0.00
ACETON 0.00	MTHACT 0.00	PRPACD 0.00	NTPRP1 0.00	NTPRP2 0.00	PROPAN 0.00	NPRPNL 0.00
IPRPNL 0.00	TRMTAM 0.00	BUTENE 0.00	BTNON2 0.00	TTRHFR 0.00	DIOANE 0.00	ETHACT 0.00
NBTANE 0.00	IBTANE 0.00	NBTNOL 0.00	IBTNOL 0.00	SBTNOL 0.00	TBTNOL 0.00	DETETH 0.00
ETHCEL 0.00	DETAMN 0.00	FRFRAL 0.00	PYRDIN 0.00	ISOPRN 0.00	CYCPTN 0.00	ISOPTN 0.00
NORPTN 0.00	CHLBZN 0.00	NTRBZN 0.00	BENZEN 0.00	PHENOL 0.00	ANILIN 0.00	CYHXON 0.00

```
*****************************************************************************************
```

* * ASSOCIATION AND SOLVATION PARAMETERS * *

```
*****************************************************************************************
```

MTCYHX 0.00

CYHXAN 0.00	HEXENE 0.00	MTCYPN 0.00	CYHXNL 0.00	MISBKT 0.00	NBTACT 0.00	NHEXAN 0.00
DMBT23 0.00	TRETAM 0.00	TOLUEN 0.00				

NHEPTN 0.00

HYDRGN 0.00	NITRGN 0.00	OXYGEN 0.00	CRBMOX 0.00	ARGON 0.00	METHAN 0.00	CRBTCL 0.00
CLRFRM 0.00	DCLMTN 0.00	FRMACD 0.00	NTRMTN 0.00	MTHNOL 0.00	CRBDOX 0.00	CRBDSF 0.00
TCLETN 0.00	ACTYLN 0.00	ACTNTR 0.00	ETHLYN 0.00	DCLETN 0.00	ACTALD 0.00	ACTACD 0.00
ETHIOD 0.00	NTRETH 0.00	ETHANE 0.00	ETHNOL 0.00	DMTAMN 0.00	ETHGLY 0.00	PRPYLN 0.00
ACETON 0.00	MTHACT 0.00	PRPACD 0.00	NTPRP1 0.00	NTPRP2 0.00	PROPAN 0.00	NPRPNL 0.00
IPRPNL 0.00	TRMTAM 0.00	BUTENE 0.00	BTNON2 0.00	TTRHFR 0.00	DIOANE 0.00	ETHACT 0.00
NBTANE 0.00	IBTANE 0.00	NBTNOL 0.00	IBTNOL 0.00	SBTNOL 0.00	TBTNOL 0.00	DETETH 0.00
ETHCEL 0.00	DETAMN 0.00	FRFRAL 0.00	PYRDIN 0.00	ISOPRN 0.00	CYCPTN 0.00	ISOPTN 0.00
NORPTN 0.00	CHLBZN 0.00	NTRBZN 0.00	BENZEN 0.00	PHENOL 0.00	ANILIN 0.00	CYHXON 0.00
CYHXAN 0.00	HEXENE 0.00	MTCYPN 0.00	CYHXNL 0.00	MISBKT 0.00	NBTACT 0.00	NHEXAN 0.00
DMBT23 0.00	TRETAM 0.00	TOLUEN 0.00	MTCYHX 0.00			

STYREN 0.00

HYDRGN 0.00	NITRGN 0.00	OXYGEN 0.00	CRBMOX 0.00	ARGON 0.00	METHAN 0.00	CRBTCL 0.25
CLRFRM 0.10	DCLMTN 0.00	FRMACD 0.40	NTRMTN 0.50	MTHNOL 0.00	CRBDOX 0.00	CRBDSF 0.00
TCLETN 0.00	ACTYLN 0.30	ACTNTR 0.70	ETHLYN 0.10	DCLETN 0.00	ACTALD 0.40	ACTACD 0.40
ETHIOD 0.00	NTRETH 0.50	ETHANE 0.00	ETHNOL 0.00	DMTAMN 0.00	ETHGLY 0.00	PRPYLN 0.10
ACETON 0.50	MTHACT 0.00	PRPACD 0.40	NTPRP1 0.00	NTPRP2 0.50	PROPAN 0.00	NPRPNL 0.00
IPRPNL 0.00	TRMTAM 0.00	BUTENE 0.10	BTNON2 0.50	TTRHFR 0.30	DIOANE 0.30	ETHACT 0.60
NBTANE 0.00	IBTANE 0.00	NBTNOL 0.00	IBTNOL 0.00	SBTNOL 0.00	TBTNOL 0.00	DETETH 0.30
ETHCEL 0.30	DETAMN 0.00	FRFRAL 0.40	PYRDIN 0.00	ISOPRN 0.10	CYCPTN 0.00	ISOPTN 0.00
NORPTN 0.00	CHLBZN 0.00	NTRBZN 0.50	BENZEN 0.00	PHENOL 0.00	ANILIN 0.00	CYHXON 0.50
CYHXAN 0.00	HEXENE 0.10	MTCYPN 0.00	CYHXNL 0.00	MISBKT 0.00	NBTACT 0.60	NHEXAN 0.00
DMBT23 0.00	TRETAM 0.00	TOLUEN 0.00	MTCYHX 0.00	NHEPTN 0.00		

ETHBZN 0.00

HYDRGN 0.00	NITRGN 0.00	OXYGEN 0.00	CRBMOX 0.00	ARGON 0.00	METHAN 0.00	CRBTCL 0.25
CLRFRM 0.10	DCLMTN 0.00	FRMACD 0.40	NTRMTN 0.50	MTHNOL 0.00	CRBDOX 0.00	CRBDSF 0.00
TCLETN 0.00	ACTYLN 0.30	ACTNTR 0.70	ETHLYN 0.10	DCLETN 0.10	ACTALD 0.40	ACTACD 0.40
ETHIOD 0.00	NTRETH 0.50	ETHANE 0.00	ETHNOL 0.00	DMTAMN 0.00	ETHGLY 0.00	PRPYLN 0.10
ACETON 0.50	MTHACT 0.60	PRPACD 0.10	NTPRP1 0.00	NTPRP2 0.50	PROPAN 0.00	NPRPNL 0.00
IPRPNL 0.00	TRMTAM 0.00	BUTENE 0.40	BTNON2 0.50	TTRHFR 0.30	DIOANE 0.30	ETHACT 0.60
NBTANE 0.00	IBTANE 0.00	NBTNOL 0.00	IBTNOL 0.00	SBTNOL 0.00	TBTNOL 0.00	DETETH 0.30
ETHCEL 0.30	DETAMN 0.00	FRFRAL 0.40	PYRDIN 0.00	ISOPRN 0.10	CYCPTN 0.00	ISOPTN 0.00
NORPTN 0.00	CHLBZN 0.00	NTRBZN 0.50	BENZEN 0.00	PHENOL 0.00	ANILIN 0.00	CYHXON 0.50
CYHXAN 0.00	HEXENE 0.10	MTCYPN 0.00	CYHXNL 0.00	MISBKT 0.50	NBTACT 0.60	NHEXAN 0.00
DMBT23 0.00	TRETAM 0.00	TOLUEN 0.00	MTCYHX 0.00	NHEPTN 0.00	STYREN 0.00	

MXYLEN 0.00

HYDRGN 0.00	NITRGN 0.00	OXYGEN 0.00	CRBMOX 0.00	ARGON 0.00	METHAN 0.00	CRBTCL 0.25

```
*****************************************************************************************
```

* * ASSOCIATION AND SOLVATION PARAMETERS * *

```
*************************************************************************

MXYLEN 0.00
  CLRFRM 0.10   DCLMTN 0.00   FRMACD 0.40   NTRMTN 0.50   MTHNOL 0.00   CRBDOX 0.00   CRBDSF 0.00
  TCLETN 0.00   ACTYLN 0.30   ACTNTR 0.70   ETHLYN 0.10   DCLETN 0.00   ACTALD 0.40   ACTALD 0.40
  ETHIOD 0.00   NTRETH 0.50   ETHANE 0.00   ETHNOL 0.00   DMTAMN 0.00   ETHGLY 0.00   PRPYLN 0.10
  ACETON 0.50   MTHACT 0.60   PRPACD 0.40   NTPRP1 0.50   NTPRP2 0.30   PROPAN 0.00   NPRPNL 0.00
  IPRPNL 0.00   TRMTAM 0.00   BUTENE 0.00   BTNON2 0.50   TTRHFR 0.30   DIOANE 0.30   ETHACT 0.60
  NBTANE 0.00   IBTANE 0.00   NBTNOL 0.00   IBTNOL 0.00   SBTNOL 0.00   TBTNOL 0.00   DETETH 0.30
  ETHCEL 0.30   DETAMN 0.00   FRFRAL 0.40   PYRDIN 0.00   ISOPRN 0.10   CYCPTN 0.00   ISOPTN 0.00
  NORPTN 0.00   CHLBZN 0.00   NTRBZN 0.50   BENZEN 0.00   PHENOL 0.00   ANILIN 0.00   CYHXON 0.50
  CYHXAN 0.00   HEXENE 0.10   MTCYPN 0.00   CYHXNL 0.00   MISBKT 0.50   NBTACT 0.60   NHEXAN 0.60
  DMBT23 0.00   TRETAM 0.00   TOLUEN 0.00   MTCYHX 0.00   NHEPTN 0.00   STYREN 0.00   ETHBZN 0.00

OXYLEN 0.00
  HYDRGN 0.00   NITRGN 0.00   OXYGEN 0.00   CRBMOX 0.00   ARGON  0.00   METHAN 0.00   CRBTCL 0.25
  CLRFRM 0.10   DCLMTN 0.10   FRMACD 0.40   NTRMTN 0.00   MTHNOL 0.00   CRBDOX 0.00   CRBDSF 0.00
  TCLETN 0.00   ACTYLN 0.30   ACTNTR 0.70   ETHLYN 0.10   DCLETN 0.00   ACTALD 0.40   ACTACD 0.40
  ETHIOD 0.00   NTRETH 0.50   ETHANE 0.00   ETHNOL 0.00   DMTAMN 0.00   ETHGLY 0.00   PRPYLN 0.10
  ACETON 0.50   MTHACT 0.60   PRPACD 0.40   NTPRP1 0.50   NTPRP2 0.50   PROPAN 0.00   NPRPNL 0.00
  IPRPNL 0.00   TRMTAM 0.00   BUTENE 0.10   BTNON2 0.50   TTRHFR 0.30   DIOANE 0.30   ETHACT 0.60
  NBTANE 0.00   IBTANE 0.00   NBTNOL 0.00   IBTNOL 0.00   SBTNOL 0.00   TBTNOL 0.00   DETETH 0.30
  ETHCEL 0.30   DETAMN 0.00   FRFRAL 0.00   PYRDIN 0.00   ISOPRN 0.10   CYCPTN 0.00   ISOPTN 0.00
  NORPTN 0.00   CHLBZN 0.00   NTRBZN 0.50   BENZEN 0.00   PHENOL 0.00   ANILIN 0.00   CYHXON 0.50
  CYHXAN 0.00   HEXENE 0.10   MTCYPN 0.00   CYHXNL 0.00   MISBKT 0.50   NBTACT 0.60   NHEXAN 0.00
  DMBT23 0.00   TRETAM 0.00   TOLUEN 0.00   MTCYHX 0.00   NHEPTN 0.00   STYREN 0.00   ETHBZN 0.00
  MXYLEN 0.00

PXYLEN 0.00
  HYDRGN 0.00   NITRGN 0.00   OXYGEN 0.00   CRBMOX 0.00   ARGON  0.00   METHAN 0.00   CRBTCL 0.25
  CLRFRM 0.10   DCLMTN 0.10   FRMACD 0.40   NTRMTN 0.50   MTHNOL 0.00   CRBDOX 0.00   CRBDSF 0.00
  TCLETN 0.00   ACTYLN 0.30   ACTNTR 0.70   ETHLYN 0.10   DCLETN 0.00   ACTALD 0.40   ACTACD 0.40
  ETHIOD 0.00   NTRETH 0.50   ETHANE 0.00   ETHNOL 0.00   DMTAMN 0.00   ETHGLY 0.00   PRPYLN 0.10
  ACETON 0.50   MTHACT 0.60   PRPACD 0.40   NTPRP1 0.50   NTPRP2 0.50   PROPAN 0.00   NPRPNL 0.00
  IPRPNL 0.00   TRMTAM 0.00   BUTENE 0.10   BTNON2 0.50   TTRHFR 0.30   DIOANE 0.30   ETHACT 0.60
  NBTANE 0.00   IBTANE 0.00   NBTNOL 0.00   IBTNOL 0.00   SBTNOL 0.00   TBTNOL 0.00   DETETH 0.30
  ETHCEL 0.30   DETAMN 0.00   FRFRAL 0.40   PYRDIN 0.00   ISOPRN 0.10   CYCPTN 0.00   ISOPTN 0.00
  NORPTN 0.00   CHLBZN 0.00   NTRBZN 0.00   BENZEN 0.00   PHENOL 0.00   ANILIN 0.00   CYHXON 0.50
  CYHXAN 0.00   HEXENE 0.10   MTCYPN 0.00   CYHXNL 0.00   MISBKT 0.50   NBTACT 0.60   NHEXAN 0.00
  DMBT23 0.00   TRETAM 0.00   TOLUEN 0.00   MTCYHX 0.00   NHEPTN 0.00   STYREN 0.00   ETHBZN 0.00
  MXYLEN 0.00

NOCTAN 0.00
  HYDRGN 0.00   NITRGN 0.00   OXYGEN 0.00   CRBMOX 0.00   ARGON  0.00   METHAN 0.00   CRBTCL 0.00
  CLRFRM 0.00   DCLMTN 0.00   FRMACD 0.00   NTRMTN 0.00   MTHNOL 0.00   CRBDOX 0.00   CRBDSF 0.00
  TCLETN 0.00   ACTYLN 0.00   ACTNTR 0.00   ETHLYN 0.00   DCLETN 0.00   ACTALD 0.00   ACTACD 0.00

*************************************************************************
```

* * ASSOCIATION AND SOLVATION PARAMETERS * *

NOCTAN 0.00

ETHIOD 0.00	NTRETH 0.00	ETHANE 0.00	ETHNOL 0.00	DMTAMN 0.00	ETHGLY 0.00	PRPYLN 0.00
ACETON 0.00	MTHACT 0.00	PRPACD 0.00	NTPRP1 0.00	NTPRP2 0.00	PROPAN 0.00	NPRPNL 0.00
IPRPNL 0.00	TRNTAM 0.00	BUTENE 0.00	BTNON2 0.00	TTRHFR 0.00	DIOANE 0.00	ETHACT 0.00
NBTANE 0.00	IBTANE 0.00	NBTNOL 0.00	IBTNOL 0.00	SBTNOL 0.00	TBTNOL 0.00	DETETH 0.00
ETHCEL 0.00	DETAMN 0.00	FRFRAL 0.00	PYRDIN 0.00	ISOPRN 0.00	ANILIN 0.00	ISOPTN 0.00
NORPTN 0.00	CHLBZN 0.00	NTRBZN 0.00	BENZEN 0.00	PHENOL 0.00	NBTACT 0.00	CYHXON 0.00
CYHXAN 0.00	HEXENE 0.00	MTCYPN 0.00	CYHXNL 0.00	MISBKT 0.00	STYREN 0.00	NHEXAN 0.00
DMBT23 0.00	TRETAM 0.00	TOLUEN 0.00	MTCYHX 0.00	NHEPTN 0.00		ETHBZN 0.00
MXYLEN 0.00	OXYLEN 0.00	PXYLEN 0.00				

TMP224 0.00

HYDRGN 0.00	NITRGN 0.00	OXYGEN 0.00	CRBMOX 0.00	ARGON 0.00	METHAN 0.00	CRBTCL 0.00
CLRFRM 0.00	DCLMTN 0.00	FRMACD 0.00	NTRMTN 0.00	MTHNOL 0.00	CRBDOX 0.00	CRBDSF 0.00
TCLETN 0.00	ACTYLN 0.00	ACTNTR 0.00	ETHLYN 0.00	DCLETN 0.00	ACTALD 0.00	ACTACD 0.00
ETHIOD 0.00	NTRETH 0.00	ETHANE 0.00	ETHNOL 0.00	DMTAMN 0.00	ETHGLY 0.00	PRPYLN 0.00
ACETON 0.00	MTHACT 0.00	PRPACD 0.00	NTPRP1 0.00	NTPRP2 0.00	PROPAN 0.00	NPRPNL 0.00
IPRPNL 0.00	TRMTAM 0.00	BUTENE 0.00	BTNON2 0.00	TTRHFR 0.00	DIOANE 0.00	ETHACT 0.00
NBTANE 0.00	IBTANE 0.00	NBTNOL 0.00	IBTNOL 0.00	SBTNOL 0.00	TBTNOL 0.00	DETETH 0.00
ETHCEL 0.00	DETAMN 0.00	FRFRAL 0.00	PYRDIN 0.00	ISOPRN 0.00	CYCPTN 0.00	ISOPTN 0.00
NORPTN 0.00	CHLBZN 0.00	NTRBZN 0.00	BENZEN 0.00	PHENOL 0.00	ANILIN 0.00	CYHXON 0.00
CYHXAN 0.00	HEXENE 0.00	MTCYPN 0.00	CYHXNL 0.00	MISBKT 0.00	NBT ACT 0.00	NHEXAN 0.00
DMBT23 0.00	TRETAM 0.00	TOLUEN 0.00	MTCYHX 0.00	NHEPTN 0.00	STYREN 0.00	ETHBZN 0.00
MXYLEN 0.00	OXYLEN 0.00	PXYLEN 0.00	NOCTAN 0.00			

NDECAN 0.00

HYDRGN 0.00	NITRGN 0.00	OXYGEN 0.00	CRBMOX 0.00	ARGON 0.00	METHAN 0.00	CRBTCL 0.00
CLRFRM 0.00	DCLMTN 0.00	FRMACD 0.00	NTRMTN 0.00	MTHNOL 0.00	CRBDOX 0.00	CRBDSF 0.00
TCLETN 0.00	ACTYLN 0.00	ACTNTR 0.00	ETHLYN 0.00	DCLETN 0.00	ACTALD 0.00	ACTACD 0.00
ETHIOD 0.00	NTRETH 0.00	ETHANE 0.00	ETHNOL 0.00	DMTAMN 0.00	ETHGLY 0.00	PRPYLN 0.00
AGETON 0.00	MTHACT 0.00	PRPACD 0.00	NTPRP1 0.00	NTPRP2 0.00	PROPAN 0.00	NPRPNL 0.00
IPRPNL 0.00	TRMTAM 0.00	BUTENE 0.00	BTNON2 0.00	TTRHFR 0.00	DIOANE 0.00	ETHACT 0.00
NBTANE 0.00	IBTANE 0.00	NBTNOL 0.00	IBTNOL 0.00	SBTNOL 0.00	TBTNOL 0.00	DETETH 0.00
ETHCEL 0.00	DETAMN 0.00	FRFRAL 0.00	PYRDIN 0.00	ISOPRN 0.00	CYCPTN 0.00	ISOPTN 0.00
NORPTN 0.00	CHLBZN 0.00	NTRBZN 0.00	BENZEN 0.00	PHENOL 0.00	ANILIN 0.00	CYHXON 0.00
CYHXAN 0.00	HEXENE 0.00	MTCYPN 0.00	CYHXNL 0.00	MISBKT 0.00	NBTACT 0.00	NHEXAN 0.00
DMBT23 0.00	TRETAM 0.00	TOLUEN 0.00	MTCYHX 0.00	NHEPTN 0.00	STYREN 0.00	ETHBZN 0.00
MXYLEN 0.00	OXYLEN 0.00	PXYLEN 0.00	NOCTAN 0.00	TMP224 0.00		

NHXDCN 0.00

HYDRGN 0.00	NITRGN 0.00	OXYGEN 0.00	CRBMOX 0.00	ARGON 0.00	METHAN 0.00	CRBTCL 0.00
CLRFRM 0.00	DCLMTN 0.00	FRMACD 0.00	NTRMTN 0.00	MTHNOL 0.00	CRBDOX 0.00	CRBDSF 0.00
TCLETN 0.00	ACTYLN 0.00	ACTNTR 0.00	ETHLYN 0.00	DCLETN 0.00	ACTALD 0.00	ACTACD 0.00
ETHIOD 0.00	NTRETH 0.00	ETHANE 0.00	ETHNOL 0.00	DMTAMN 0.00	ETHGLY 0.00	PRPYLN 0.00

175

NHXDCN 0.00

ACETON 0.00	MTHACT 0.00	PRPACD 0.00	NTPRP1 0.00	NTPRP2 0.00	PROPAN 0.00	NPRPNL 0.00
IPRPNL 0.00	TRMTAM 0.00	BUTENE 0.00	BTNON2 0.00	TTRHFR 0.00	DIOANE 0.00	ETHACT 0.00
NBTANE 0.00	IBTANE 0.00	NBTNOL 0.00	IBTNOL 0.00	SBTNOL 0.00	TBTNOL 0.00	DETETH 0.00
ETHCEL 0.00	DETAMN 0.00	FRFRAL 0.00	PYRDIN 0.00	ISOPRN 0.00	CYCPTN 0.00	ISOPTN 0.00
NORPTN 0.00	CHLBZN 0.00	NTRBZN 0.00	BENZEN 0.00	PHENOL 0.00	ANILIN 0.00	CYHXON 0.00
CYHXAN 0.00	HEXENE 0.00	MTCYPN 0.00	CYHXNL 0.00	MISBKT 0.00	NBTACT 0.00	NHEXAN 0.00
DMBT23 0.00	TRETAM 0.00	TOLUEN 0.00	MTCYHX 0.00	NHEPTN 0.00	STYREN 0.00	ETHBZN 0.00
MXYLEN 0.00	OXYLEN 0.00	PXYLEN 0.00	NOCTAN 0.00	TMP224 0.00	NDECAN 0.00	

HYDCHL 0.00

HYDRGN 0.00	NITRGN 0.00	OXYGEN 0.00	CRBMOX 0.00	ARGON 0.00	METHAN 0.00	CRBTCL 0.00
CLRFRM 0.00	DCLMTN 0.00	FRMACD 1.00	NTRMTN 1.20	MTHNOL 1.30	CRBDOX 0.00	CRBDSF 0.00
TCLETN 0.00	ACTYLN 0.00	ACTNTR 0.50	ETHLYN 0.00	DCLETN 0.00	ACTALD 0.75	ACTACD 1.00
ETHIOD 0.00	NTRETH 1.20	ETHANE 0.00	ETHNOL 1.38	DMTAMN 2.20	ETHGLY 1.38	PRPYLN 0.00
ACETON 0.50	MTHACT 0.50	PRPACD 1.00	NTPRP1 1.20	NTPRP2 1.20	PROPAN 0.00	NPRPNL 1.38
IPRPNL 1.38	TRMTAM 2.20	BUTENE 0.00	BTNON2 0.50	TTRHFR 0.00	DIOANE 0.00	ETHACT 0.50
NBTANE 0.00	IBTANE 0.00	NBTNOL 1.30	IBTNOL 1.30	SBTNOL 1.38	TBTNOL 1.38	DETETH 0.00
ETHCEL 1.38	DETAMN 2.20	FRFRAL 0.75	PYRDIN 1.85	ISOPRN 0.00	CYCPTN 0.00	ISOPTN 0.00
NORPTN 0.00	CHLBZN 0.00	NTRBZN 1.20	BENZEN 0.00	PHENOL 1.38	ANILIN 2.20	CYHXON 0.50
CYHXAN 0.00	HEXENE 0.00	MTCYPN 0.00	CYHXNL 1.38	MISBKT 0.50	NBTACT 0.50	NHEXAN 0.00
DMBT23 0.00	TRETAM 2.20	TOLUEN 0.00	MTCYHX 0.00	NHEPTN 0.00	STYREN 0.00	ETHBZN 0.00
MXYLEN 0.00	OXYLEN 0.00	PXYLEN 0.00	NOCTAN 0.00	TMP224 0.00	NDECAN 0.00	NHXDCN 0.00

WATER 1.70

HYDRGN 0.00	NITRGN 0.00	OXYGEN 0.00	CRBMOX 0.00	ARGON 0.00	METHAN 0.00	CRBTCL 0.00
CLRFRM 0.00	DCLMTN 0.00	FRMACD 2.50	NTRMTN 1.20	MTHNOL 1.55	CRBDOX 0.30	CRBDSF 0.15
TCLETN 0.00	ACTYLN 0.30	ACTNTR 1.50	ETHLYN 0.00	DCLETN 0.00	ACTALD 0.80	ACTACD 2.50
ETHIOD 0.00	NTRETH 1.20	ETHANE 0.00	ETHNOL 1.55	DMTAMN 0.20	ETHGLY 1.55	PRPYLN 0.00
ACETON 1.00	MTHACT 1.30	PRPACD 2.50	NTPRP1 1.20	NTPRP2 1.20	PROPAN 0.00	NPRPNL 1.55
IPRPNL 1.55	TRMTAM 0.20	BUTENE 0.00	BTNON2 1.00	TTRHFR 0.50	DIOANE 0.50	ETHACT 1.30
NBTANE 0.00	IBTANE 0.00	NBTNOL 1.55	IBTNOL 1.55	SBTNOL 1.55	TBTNOL 1.55	DETETH 0.50
ETHCEL 1.55	DETAMN 0.20	FRFRAL 0.80	PYRDIN 0.40	ISOPRN 1.55	CYCPTN 0.00	ISOPTN 0.00
NORPTN 0.00	CHLBZN 0.00	NTRBZN 1.20	BENZEN 0.00	PHENOL 1.55	ANILIN 0.20	CYHXON 1.00
CYHXAN 0.00	HEXENE 0.00	MTCYPN 0.00	CYHXNL 1.55	MISBKT 1.00	NBTACT 1.30	NHEXAN 0.00
DMBT23 0.00	TRETAM 0.20	TOLUEN 0.00	MTCYHX 0.00	NHEPTN 0.00	STYREN 0.00	ETHBZN 0.00
MXYLEN 0.00	OXYLEN 0.00	PXYLEN 0.00	NOCTAN 0.00	TMP224 0.00	NDECAN 0.00	NHXDCN 0.00
HYDCHL 1.38						

HYDSUL 0.00

HYDRGN 0.00	NITRGN 0.00	OXYGEN 0.00	CRBMOX 0.00	ARGON 0.00	METHAN 0.00	CRBTCL 0.00
CLRFRM 0.00	DCLMTN 0.00	FRMACD 0.50	NTRMTN 0.60	MTHNOL 0.70	CRBDOX 0.10	CRBDSF 0.00
TCLETN 0.00	ACTYLN 0.00	ACTNTR 0.70	ETHLYN 0.70	DCLETN 0.00	ACTALD 0.15	ACTACD 0.50
ETHIOD 0.00	NTRETH 0.60	ETHANE 0.00	ETHNOL 0.70	DMTAMN 0.15	ETHGLY 0.70	PRPYLN 0.00

HYDSUL 0.00

ACETON 0.20	MTHACT 0.30	PRPACD 0.50	NTPRP1 0.60	PROPAN 0.00	NPRPNL 0.70
IPRPNL 0.70	TRMTAM 0.15	BUTENE 0.00	BTNON2 0.20	DIOANE 0.10	ETHACT 0.30
NBTANE 0.00	IBTANE 0.00	NBTNOL 0.70	IBTNOL 0.70	TBTNOL 0.70	DETETH 0.10
ETHCEL 0.70	DETAMN 0.15	FRFRAL 0.15	PYRDIN 0.10	CYCPTN 0.00	ISOPTN 0.00
NCRPTN 0.00	CHLBZN 0.05	NTRBZN 0.60	BENZEN 0.05	ANILIN 0.15	CYHXON 0.20
CYHXAN 0.00	HEXENE 0.00	MTCYPN 0.00	CYHXNL 0.70	MISBKT 0.20	NHEXAN 0.00
DMBT23 0.00	TRETAM 0.15	TOLUEN 0.05	MTCYHX 0.00	STYREN 0.05	ETHBZN 0.05
MXYLEN 0.05	OXYLEN 0.00	PXYLEN 0.05	NOCTAN 0.00	NDECAN 0.00	NHXDCN 0.00
HYDCHL 0.50	WATER 0.70			TMP224 0.00	

AMMNIA 0.00

HYDRGN 0.00	NITRGN 0.00	OXYGEN 0.00	CRBMOX 0.00	ARGON 0.00	CRBTCL 0.00
CLRFRM 0.30	DCLMTN 0.00	FRMACD 2.20	NTRMTN 1.40	MTHNOL 0.20	CRBDSF 0.10
TCLETN 0.20	ACTYLN 0.50	ACTNTR 1.40	ETHLYN 0.20	DCLETN 0.00	ACTACD 2.20
ETHIOD 0.00	NTRETH 1.40	ETHANE 0.00	ETHNOL 0.20	DMTAMN 0.20	PRPYLN 0.20
ACETON 1.00	MTHACT 1.10	PRPACD 2.20	NTPRP1 1.40	NTPRP2 1.40	NPRPNL 0.20
IPRPNL 0.20	TRMTAM 0.20	BUTENE 0.20	BTNON2 1.00	TTRHFR 0.70	ETHACT 1.10
NBTANE 0.00	IBTANE 0.00	NBTNOL 0.20	IBTNOL 0.20	SBTNOL 0.20	DETETH 0.70
ETHCEL 0.70	DETAMN 0.20	FRFRAL 0.85	PYRDIN 0.20	ISOPRN 0.20	ISOPTN 0.00
NORPTN 0.00	CHLBZN 0.00	NTRBZN 1.40	BENZEN 0.00	PHENOL 0.00	CYHXON 1.00
CYHXAN 0.00	HEXENE 0.20	MTCYPN 0.00	CYHXNL 0.20	MISBKT 0.20	NHEXAN 0.00
DMBT23 0.00	TRETAM 0.20	TOLUEN 0.00	MTCYHX 0.00	NHEPTN 0.00	ETHBZN 0.00
MXYLEN 0.30	OXYLEN 0.00	PXYLEN 0.00	NOCTAN 0.00	TMP224 0.00	NHXDCN 0.00
HYDCHL 2.20	WATER 0.20	HYDSUL 0.15			

SULDIO 0.00

HYDRGN 0.00	NITRGN 0.00	OXYGEN 0.00	CRBMOX 0.00	ARGON 0.00	CRBTCL 0.00
CLRFRM 0.00	DCLMTN 0.00	FRMACD 0.90	NTRMTN 1.00	MTHNOL 1.50	CRBDOX 0.00
TCLETN 0.20	ACTYLN 0.50	ACTNTR 0.80	ETHLYN 0.20	DCLETN 0.00	ACTACD 0.90
ETHIOD 0.00	NTRETH 1.00	ETHANE 0.00	ETHNOL 1.50	DMTAMN 1.50	PRPYLN 0.20
ACETON 0.70	MTHACT 0.80	PRPACD 0.90	NTPRP1 1.00	NTPRP2 1.00	NPRPNL 1.50
IPRPNL 1.50	TRMTAM 1.50	BUTENE 0.20	BTNON2 0.70	TTRHFR 0.58	ETHACT 0.80
NBTANE 0.00	IBTANE 0.00	NBTNOL 1.50	IBTNOL 1.50	SBTNOL 1.50	DETETH 0.58
ETHCEL 1.50	DETAMN 1.50	FRFRAL 0.70	PYRDIN 1.00	ISOPRN 0.20	ISOPTN 0.00
NORPTN 0.00	CHLBZN 0.30	NTRBZN 1.00	BENZEN 0.30	PHENOL 1.50	CYHXON 0.70
CYHXAN 0.00	HEXENE 0.20	MTCYPN 0.00	CYHXNL 1.50	MISBKT 0.70	NHEXAN 0.00
DMBT23 0.00	TRETAM 1.50	TOLUEN 0.30	MTCYHX 0.00	NHEPTN 0.00	ETHBZN 0.30
MXYLEN 0.30	OXYLEN 0.30	PXYLEN 0.30	NOCTAN 0.00	TMP224 0.00	NHXDCN 0.00
HYDCHL 0.00	WATER 1.50	HYDSUL 0.50	AMMNIA 1.50		

ANISOL 0.00

HYDRGN 0.00	NITRGN 0.00	OXYGEN 0.00	CRBMOX 0.00	ARGON 0.00	CRBTCL 0.25
CLRFRM 0.95	DCLMTN 0.35	FRMACD 1.20	NTRMTN 1.25	MTHNOL 0.50	CRBDSF 0.00

ANISOL 0.00

TCLETN	0.10	ACTYLN	0.30	ACTNTR	1.00	ETHLYN	0.10	DCLETN	0.00	ACTALD	0.40	ACTACD	1.20
ETHIOD	0.00	NTRETH	1.25	ETHANE	0.00	ETHNOL	0.50	DMTAMN	0.70	ETHGLY	0.50	PRPYLN	0.10
ACETON	0.50	MTHACT	0.60	PRPACD	1.20	BTNON2	0.50	NTPRP2	1.25	PROPAN	0.00	NPRPNL	0.50
IPRPNL	0.50	TRMTAM	0.70	BUTENE	0.10	NTPRP1	1.25	TTRHFR	0.00	DIOANE	0.00	ETHACT	0.60
NBTANE	0.00	IBTANE	0.00	NBTNOL	0.50	IBTNOL	0.50	SBTNOL	0.50	TBTNOL	0.50	DETETH	0.00
ETHCEL	0.50	DETAMN	0.70	FRFRAL	0.40	PYRDIN	0.00	ISOPRN	0.10	CYCPTN	0.00	ISOPTN	0.00
NQRPTN	0.00	CHLBZN	0.30	NTRBZN	1.25	BENZEN	0.30	PHENOL	0.50	ANILIN	0.70	CYHXON	0.50
CYHXAN	0.00	HEXENE	0.10	MTCYPN	0.00	CYHXNL	0.00	MISBKT	0.50	NBTACT	0.60	NHEXAN	0.00
DMBT23	0.00	TRETAM	0.70	TOLUEN	0.30	MTCYHX	0.30	NHEPTN	0.00	STYREN	0.30	ETHBZN	0.30
MXYLEN	0.30	DXYLEN	0.30	PXYLEN	0.30	NOCTAN	0.00	TMP224	0.00	NDECAN	0.00	NHXDCN	0.00
HYDCHL	0.00	WATER	0.50	HYDSUL	0.10	AMMNIA	0.70	SULDIO	0.58				

ACRNTR 1.65

HYDRGN	0.00	NITRGN	0.00	OXYGEN	0.00	CRBMOX	0.00	ARGON	0.00	METHAN	0.00	CRBTCL	0.30
CLRFRM	1.80	DCLMTN	0.60	FRMACD	0.40	NTRMTN	2.00	MTHNOL	1.50	CRBDOX	0.00	CRBDSF	0.00
TCLETN	0.20	ACTYLN	0.50	ACTNTR	1.65	ETHLYN	0.20	DCLETN	0.00	ACTALD	2.38	ACTACD	2.50
ETHIOD	0.00	NTRETH	2.00	ETHANE	0.00	ETHNOL	1.50	DMTAMN	1.40	ETHGLY	0.00	PRPYLN	0.20
ACETON	1.30	MTHACT	1.40	PRPACD	2.50	NTPRP1	2.00	NTPRP2	2.00	PROPAN	0.00	NPRPNL	1.50
IPRPNL	1.50	TRMTAM	1.40	BUTENE	0.20	BTNON2	1.30	TTRHFR	1.00	DIOANE	1.00	ETHACT	1.40
NBTANE	0.00	IBTANE	0.00	NBTNOL	1.50	IBTNOL	1.50	SBTNOL	1.50	TBTNOL	1.50	DETETH	1.00
ETHCEL	1.50	DETAMN	1.40	FRFRAL	2.38	PYRDIN	1.30	ISODRN	0.20	CYCPTN	0.00	ISOPTN	0.00
NORPTN	0.00	CHLBZN	0.70	NTRBZN	2.00	BENZEN	0.70	PHENOL	1.50	ANILIN	1.40	CYHXON	1.30
CYHXAN	0.00	HEXENE	0.20	MTCYPN	0.00	CYHXNL	1.50	MISBKT	1.30	NBTACT	1.40	NHEXAN	0.00
DMBT23	0.00	TRETAM	1.40	TOLUEN	0.70	MTCYHX	0.70	NHEPTN	0.00	STYREN	0.70	ETHBZN	0.70
MXYLEN	0.70	DXYLEN	0.70	PXYLEN	0.70	NOCTAN	0.70	TMP224	0.00	NDECAN	0.00	NHXDCN	0.00
HYDCHL	0.50	WATER	1.50	HYDSUL	0.70	AMMNIA	1.40	SULDIO	0.80	ANISOL	1.00		

VNLACT 0.50

HYDRGN	0.00	NITRGN	0.00	OXYGEN	0.00	CRBMOX	0.00	ARGON	0.00	METHAN	0.00	CRBTCL	0.25
CLRFRM	1.55	DCLMTN	0.50	FRMACD	2.00	NTRMTN	1.80	MTHNOL	1.80	CRBDOX	0.00	CRBDSF	0.00
TCLETN	0.00	ACTYLN	0.30	ACTNTR	1.40	ETHLYN	0.10	DCLETN	0.10	ACTALD	0.75	ACTACD	2.00
ETHIOD	0.00	NTRETH	1.80	ETHANE	0.00	ETHNOL	1.30	DMTAMN	1.10	ETHGLY	1.30	PRPYLN	0.10
ACETON	1.10	MTHACT	0.60	PRPACD	2.00	NTPRP1	1.80	NTPRP2	1.80	PROPAN	0.00	NPRPNL	1.30
IPRPNL	1.30	TRMTAM	1.10	BUTENE	0.10	BTNON2	1.10	TTRHFR	0.30	DIOANE	0.30	ETHACT	0.60
NBTANE	0.00	IBTANE	0.00	NBTNOL	1.30	IBTNOL	1.30	SBTNOL	1.30	TBTNOL	1.30	DETETH	0.30
ETHCEL	1.30	DETAMN	1.10	FRFRAL	0.75	PYRDIN	0.30	ISOPRN	0.10	CYCPTN	0.00	ISOPTN	0.00
NORPTN	0.00	CHLBZN	0.60	NTRBZN	1.80	BENZEN	0.60	PHENOL	1.30	ANILIN	1.10	CYHXON	1.10
CYHXAN	0.00	HEXENE	0.10	MTCYPN	0.00	CYHXNL	1.30	MISBKT	1.10	NBTACT	0.60	NHEXAN	0.00
DMBT23	0.00	TRETAM	1.10	TOLUEN	0.60	MTCYHX	0.00	NHEPTN	0.00	STYREN	0.60	ETHBZN	0.60
MXYLEN	0.60	DXYLEN	0.60	PXYLEN	0.60	NOCTAN	0.00	TMP224	0.00	NDECAN	0.00	NHXDCN	0.00
HYDCHL	0.50	WATER	1.30	HYDSUL	0.30	AMMNIA	1.10	SULDIO	0.80	ANISOL	0.60	ACRNTR	1.40

Appendix C-5

Selected UNIQUAC Binary Parameters.
Characteristic Binary Parameters for
Noncondensable-Condensable Interactions

* * INTERACTION PARAMETERS FOR LIQUID-PHASE MIXTURES * *

INDICES		COMPONENT NAMES		BINARY PARAMETERS	
(1)	(2)	(1)	(2)	A12	A21
1	12	*HYDROGEN	METHANOL	3.94	218.00
1	29	*HYDROGEN	ACETONE	3.23	218.00
1	60	*HYDROGEN	BENZENE	1.97	810.00
1	70	*HYDROGEN	N-HEXANE	2.99	-4.00
1	78	*HYDROGEN	M-XYLENE	2.09	765.00
1	81	*HYDROGEN	N-OCTANE	-.05	1646.00
1	86	*HYDROGEN	WATER	9.97	-480.00
1	88	*HYDROGEN	AMMONIA	.68	1344.00
2	43	*NITROGEN	N-BUTANE	-1.67	814.00
2	70	*NITROGEN	N-HEXANE	.07	336.00
2	86	*NITROGEN	WATER	3.76	913.00
2	88	*NITROGEN	AMMONIA	-3.35	2099.00
4	12	*CARBON MONOXIDE	METHANOL	.83	552.00
4	29	*CARBON MONOXIDE	ACETONE	.48	408.00
4	60	*CARBON MONOXIDE	BENZENE	-1.04	1062.00
4	81	*CARBON MONOXIDE	N-OCTANE	-3.38	1878.00
6	29	*METHANE	ACETONE	-.26	471.00
6	34	*METHANE	PROPANE	-2.01	604.00
6	43	*METHANE	N-BUTANE	-2.05	731.00
6	57	*METHANE	N-PENTANE	-1.32	509.00

* * INTERACTION PARAMETERS FOR LIQUID-PHASE MIXTURES * *

INDICES (1)	(2)	COMPONENT NAMES (1)	(2)	BINARY PARAMETERS A12	A21
6	60	METHANE	BENZENE	-.38	473.00
6	70	METHANE	N-HEXANE	-.06	94.00
6	83	METHANE	N-DECANE	-.24	150.00
7	17	CARBON TETRACHLORIDE	ACETONITRILE	458.86	-40.13
7	25	CARBON TETRACHLORIDE	ETHANOL	1192.49	-135.27
7	52	CARBON TETRACHLORIDE	FURFURAL	476.85	-100.42
7	60	CARBON TETRACHLORIDE	BENZENE	-37.52	43.39
7	64	CARBON TETRACHLORIDE	CYCLOHEXANE	98.18	-84.82
7	66	CARBON TETRACHLORIDE	METHYLCYCLOPENTANE	161.95	-129.21
7	73	CARBON TETRACHLORIDE	TOLUENE	-168.53	203.67
7	75	CARBON TETRACHLORIDE	N-HEPTANE	98.30	-59.13
8	7	CHLOROFORM	CARBON TETRACHLORIDE	-14.74	78.19
8	10	CHLOROFORM	FORMIC ACID	461.38	90.36
8	12	CHLOROFORM	METHANOL	926.31	-143.50
9	21	CHLOROFORM	ACETIC ACID	346.43	-98.44
8	25	CHLOROFORM	ETHANOL	888.68	-203.03
8	42	CHLOROFORM	ETHYL ACETATE	24.16	-119.49
8	60	CHLOROFORM	BENZENE	4.98	-50.53
10	21	FORMIC ACID	ACETIC ACID	-144.53	241.54
11	7	NITRO-METHANE	CARBON TETRACHLORIDE	.59	398.40

* * INTERACTION PARAMETERS FOR LIQUID-PHASE MIXTURES * *

****

INDICES		COMPONENT NAMES		BINARY PARAMETERS	
(1)	(2)	(1)	(2)	A12	A21

****

(1)	(2)	(1)	(2)	A12	A21
11	60	NITRO-METHANE	BENZENE	73.79	82.20
12	7	METHANOL	CARBON TETRACHLORIDE	-29.64	1127.95
12	25	METHANOL	ETHANOL	660.19	-292.39
12	39	METHANOL	2-BUTANONE	-154.22	803.49
12	42	METHANOL	ETHYL ACETATE	-107.54	579.61
12	51	METHANOL	DIETHYL AMINE	-374.83	676.42
12	60	METHANOL	BENZENE	-86.89	1284.21
12	64	METHANOL	CYCLOHEXANE	6.02	1364.12
12	68	METHANOL	METHYLISOBUTYLKETONE	-105.94	688.03
12	70	METHANOL	N-HEXANE	-2.66	1636.05
12	71	METHANOL	2,3-DIMETHYL BUTANE	-7.18	1453.90
12	72	METHANOL	TRIETHYLAMINE	-196.66	664.29
12	73	METHANOL	TOLUENE	-43.91	951.82
12	75	METHANOL	N-HEPTANE	2.49	1419.32
12	86	METHANOL	WATER	-50.82	148.27
12	90	METHANOL	ANISOLE	-48.39	782.29
14	12	CARBON DISULFIDE	METHANOL	1166.49	138.15
14	29	CARBON DISULFIDE	ACETONE	157.31	103.91
17	60	ACETONITRILE	BENZENE	60.28	89.57
17	75	ACETONITRILE	N-HEPTANE	23.71	545.79

****

* * INTERACTION PARAMETERS FOR LIQUID-PHASE MIXTURES * *

INDICES (1)	(2)	COMPONENT NAMES (1)	(2)	BINARY PARAMETERS A12	A21
20	21	ACETALDEHYDE	ACETIC ACID	458.43	-212.77
20	92	ACETALDEHYDE	VINYL ACETATE	-117.74	243.51
21	73	ACETIC ACID	TOLUENE	-67.91	298.09
21	75	ACETIC ACID	N-HEPTANE	-8.49	342.57
23	7	NITRO-ETHANE	CARBON TETRACHLORIDE	73.06	295.06
23	35	NITRO-ETHANE	N-PROPANOL	574.22	-94.39
23	60	NITRO-ETHANE	BENZENE	16.61	46.05
24	75	ETHANE	N-HEPTANE	264.89	-112.61
25	17	ETHANOL	ACETONITRILE	58.72	430.51
25	19	ETHANOL	1,2-DICHLOROETHANE	-105.66	929.71
25	21	ETHANOL	ACETIC ACID	-210.53	244.67
25	35	ETHANOL	N-PROPANOL	210.95	-67.70
25	60	ETHANOL	BENZENE	-128.88	997.41
25	64	ETHANOL	CYCLOHEXANE	-113.70	1269.49
25	66	ETHANOL	METHYLCYCLOPENTANE	-118.27	1393.93
25	70	ETHANOL	N-HEXANE	-108.93	1441.57
25	73	ETHANOL	TOLUENE	-132.12	1083.75
25	74	ETHANOL	METHYL CYCLOHEXANE	-117.57	1340.56
25	75	ETHANOL	N-HEPTANE	-117.60	1224.27
25	81	ETHANOL	N-OCTANE	-109.08	1385.91

** INTERACTION PARAMETERS FOR LIQUID-PHASE MIXTURES **

INDICES (1)	(2)	COMPONENT NAMES (1)	(2)	BINARY PARAMETERS A12	A21
25	82	ETHANOL	224-TRIMETHYLPENTANE	-120.42	1449.61
25	83	ETHANOL	N-DECANE	-127.48	1254.65
25	86	ETHANOL	WATER	-64.56	380.68
29	7	ACETONE	CARBON TETRACHLORIDE	-92.32	246.68
29	8	ACETONE	CHLOROFORM	-171.71	93.96
29	17	ACETONE	ACETONITRILE	-176.38	261.53
29	12	ACETONE	METHANOL	359.10	-96.90
29	21	ACETONE	ACETIC ACID	461.81	-262.30
29	25	ACETONE	ETHANOL	404.49	-131.25
29	52	ACETONE	FURFURAL	-101.30	195.63
29	60	ACETONE	BENZENE	-108.79	174.00
29	70	ACETONE	N-HEXANE	-33.08	261.51
29	86	ACETONE	WATER	530.99	-100.71
29	92	ACETONE	VINYL ACETATE	-92.48	110.60
30	8	METHYL ACETATE	CHLOROFORM	-187.87	121.17
30	12	METHYL ACETATE	METHANOL	516.12	-95.31
30	25	METHYL ACETATE	ETHANOL	506.85	-120.60
30	60	METHYL ACETATE	BENZENE	203.46	-143.89
31	68	PROPIONIC ACID	METHYLISOBUTYLKETONE	-78.49	136.46
31	81	PROPIONIC ACID	N-OCTANE	-193.20	556.12

* * INTERACTION PARAMETERS FOR LIQUID-PHASE MIXTURES * *

INDICES		COMPONENT NAMES		BINARY PARAMETERS	
(1)	(2)	(1)	(2)	A12	A21
32	7	1-NITROPROPANE	CARBON TETRACHLORIDE	-95.68	267.13
32	60	1-NITROPROPANE	BENZENE	535.16	-246.81
33	7	2-NITROPROPANE	CARBON TETRACHLORIDE	-134.32	307.25
33	60	2-NITROPROPANE	BENZENE	794.91	-344.15
33	70	2-NITROPROPANE	N-HEXANE	-32.95	252.20
35	7	N-PROPANOL	CARBON TETRACHLORIDE	-166.93	1336.03
35	21	N-PROPANOL	ACETIC ACID	299.33	445.77
35	60	N-PROPANOL	BENZENE	-155.10	928.50
35	70	N-PROPANOL	N-HEXANE	-144.11	1326.05
35	73	N-PROPANOL	TOLUENE	-195.40	818.34
35	75	N-PROPANOL	N-HEPTANE	-160.43	1306.22
35	86	N-PROPANOL	WATER	78.37	583.03
36	42	I-PROPANOL	ETHYL ACETATE	-190.57	522.07
36	60	I-PROPANOL	BENZENE	-145.52	854.75
36	75	I-PROPANOL	N-HEPTANE	-162.54	1295.60
36	82	I-PROPANOL	224-TRIMETHYLPENTANE	-198.06	904.00
39	73	2-BUTANONE	TOLUENE	-82.85	123.57
39	75	2-BUTANONE	N-HEPTANE	-75.13	242.53
39	86	2-BUTANONE	WATER	509.67	-11.75
41	60	DIOXANE	BENZENE	363.78	-197.65

* * INTERACTION PARAMETERS FOR LIQUID-PHASE MIXTURES * *

INDICES (1)	(2)	COMPONENT NAMES (1)	(2)	BINARY PARAMETERS A12	A21
42	21	ETHYL ACETATE	ACETIC ACID	-214.39	63.04
42	25	ETHYL ACETATE	ETHANOL	571.73	-167.61
42	35	ETHYL ACETATE	N-PROPANOL	539.64	-190.31
42	52	ETHYL ACETATE	FURFURAL	-19.15	48.52
42	60	ETHYL ACETATE	BENZENE	233.81	-181.49
42	73	ETHYL ACETATE	TOLUENE	309.41	-214.26
42	77	ETHYL ACETATE	ETHYL BENZENE	137.11	-105.50
42	86	ETHYL ACETATE	WATER	1000.87	-174.08
45	7	N-BUTANOL	CARBON TETRACHLORIDE	-188.77	1248.17
45	21	N-BUTANOL	ACETIC ACID	-296.30	546.68
45	70	N-BUTANOL	N-HEXANE	-159.24	1370.74
45	75	N-BUTANOL	N-HEPTANE	-251.11	1291.98
45	81	N-BUTANOL	N-OCTANE	-236.21	1098.91
46	60	I-BUTANOL	BENZENE	-162.39	861.06
47	50	SEC-BUTANOL	BENZENE	784.99	-158.83
48	60	TERT-BUTANOL	BENZENE	744.89	-165.40
49	25	DI-ETHYL ETHER	ETHANOL	733.67	-179.02
52	60	FURFURAL	BENZENE	71.00	12.00
52	64	FURFURAL	CYCLOHEXANE	41.17	354.83
55	60	CYCLOPENTANE	BENZENE	15.19	33.15

* * INTERACTION PARAMETERS FOR LIQUID-PHASE MIXTURES * *

INDICES (1)	(2)	COMPONENT NAMES (1)	(2)	BINARY PARAMETERS A12	A21
57	29	N-PENTANE	ACETONE	266.31	-22.83
60	39	BENZENE	2-BUTANONE	350.47	-226.16
60	45	BENZENE	N-BUTANOL	928.90	-181.24
60	70	BENZENE	N-HEXANE	-77.13	132.43
60	73	BENZENE	TOLUENE	-220.57	330.69
62	66	ANILINE	METHYLCYCLOPENTANE	54.36	228.71
60	86	BENZENE	WATER	2057.42	115.13
64	11	CYCLOHEXANE	NITRO-METHANE	517.19	105.01
64	35	CYCLOHEXANE	N-PROPANOL	1284.75	-173.42
64	45	CYCLOHEXANE	N-BUTANOL	1393.11	-196.90
64	60	CYCLOHEXANE	BENZENE	-32.57	88.26
64	66	CYCLOHEXANE	METHYLCYCLOPENTANE	144.37	-118.82
64	70	CYCLOHEXANE	N-HEXANE	172.73	-145.56
64	73	CYCLOHEXANE	TOLUENE	83.67	-44.04
65	25	HEXENE-1	ETHANOL	1222.95	-107.26
66	60	METHYLCYCLOPENTANE	BENZENE	56.47	-6.47
66	73	METHYLCYCLOPENTANE	TOLUENE	89.77	-48.05
70	23	N-HEXANE	NITRO-ETHANE	230.64	-5.86
70	62	N-HEXANE	ANILINE	283.76	34.92
70	66	N-HEXANE	METHYLCYCLOPENTANE	-138.84	152.13

* * INTERACTION PARAMETERS FOR LIQUID-PHASE MIXTURES * *

INDICES (1)	(2)	COMPONENT NAMES (1)	(2)	BINARY PARAMETERS A12	A21
70	73	N-HEXANE	TOLUENE	34.27	4.30
73	52	TOLUENE	FURFURAL	74.87	244.12
74	73	METHYL CYCLOHEXANE	TOLUENE	210.35	-134.19
75	60	N-HEPTANE	BENZENE	245.42	-135.93
75	60	N-HEPTANE	BENZENE	122.21	-55.81
75	73	N-HEPTANE	TOLUENE	108.24	-72.96
81	23	N-OCTANE	NITRO-ETHANE	333.48	-30.98
81	36	N-OCTANE	1-PROPANOL	1107.44	-166.18
82	23	224-TRIMETHYLPENTANE	NITRO-ETHANE	236.48	10.66
82	52	224-TRIMETHYLPENTANE	FURFURAL	410.08	-4.98
82	60	224-TRIMETHYLPENTANE	BENZENE	80.91	-27.13
82	64	224-TRIMETHYLPENTANE	CYCLOHEXANE	141.01	-112.66
82	73	224-TRIMETHYLPENTANE	TOLUENE	141.11	-94.60
83	35	N-DECANE	N-PROPANOL	1137.20	-201.82
83	45	N-DECANE	N-BUTANOL	1430.77	-259.67
83	36	N-DECANE	1-PROPANOL	1074.76	-207.27
86	13	WATER	FORMIC ACID	-508.85	1019.29
86	17	WATER	ACETONITRILE	122.02	122.07
86	21	WATER	ACETIC ACID	-299.90	530.94
86	23	WATER	NITRO-ETHANE	138.44	920.08

* * INTERACTION PARAMETERS FOR LIQUID-PHASE MIXTURES * *

****

| INDICES | | COMPONENT NAMES | | BINARY PARAMETERS | |
(1)	(2)	(1)	(2)	A12	A21

****

86	31	WATER	PROPIONIC ACID	-188.72	540.10
86	41	WATER	DIOXANE	-328.70	927.26
86	45	WATER	N-BUTANOL	1097.59	38.69
86	73	WATER	TOLUENE	305.71	1371.36
86	91	WATER	ACRYLONITRILE	155.78	471.21
91	17	ACRYLONITRILE	ACETONITRILE	183.65	-142.35
92	21	VINYL ACETATE	ACETIC ACID	330.03	-124.12
92	86	VINYL ACETATE	WATER	1557.23	131.36

****

NOTE-- * INDICATES A NONCONDENSABLE COMPONENT WITH A12 = DELTA-0 AND A21 = DELTA-1 (SEE EQUATION 4-21)

188

Appendix C-6

UNIQUAC Parameters for Condensable Binary Systems and Data References

** UNIQUAC BINARY INTERACTION PARAMETERS **

COMPONENTS 1	2	TYPE	NO. OF POINTS	TEMP. K	UNIQUAC PARAMETERS,K A_{12}	UNIQUAC PARAMETERS,K A_{21}	VARIANCE OF FIT	DATA REFERENCE
ACETONE	ACETIC ACID	VL	19	333-385	461.81	-262.30	460.84	YORK,1942
	ACETONITRILE	VL	9	330-353	-176.38	261.53	39.81	PRATT,1947
	BENZENE	VL	13	329-350	-115.95	253.47	132.48	SODAY,1930
	BENZENE	VL	12	313	-63.21	109.30	16.40	KRAUS,1971
	BENZENE	VL	12	303	-90.40	147.12	7.02	KRAUS,1971
	BENZENE	VL	11	318	-108.79	174.00	.84	BROWN,1957
	CHLOROFORM	VL	11	330-337	-147.83	70.60	5.24	REINDERS,1947
	CHLOROFORM	VL	31	308	-140.77	28.38	9.17	ZAWIDZKI,1900
	CHLOROFORM	VL	29	323	-171.71	93.96	6.75	SEVERNS,1955
	CARBON DISULFIDE	VL	34	308	103.91	157.31	7.33	ZAWIDZKI,1900
	CARBON TETRACHLORIDE	VL	12	318	-92.32	246.65	3.27	BROWN,1957
	ETHANOL	VL	14	305	414.46	-119.36	8.71	GORDON,1946
	ETHANOL	VL	14	321	404.49	-131.25	3.81	GORDON,1946
	ETHANOL	VL	14	313	398.80	-122.61	4.00	GORDON,1946
	FURFURAL	VL	12	331-373	-101.30	195.63	43.86	MYLES,1961
	METHANOL	VL	12	328-336	295.36	-63.85	15.27	AMER,1956
	METHANOL	VL	9	327-335	346.16	-66.71	38.83	OTHMER,1928
	METHANOL	VL	8	328-335	366.38	-111.12	50.47	GRISWOLD,1949
	METHANOL	VL	23	328-337	359.10	-96.90	6.77	UCHIDA,1950
	METHANOL	VL	25	323	379.31	-108.42	11.58	SEVERNS,1955
	N-HEXANE	VL	14	318	-33.08	261.51	4.38	SCHAEFER,1958
	N-PENTANE	VL	11	305-322	-22.83	266.31	212.14	LO,1962
	VINYL ACETATE	VL	8	330-342	-82.48	110.60	2.44	SERAFIMOV,1964
	WATER	VL	10	330-362	490.50	-67.18	36.32	OTHMER,1945
	WATER	VL	13	298	636.17	-129.35	22.07	BEARE,1930
	WATER	VL	12	309-339	572.52	-117.85	23.25	OTHMER,1945
	WATER	VL	10	295-321	603.58	-134.20	69.05	OTHMER,1945
	WATER	VL	13	318-344	530.98	-100.94	62.97	OTHMER,1945
	WATER	VL	21	329-365	466.36	-56.91	71.35	BRUNJES,1943
	WATER	VL	13	331-368	530.99	-100.71	47.43	REINDERS,1947
ACRYLONITRILE	ACETONITRILE	VL	7	346-352	573.80	-263.75	770.21	BLACKFORD,1965
	ACETONITRILE	VL	8	351-354	80.53	-49.84	34.04	PROKHOROVA,1964
	WATER	MS	1	298	575.58	141.17		VOLPICELLI,1968
ACETIC ACID	ACETONE	VL	19	333-385	-262.30	461.81	460.94	YORK,1942
	ACETALDEHYDE	VL	13	305-380	-212.77	458.43	125.65	FRIED,1963
	CHLOROFORM	VL	13	338-379	-98.44	345.43	385.19	CONTI,1960
	ETHYL ACETATE	VL	6	353-381	-214.39	426.54	63.04	GARNER,1954
	ETHANOL	VL	16	351-385	244.67	-210.53	32.68	RIUS,1959
	FORMIC ACID	VL	12	374-387	241.64	-144.58	12.84	CONTI,1960
	N-BUTANOL	VL	18	388-393	546.68	-296.30	10.54	RIUS,1959
	N-HEPTANE	VL	15	303	-8.49	342.57	76.23	MARKUZIN,1971
	N-HEPTANE	VL	14	313	-1.74	339.28	80.87	MARKUZIN,1971
	N-HEPTANE	VL	13	293	8.24	311.32	93.50	MARKUZIN,1971
	N-PROPANOL	VL	14	370-387	445.77	299.33	15.97	RIUS,1959
	TOLUENE	VL	15	303	-67.91	298.09	25.58	MARKUZIN,1971

COMPONENTS 1	2	TYPE	NO. OF POINTS	TEMP. K	UNIQUAC PARAMETERS, K A12	A21	VARIANCE OF FIT	DATA REFERENCE
ACETIC ACID	VINYL ACETATE	VL	14	348-386	-140.18	360.55	142.93	CAPKOVA,1963
	WATER	VL	14	317-326	570.35	-277.77	8.88	ITO,1963
	WATER	VL	23	373-386	518.98	-296.05	68.65	SEBASTIANI,1967
	WATER	VL	10	293	555.01	-258.51	1.18	LAZEVA,1973
	WATER	VL	16	373-389	530.94	-299.90	23.34	ITO,1963
ACETALDEHYDE	ACETIC ACID	VL	15	339-350	602.74	-291.10	14.56	ITO,1963
ACETONITRILE	VINYL ACETATE	VL	13	305-380	458.43	-212.77	125.65	FRIED,1963
	ACETONE	VL	9	302-339	-117.74	243.51	96.59	FRIED,1963
	ACRYLONITRILE	VL	9	330-353	261.53	-176.38	39.81	PRATT,1947
	ACRYLONITRILE	VL	8	351-354	-49.84	80.53	34.04	PROKHOROVA,1964
	BENZENE	VL	7	346-352	-263.76	573.80	770.21	BLACKFORD,1965
	BENZENE	VL	12	318	-40.70	229.79	2.68	BROWN,1955A
	CARBON TETRACHLORIDE	VL	11	318	-51.54	247.13	9.74	PALMER,1972
	ETHANOL	VL	13	318	-40.18	458.86	16.91	BROWN,1954B
	N-HEPTANE	MS	8	293	430.51	68.72	16.25	VIERK,1950
	WATER	MS	1	318	23.71	545.79	74.17	PALMER,1972
	WATER	VL	13	349-363	301.93	83.42	412.28	MASLAN,1956
	WATER	VL	11	349-359	388.46	23.81	45.35	BLACKFORD,1965
ANILINE	METHYLCYCLOPENTANE	MS	1	298	54.36	228.71		DARWENT,1943
	N-HEXANE	MS	1	298	34.82	283.76		DARWENT,1943
ANISOLE	METHANOL	VL	14	338-401	782.28	-48.39	87.31	NAKANISHI,1958
BENZENE	ACETONE	VL	12	303	147.12	-90.40	7.02	KRAUS,1971
	ACETONE	VL	12	313	109.30	-63.21	16.40	KRAUS,1971
	ACETONE	VL	13	329-350	253.47	-115.95	132.48	SODAY,1930
	ACETONE	VL	11	318	174.00	-108.79	.84	BROWN,1957
	ACETONITRILE	VL	12	318	229.79	-40.70	2.68	BROWN,1955A
	ACETONITRILE	VL	11	318	247.13	-51.54	9.74	PALMER,1972
	2-BUTANONE	VL	15	323	387.60	-242.14	26.15	KRAUS,1971
	2-BUTANONE	VL	15	313	409.15	-245.63	35.18	KRAUS,1971
	2-BUTANONE	VL	22	352	185.55	-144.21	2.18	STEINHAUSER,1949
	CHLOROFORM	VL	15	333	350.47	-226.16	6.72	KRAUS,1971
	CHLOROFORM	VL	19	323	-50.53	4.98	2.83	NAGATA,1970
	CARBON TETRACHLORIDE	VL	14	336-352	-121.79	83.58	8.33	REINDERS,1940
	CYCLOPENTANE	VL	8	313	43.39	-37.52	.10	SCATCHARD,1946
	CYCLOHEXANE	VL	32	322-352	33.15	15.19	4.13	MYERS,1956
	CYCLOHEXANE	VL	23	350-353	4.29	45.93	31.72	RICHARDS,1944
	CYCLOHEXANE	VL	12	343	-16.69	69.48	2.01	DIAZ PENA,1970A
	CYCLOHEXANE	VL	10	283	4.33	65.28	.22	BOUBLIK,1963
	CYCLOHEXANE	VL	30	350-352	-16.85	66.85	1.57	RIDGWAY,1967
	CYCLOHEXANE	VL	7	313	-11.20	70.13	.72	SCATCHARD,1939
	DIOXANE	VL	9	351	-85.00	192.72	334.64	THORTON,1951A
	ETHYL ACETATE	VL	5	298	-197.65	363.78	19.62	TEAGUE,1943
	ETHYL ACETATE	VL	12	343	-192.00	248.71	1.39	LINEK,1972
	ETHYL ACETATE	VL	13	323	-181.49	233.81	1.37	LINEK,1972

* * UNIQUAC BINARY INTERACTION PARAMETERS * *

1 COMPONENTS 2	TYPE	NO. OF POINTS	TEMP. K	UNIQUAC PARAMETERS,K A12	A21	VARIANCE OF FIT	DATA REFERENCE
BENZENE	ETHYL ACETATE	VL	9	333	-166.35	206.66	1.36	LINEK,1972
	ETHANOL	VL	8	340-351	947.20	-138.90	9.12	WEHE,1955
	ETHANOL	VL	10	324-329	920.39	-123.54	7.90	NIELSEN,1959
	ETHANOL	VL	12	318	997.41	-128.88	1.64	BROWN,1954A
	FURFURAL	VL	10	355-427	-85.00	192.72	334.64	THORTON,1951A
	1-BUTANOL	VL	10	318	861.06	-162.39	1.58	BROWN,1969
	1-PROPANOL	VL	12	318	854.75	-145.52	2.16	BROWN,1956
	METHYLCYCLOPENTANE	VL	10	333	-76.06	147.22	24.77	BEYER,1965
	METHYLCYCLOPENTANE	VL	33	344-352	-6.47	56.47	.78	MYERS,1956
	METHYL ACETATE	VL	13	330-350	-207.37	278.33	27.43	HUDSON,1969
	METHYL ACETATE	VL	17	323	-143.88	203.46	5.70	NAGATA,1970
	METHANOL	VL	9	328	972.09	-56.35	9.49	SCATCHARD,1946
	METHANOL	VL	20	330-348	849.76	-36.27	26.30	STRUBL,1972
	METHANOL	VL	10	331-343	912.41	-50.58	11.52	HUDSON,1969
	METHANOL	VL	9	328	972.09	-56.35	9.49	SCATCHARD,1946
	METHANOL	VL	25	318	902.56	-38.37	16.03	STRUBL,1972
	METHANOL	VL	13	313	1254.64	-67.50	51.42	LEE,1931
	METHANOL	VL	9	308	988.60	-56.05	3.62	SCATCHARD,1946
	N-BUTANOL	VL	5	298	754.21	-135.92	19.75	ALLEN,1939
	N-BUTANOL	VL	9	318	928.90	-181.24	1.32	BROWN,1959
	N-HEPTANE	VL	15	318	19.07	31.35	1.77	PALMER,1972
	N-HEPTANE	VL	9	333-347	-32.03	87.50	2.84	NIELSEN,1959
	N-HEPTANE	VL	12	333	-21.60	68.52	4.30	BROWN,1951
	N-HEXANE	VL	12	341-350	-20.14	62.19	5.18	PRAHBU,1963B
	N-HEXANE	VL	24	341-352	-77.13	132.43	2.87	RIDGWAY,1967
	N-HEXANE	VL	10	298	27.92	31.94	4.36	SMITH,1970
	N-HEXANE	VL	10	333	-87.13	155.51	38.30	BEYER,1965
	N-PROPANOL	VL	11	318	928.50	-155.10	14.47	BROWN,1959
	1-NITROPROPANE	VL	8	298	-246.81	535.16	41.27	SAUNDERS,1961
	2-NITROPROPANE	VL	7	298	-344.15	794.91	387.65	SOUNDERS,1961
	NITRO-ETHANE	VL	9	298	46.05	16.61	11.52	SAUNDERS,1961
	NITRO-METHANE	VL	10	298	134.43	29.71	11.30	SANDERS,1961
	NITRO-METHANE	VL	12	318	163.25	8.84	1.80	BROWN,1955B
	SEC-BUTANOL	VL	10	318	-168.83	784.99	2.77	BROWN,1969
	TERT-BUTANOL	VL	11	318	-165.40	744.99	7.88	BROWN,1969
	224-TRIMETHYLPENTANE	VL	11	353-368	-35.12	91.65	2.19	CHU,1956
	TOLUENE	VL	10	354-378	-220.57	330.69	22.04	ROSANOFF,1914
	WATER	MS	1	298	2057.42	115.13		STEPHEN,1964
2-BUTANONE	BENZENE	VL	15	313	-245.63	409.15	35.18	KRAUS,1971
	BENZENE	VL	15	323	-242.14	387.60	26.15	KRAUS,1971
	BENZENE	VL	15	333	-226.16	350.47	6.72	KRAUS,1971
	BENZENE	VL	22	352	-144.21	185.55	2.18	STEINHAUSER,1949
	METHANOL	VL	10	335-340	803.49	-164.22	580.14	HILL,1952
	N-HEPTANE	VL	17	350-369	-75.13	242.53	37.03	STEINHAUSER,1949
	TOLUENE	VL	23	352-383	-82.85	123.57	5.51	STEINHAUSER,1949

COMPONENTS 1	COMPONENTS 2	TYPE	NO. OF POINTS	TEMP. K	UNIQUAC PARAMETERS, A12	A21	VARIANCE OF FIT	DATA REFERENCE
2-BUTANONE	WATER	VL	5	325-330	546.76	-8.79	67.39	OTHMER,1945
	WATER	VL	8	335-337	547.14	-8.83	25.60	OTHMER,1945
	WATER	VL	7	329-337	486.97	-6.76	82.57	ALTSYBEEVA,1964
	WATER	VL	7	340-347	503.58	-2.58	211.29	ALTSYBEEVA,1946
	WATER	VL	5	313	519.97	-6.36	104.54	OTHMER,1945
	WATER	VL	27	346-370	622.84	-55.89	133.11	OTHMER,1952
	WATER	VL	8	333	509.67	-11.75	22.45	ALTSYBEEVA,1964
	WATER	VL	7	346-363	542.99	-15.70	153.66	ALTSYBEEVA,1964
	WATER	VL	16	346-371	592.04	-46.55	81.34	ELLIS,1960
	WATER	VL	8	346-348	593.57	-23.29	18.46	OTHMER,1945
	WATER	VL	7	313-325	503.39	-23.39	129.40	ALTSYBEEVA,1964
CHLOROFORM	ACETONE	VL	29	323	93.96	-171.71	6.75	SEVERNS,1955
	ACETONE	VL	11	330-337	70.60	-147.83	5.24	REINDERS,1947
	ACETONE	VL	31	308	28.38	-140.77	9.17	ZAWIDZKI,1900
	ACETIC ACID	VL	13	338-379	346.43	-98.44	385.19	CONTI,1960
	BENZENE	VL	19	323	4.98	-50.53	2.83	NAGATA,1970
	BENZENE	VL	14	336-352	83.58	-121.79	8.33	REINDERS,1940
	CARBON TETRACHLORIDE	VL	22	313	-14.74	38.19	.27	MC GLASHAN,1954
	CARBON TETRACHLORIDE	VL	20	328	-.10	20.02	8.22	MC GLASHAN,1954
	CARBON TETRACHLORIDE	VL	29	298	-47.33	76.09	.43	MC GLASHAN,1954
	ETHYL ACETATE	VL	18	336-350	24.16	-119.49	3.60	NAGATA,1962A
	ETHANOL	VL	23	328	888.68	-203.03	17.63	SCATCHARD,1938
	ETHANOL	VL	8	328	1315.02	-235.47	4.75	RODECK,1956
	FORMIC ACID	VL	13	332-352	461.38	90.36	172.89	CONTI,1960
	METHYL ACETATE	VL	16	323	121.17	-187.87	3.66	NAGATA,1970
	METHANOL	VL	22	326-336	926.31	-143.50	7.44	NAGATA,1962B
	METHANOL	VL	19	322	1007.23	-141.01	20.80	KIREEV,1941
	METHANOL	VL	9	326-337	1001.00	-154.69	27.38	TYREV,1912
	METHANOL	VL	19	308	1023.77	-150.65	25.57	KIREEV,1941
	METHANOL	VL	25	323	1018.86	-152.44	16.01	SEVERNS,1955
CARBON DISULFIDE	ACETONE	VL	34	308	157.31	103.91	7.33	ZAWIDZKI,1900
	METHANOL	VL	18	310-331	1166.49	138.15	90.65	IIND,1970
CARBON TETRACHLORIDE	ACETONE	VL	12	318	246.68	-92.32	3.27	BROWN,1957
	ACETONITRILE	VL	8	318	458.86	-40.18	16.91	BROWN,1954B
	BENZENE	VL	29	298	-37.52	43.39	.10	SCATCHARD,1945
	CHLOROFORM	VL	20	328	76.09	-47.33	.43	MC GLASHAN,1954
	CHLOROFORM	VL	22	313	20.02	-.10	8.22	MC GLASHAN,1954
	CHLOROFORM	VL	18	313	38.19	-14.74	.27	MC GLASHAN,1954
	CYCLOHEXANE	VL	12	349-353	124.50	-101.32	1.03	RODGER,1969
	CYCLOHEXANE	VL	9	343	98.18	-84.82	.04	BROWN,1950
	CYCLOHEXANE	VL	10	313	-85.18	102.03	.08	SCATCHARD,1939
	CYCLOHEXANE	VL	9	333	-16.01	18.00	.80	DVORAK,1963
	CYCLOHEXANE	VL	9	343	80.65	-71.18	.05	SCATCHARD,1939
	ETHANOL	VL	15	338	1192.49	-135.27	4.28	BARKER,1953
	ETHANOL	VL	13	318	1287.98	-128.51	2.85	BARKER,1953

* * UNIQUAC BINARY INTERACTION PARAMETERS * *

COMPONENTS 1	COMPONENTS 2	TYPE	NO. OF POINTS	TEMP. K	UNIQUAC PARAMETERS, K A12	UNIQUAC PARAMETERS, K A21	VARIANCE OF FIT	DATA REFERENCE
CARBON TETRACHLORIDE	ETHANOL	VL	9	337-347	1583.69	-143.68	53.84	TYRER,1912
	FURFURAL	VL	8	351-397	476.85	-100.42	47.57	MYLES,1961
	METHYLCYCLOPENTANE	VL	18	345-349	161.96	-129.21	.49	RODGER,1969
	METHANOL	VL	9	328	1127.95	-29.64	26.50	SCATCHARD,1946
	N-BUTANOL	VL	21	308	1248.17	-188.77	1.22	PARASKEVOPOULOS,1962
	N-BUTANOL	VL	13	323	1117.85	-201.94	106.98	SMITH,1929
	N-HEPTANE	VL	13	323	88.30	-59.13	24.29	SMITH,1929
	N-PROPANOL	VL	20	308	1336.03	-166.93	4.14	PARASKEVOPOULOS,1962
	N-PROPANOL	VL	11	343	1227.39	-194.57	14.01	PAPOUSEK,1959
	N-PROPANOL	VL	9	346-363	1075.47	-186.06	10.13	CAPLEY,1949
	1-NITROPROPANE	VL	21	298	267.13	-95.68	13.28	SAUNDERS,1961
	2-NITROPROPANE	VL	13	298	307.25	-134.32	22.28	SAUNDERS,1951
	NITRO-ETHANE	VL	17	298	295.06	73.06	12.97	SAUNDERS,1961
	NITRO-METHANE	VL	12	318	398.40	.59	18.55	BROWN,1955B
	TOLUENE	VL	17	352-382	-168.53	203.67	5.47	RODGER,1969
CYCLOPENTANE	BENZENE	VL	32	322-352	15.19	-33.15	4.13	MYERS,1956
CYCLOHEXANE	BENZENE	VL	7	313	70.13	-11.20	.72	SCATCHARD,1939
	BENZENE	VL	30	350-352	66.65	-16.85	1.57	RIDGWAY,1967
	BENZENE	VL	12	343	69.48	-16.69	2.01	DIAZ PENA,1970A
	BENZENE	VL	9	351	192.72	-85.00	334.64	THORTON,1951A
	BENZENE	VL	10	283	65.28	4.33	.22	BOUBLIK,1963
	BENZENE	VL	23	350-353	45.93	4.29	31.72	RICHARDS,1944
	CARBON TETRACHLORIDE	VL	12	343	-84.82	98.18	.04	BROWN,1950
	CARBON TETRACHLORIDE	VL	9	343	-71.18	80.65	.05	SCATCHARD,1939
	CARBON TETRACHLORIDE	VL	9	313	102.03	-85.18	.08	SCATCHAPD,1939
	CARBON TETRACHLORIDE	VL	18	349-353	-101.32	124.50	1.03	RODGER,1969
	CARBON TETRACHLORIDE	VL	10	333	18.00	-16.01	.80	DVORAK,1963
	ETHANOL	VL	7	293	1358.55	-79.36	1.43	SCATCHARD,1964
	ETHANOL	VL	7	323	1298.10	-101.78	.66	SCATCHARD,1964
	ETHANOL	VL	7	278	1349.20	-68.00	1.98	SCATCHARD,1964
	ETHANOL	VL	7	308	1369.41	-90.10	5.77	SCATCHARD,1954
	FURFURAL	VL	7	338	1269.49	-113.70	1.73	SCATCHAPD,1954
	FURFURAL	MS	10	354-421	265.87	44.32	198.64	THORTON,1951A
	FURFURAL	MS	1	298	354.83	41.17		HENTY,1964
	METHYLCYCLOPENTANE	VL	10	345-352	144.37	-118.82	5.19	SUSAREV,1962
	METHANOL	VL	13	328	1364.17	6.02	189.81	STRUBL,1970
	N-BUTANOL	VL	12	318	1591.50	-198.09	9.83	SMIRNOVA,1969
	N-BUTANOL	VL	15	343	1360.84	-217.82	16.54	VONKA,1971
	N-BUTANOL	VL	14	298	1299.00	-175.51	13.51	SMIRNOVA,1969
	N-BUTANOL	VL	14	323	1393.11	-196.90	5.45	VONKA,1971
	N-HEXANE	VL	16	343-352	172.73	-145.56	3.22	RIDGWAY,1967
	N-PROPANOL	VL	10	298	1328.49	-136.92	12.30	SMIRNOVA,1969
	N-PROPANOL	VL	17	328	1364.65	-168.39	34.76	STRUBL,1970
	N-PROPANOL	VL	14	338	1284.75	-173.42	8.37	STRUBL,1970
	NITROMETHANE	MS	1	298	517.19	105.01		WECK,1954

COMPONENTS 1	COMPONENTS 2	TYPE	NO. OF POINTS	TEMP. K	UNIQUAC PARAMETERS,K A12	A21	VARIANCE OF FIT	DATA REFERENCE
CYCLOHEXANE	224-TRIMETHYLPENTANE	VL	7	308	-112.66	141.01	.68	BATTINO,1966
	224-TRIMETHYLPENTANE	VL	8	318	-105.65	128.20	.96	BATTINO,1966
	TOLUENE	VL	20	354-380	63.31	-25.90	9.97	SIEG,1950
1,2-DICHLOROETHANE	TOLUENE	VL	30	354-381	83.67	-44.04	3.72	MYERS,1956
	ETHANOL	VL	10	333	787.17	-104.86	263.32	UDOVENKO,1952A
	ETHANOL	VL	10	313	929.71	-105.66	24.01	UDOVENKO,1952A
	ETHANOL	VL	10	323	858.71	-108.84	34.84	UDOVENKO,1952A
DIETHYL AMINE	METHANOL	VL	20	330-340	676.42	-374.88	20.11	NAKANISHI,1967
DI-ETHYL ETHER	ETHANOL	VL	19	283	741.58	-156.77	3.28	NAGAI,1635
	ETHANOL	VL	19	293	727.64	-159.84	1.73	NAGAI,1935
	ETHANOL	VL	19	303	738.74	-169.24	8.61	NAGAI,1935
	ETHANOL	VL	19	323	733.67	-179.02	5.87	NAGAI,1935
	ETHANOL	VL	19	313	730.44	-172.35	3.19	NAGAI,1935
	ETHANOL	VL	19	273	733.92	-149.22	1.36	NAGAI,1935
DIOXANE	BENZENE	VL	5	298	363.78	-197.65	19.62	TEAGUE,1943
2,3-DIMETHYL BUTANE	METHANOL	VL	18	360-372	927.26	-328.70	113.75	KATO,1970
ETHYL ACETATE	ACETIC ACID	VL	17	317-333	1463.90	-7.18	36.96	MILLOCK,1970
	BENZENE	VL	6	353-381	426.54	-214.39	63.04	GARNER,1954
	BENZENE	VL	12	343	248.71	-192.00	1.39	LINEK,1972
	BENZENE	VL	13	323	233.81	-181.49	1.37	LINEK,1972
	BENZENE	VL	9	333	206.66	-166.35	1.36	LINEK,1972
	CHLOROFORM	VL	18	336-350	-119.49	24.16	3.60	NAGATA,1962A
	ETHYL BENZENE	VL	14	343	183.96	-137.27	11.82	LINEK,1972
	ETHYL BENZENE	VL	14	323	224.44	-159.22	14.28	LINEK,1972
	ETHYL BENZENE	VL	14	333	137.11	-105.50	7.48	LINEK,1972
	ETHANOL	VL	15	343	571.73	-167.61	2.03	MERTL,1972
	ETHANOL	VL	14	328	574.22	-154.11	2.54	MERTL,1972
	ETHANOL	VL	14	313	594.60	-148.29	3.09	MERTL,1972
	FURFURAL	VL	18	313	647.28	-159.51	20.04	MURTI,1958
	1-PROPANOL	VL	10	353-400	-19.15	48.52	30.91	MYLES,1961
	METHANOL	VL	19	333	522.07	-190.57	6.35	MURTI,1958
	N-PROPANOL	VL	20	335-347	579.61	-107.54	16.97	NAGATA,1962S
	TOLUENE	VL	14	333	539.64	-190.31	21.17	MURTI,1958
	TOLUENE	VL	18	351-382	309.41	-214.26	5.55	CARR,1962
	WATER	VL	33	314-341	273.45	-179.88	48.03	RIVENQ,1970
	WATER	VL	8	343	1081.40	-222.02	8.26	MERTL,1972
	WATER	VL	8	328	1000.87	-174.98	6.74	MERTL,1972
	WATER	VL	8	313	877.04	-107.58	5.32	MERTL,1972
	WATER	MS	1	323	569.86	80.91		STEPHEN,1964
ETHANE	N-HEPTANE	VL	15	235-448	264.89	-112.61	4.81	KAY,1938
ETHYL BENZENE	ETHYL ACETATE	VL	14	343	-137.27	183.95	11.82	LINEK,1972
	ETHYL ACETATE	VL	14	333	-105.50	137.11	7.48	LINEK,1972
	ETHYL ACETATE	VL	14	323	-159.22	224.44	14.28	LINEK,1972
ETHANOL	ACETONE	VL	14	305	-119.36	414.46	8.71	GORDON,1946
	ACETONE	VL	14	313	-122.61	398.80	4.00	GORDON,1946

* * UNIQUAC BINARY INTERACTION PARAMETERS * *

COMPONENTS 1	COMPONENTS 2	TYPE	NO. OF POINTS	TEMP. K	UNIQUAC PARAMETERS,K A12	UNIQUAC PARAMETERS,K A21	VARIANCE OF FIT	DATA REFERENCE
ETHANOL	ACETONE	VL	14	321	-131.25	404.49	3.81	GORDON,1946
	ACETIC ACID	VL	16	351-385	-210.53	244.67	32.68	RIUS,1959
	ACETONITRILE	VL	8	293	68.72	430.51	16.25	VIERK,1950
	BENZENE	VL	12	318	-128.88	997.41	1.64	BROWN,1954A
	BENZENE	VL	10	324-329	-123.54	920.39	7.90	NIELSEN,1959
	BENZENE	VL	8	340-351	-138.90	947.20	9.12	WEHE,1955
	CHLOROFORM	VL	8	328	-235.47	1315.02	4.75	ROECK,1956
	CHLOROFORM	VL	23	328	-203.03	888.68	17.63	SCATCHARD,1939
	CARBON TETRACHLORIDE	VL	9	337-347	-143.68	1583.69	53.84	TYRER,1912
	CARBON TETRACHLORIDE	VL	15	338	-135.27	1192.49	4.28	BARKER,1953
	CARBON TETRACHLORIDE	VL	13	318	-128.51	1287.98	2.85	BARKER,1953
	CYCLOHEXANE	VL	7	293	-79.36	1358.55	1.43	SCATCHARD,1964
	CYCLOHEXANE	VL	7	323	-101.78	1298.10	.66	SCATCHARD,1964
	CYCLOHEXANE	VL	7	308	-90.10	1369.41	5.77	SCATCHARD,1964
	CYCLOHEXANE	VL	7	278	-68.00	1349.20	1.98	SCATCHARD,1964
	CYCLOHEXANE	VL	7	338	-113.70	1269.44	1.73	SCATCHARD,1964
	1,2-DICHLOROETHANE	VL	10	313	-105.65	929.71	24.01	UDOVENKO,1952A
	1,2-DICHLOROETHANE	VL	10	323	-108.84	858.71	34.84	UDOVENKO,1952A
	1,2-DICHLOROETHANE	VL	10	333	-104.86	787.17	263.32	UDOVENKO,1952A
	DI-ETHYL ETHER	VL	19	313	-172.35	730.44	3.19	NAGAI,1935
	DI-ETHYL ETHER	VL	19	323	-179.02	733.67	5.87	NAGAI,1935
	DI-ETHYL ETHER	VL	19	293	-159.84	727.64	1.73	NAGAI,1935
	DI-ETHYL ETHER	VL	19	283	-156.77	741.53	3.28	NAGAI,1635
	DI-ETHYL ETHER	VL	19	303	-169.24	738.74	8.61	NAGAI,1935
	DI-ETHYL ETHER	VL	19	273	-149.22	733.92	1.36	NAGAI,1935
	ETHYL ACETATE	VL	14	313	-148.29	594.60	3.09	MERTL,1972
	ETHYL ACETATE	VL	15	343	-167.61	571.73	2.03	MERTL,1972
	ETHYL ACETATE	VL	18	313	-159.51	647.29	20.04	WURTI,1958
	ETHYL ACETATE	VL	14	328	-154.11	574.22	2.54	MERTL,1972
	HEXENE-1	VL	8	333	-107.26	1222.95	34.96	LINDBERG,1971
	METHYL CYCLOHEXANE	VL	8	328	-117.57	1340.56	4.25	KRETSCHMER,1949B
	METHYLCYCLOPENTANE	VL	13	333-349	-118.27	1383.93	37.77	SINOR,1960
	METHYL ACETATE	VL	13	328	-120.60	506.85	3.82	NAGATA,1972
	METHYL ACETATE	VL	11	318	-130.78	573.43	4.84	NAGATA,1972
	METHANOL	VL	12	338-349	-292.39	660.19	16.27	AMER,1956
	N-DECANE	VL	13	352-433	-127.48	1254.65	65.76	ELLIS,1961
	N-HEPTANE	VL	13	313	-101.26	1435.59	25.27	DIAZ PENA,1970B
	N-HEPTANE	VL	12	323	-115.90	1350.02	111.15	SMITH,1929
	N-HEPTANE	VL	14	333	-110.15	1356.90	10.84	DIAZ PENA,1970B
	N-HEPTANE	VL	17	343	-113.40	1192.47	84.65	SMITH,1929
	N-HEPTANE	VL	18	327-336	-88.48	1180.60	121.59	KATZ,1956
	N-HEPTANE	VL	15	303	-117.60	1224.27	170.57	SMITH,1929
	N-HEXANE	VL	8	333	-107.93	1376.23	41.82	LINDBERG,1971
	N-HEXANE	VL	16	331-349	-108.93	1441.57	16.15	SINOR,1960
	N-OCTANE	VL	19	328	-109.08	1385.91	2.71	BOUBLIKOVA,1969

COMPONENTS 1	2	TYPE	NO. OF POINTS	TEMP. K	UNIQUAC PARAMETERS A12	A21	VARIANCE OF FIT	DATA REFERENCE
ETHANOL	N-OCTANE	VL	18	338	-115.85	1388.59	3.07	BOURLIKOVA,1969
	N-OCTANE	VL	19	348	-123.57	1354.92	2.35	BOURLIKOVA,1969
	N-OCTANE	VL	17	318	-103.04	1425.20	2.18	BOURLIKOVA,1969
	N-PROPANOL	VL	9	333	210.95	-67.70	7.57	UDOVENKO,1943
	224-TRIMETHYLPENTANE	VL	10	298	-103.21	1443.34	2.95	KRETSCHMER,1948
	224-TRIMETHYLPENTANE	VL	10	323	-120.42	1449.61	3.26	KRETSCHMER,1948
	TOLUENE	VL	10	308	-132.12	1083.75	1.10	KRETSCHMER,1949A
	TOLUENE	VL	11	358	-103.87	805.83	438.65	WRIGHT,1933
	TOLUENE	VL	10	328	-141.16	1009.48	2.20	KRETSCHMER,1949A
	WATER	VL	13	313	-27.38	284.81	11.40	MERTL,1972
	WATER	VL	34	351-372	-71.06	387.38	7.58	RIEDER,1949
	WATER	VL	13	328	-64.56	380.68	6.27	MERTL,1972
	WATER	VL	10	333	-129.66	561.82	154.75	UDOVENKO,1952B
	WATER	VL	13	343	-81.94	437.92	7.99	MERTL,1972
FURFURAL	ACETONE	VL	12	331-373	195.63	-101.30	43.86	MYLES,1961
	BENZENE	VL	10	355-427	192.72	-85.00	334.64	THORTON,1951A
	CARBON TETRACHLORIDE	VL	9	351-397	-100.42	476.85	47.57	MYLES,1961
	CYCLOHEXANE	VL	10	354-421	44.32	265.87	198.64	THORTON,1951A
	CYCLOHEXANE	MS	1	298	41.17	354.83		HENTY,1964
	ETHYL ACETATE	VL	10	353-400	48.52	-19.15	30.91	MYLES,1961
	224-TRIMETHYLPENTANE	MS	1	298	-4.98	410.08		HENTY,1964
	224-TRIMETHYLPENTANE	VL	10	372-399	-63.19	385.84	172.99	THORTON,1951B
	TOLUENE	VL	11	394-426	244.12	74.87	268.89	THORTON,1951B
FORMIC ACID	ACETIC ACID	VL	12	374-387	-144.58	241.64	12.84	CONTI,1960
	CHLOROFORM	VL	13	332-352	90.36	461.38	172.89	CONTI,1960
	WATER	VL	9	311-321	1047.51	-483.39	3.74	ITO,1963
	WATER	VL	12	374-380	924.01	-525.85	20.73	ITO,1963
	WATER	VL	11	335-345	1019.29	-508.85	12.92	ITO,1963
HEXENE-1	ETHANOL	VL	8	333	1222.95	-107.26	34.96	LINDBERG,1971
1-BUTANOL	BENZENE	VL	10	318	-162.39	861.06	1.58	BROWN,1969
1-PROPANOL	BENZENE	VL	12	318	-145.52	854.75	2.16	BROWN,1956
	ETHYL ACETATE	VL	19	333	-190.57	522.07	6.35	MURTI,1958
	N-DECANE	VL	7	363	-207.27	1074.76	4.44	PATCLIFF,1969
	N-HEPTANE	VL	18	333	-173.79	1243.35	15.73	VAN NESS,1967
	N-HEPTANE	VL	18	318	-162.54	1295.60	11.45	VAN NESS,1967
	N-HEPTANE	VL	18	303	-151.85	1340.62	10.02	VAN NESS,1967
	N-OCTANE	VL	19	338-370	-166.18	1107.44	118.81	PRABHU,1963A
	224-TRIMETHYLPENTANE	VL	12	350-366	-198.06	994.00	26.54	BURES,1959
METHYLISOBUTYLKETONE	METHANOL	VL	10	336-359	688.03	-105.94	302.66	HILL,1952
METHYL CYCLO-HEXANE	PROPIONIC ACID	VL	20	390-411	136.46	-78.49	17.07	WISNIAK,1975
	ETHANOL	VL	8	328	1340.56	-117.57	4.25	KRETSCHMER,1949B
METHYLCYCLOPENTANE	TOLUENE	VL	15	373	210.35	-134.19	8.04	SCHNEIDER,1961
	ANILINE	MS	1	298	228.71	54.36		DARWENT,1943
	BENZENE	VL	33	344-352	56.47	-6.47	.78	MYERS,1956
	BENZENE	VL	10	333	147.22	-76.05	24.77	BEYER,1965

COMPONENTS 1	2	TYPE	NO. OF POINTS	TEMP. K	UNIQUAC PARAMETERS, K A12	A21	VARIANCE OF FIT	DATA REFERENCE
METHYLCYCLOPENTANE	CARBON TETRACHLORIDE	VL	18	345-349	-129.21	161.96	.49	RODGER,1969
	CYCLOHEXANE	VL	10	345-352	-118.82	144.37	5.19	SUSAREV,1962
	ETHANOL	VL	13	333-349	1383.93	-118.27	37.77	SINOR,1960
	N-HEXANE	VL	9	333	162.13	-138.84	.51	BEYER,1965
	TOLUENE	VL	27	345-381	89.77	-48.05	5.10	MYERS,1956
METHYL ACETATE	BENZENE	VL	17	323	203.46	-143.88	5.70	NAGATA,1970
	BENZENE	VL	13	330-350	278.33	-207.37	27.43	HUDSON,1969
	CHLOROFORM	VL	16	323	-187.87	121.17	3.66	NAGATA,1970
	ETHANOL	VL	13	328	506.85	-120.60	3.82	NAGATA,1972
	METHANOL	VL	11	318	573.43	-130.78	4.84	NAGATA,1972
	METHANOL	VL	14	308	508.39	-83.96	3.72	NAGATA,1972
	METHANOL	VL	13	318	502.74	-81.84	8.72	NAGATA,1972
	METHANOL	VL	15	323	516.12	-95.31	2.58	SEVERNS,1955
METHANOL	ACETONE	VL	12	328-336	-63.86	295.35	15.27	AMER,1956
	ACETONE	VL	8	328-335	-111.12	366.38	50.47	GRISWOLD,1949
	ACETONE	VL	25	323	-108.42	379.31	11.58	SEVERNS,1955
	ACETONE	VL	9	327-335	-66.71	346.16	38.83	OTHMER,1928
	ACETONE	VL	23	328-337	-96.90	359.10	6.77	UCHIDA,1950
	ANISOLE	VL	14	338-401	-48.39	782.23	87.31	NAKANISHI,1968
	BENZENE	VL	25	318	-38.37	902.56	16.03	STRUBL,1972
	BENZENE	VL	9	308	-56.05	988.60	3.62	SCATCHARD,1946
	BENZENE	VL	9	328	-56.35	972.09	9.49	SCATCHARD,1946
	BENZENE	VL	10	331-343	-50.58	912.41	11.52	HUDSON,1969
	BENZENE	VL	13	313	-67.50	1254.64	51.42	LEE,1931
	BENZENE	VL	20	330-348	-36.27	849.76	26.30	STRUBL,1972
	BENZENE	VL	9	328	-56.35	972.09	9.49	SCATCHARD,1946
	2-BUTANONE	VL	10	335-340	-164.22	803.49	580.14	HILL,1952
	CHLOROFORM	VL	22	326-336	-143.50	926.31	7.44	NAGATA,19623
	CHLOROFORM	VL	19	322	-141.01	1007.23	20.80	KIREEV,1941
	CHLOROFORM	VL	9	326-337	-154.69	1001.20	27.38	TYPER,1912
	CHLOROFORM	VL	19	308	-150.65	1023.77	25.57	KIREEV,1941
	CHLOROFORM	VL	25	323	-152.44	1018.86	16.01	SEVERNS,1955
	CARBON DISULFIDE	VL	18	310-331	138.15	1166.69	90.65	IINO,1970
	CARBON TETRACHLORIDE	VL	9	328	-29.64	1127.95	26.50	SCATCHARD,1946
	CYCLOHEXANE	VL	13	328	6.02	1364.12	189.81	STRUBL,1970
	DIETHYL AMINE	VL	20	330-340	-374.88	676.42	20.11	NAKANISHI,1967
	2,3-DIMETHYL BUTANE	VL	17	317-333	-7.18	1463.90	36.96	WILLOCK,1970
	ETHYL ACETATE	VL	20	335-347	-107.54	579.61	16.97	NAGATA,19623
	ETHANOL	VL	12	338-349	660.19	-292.39	16.27	AMER,1956
	METHYL ISOBUTYLKETONE	VL	10	336-359	-105.94	688.03	302.66	HILL,1952
	METHYL ACETATE	VL	13	318	-81.84	502.74	8.72	NAGATA,1972
	METHYL ACETATE	VL	14	308	-83.96	508.39	3.72	NAGATA,1972
	METHYL ACETATE	VL	15	323	-95.31	516.12	2.58	SEVEPNS,1955
	N-HEPTANE	VL	7	332	4.84	1325.89	32.06	BENEDICT,1945
	N-HEPTANE	MS	1	306	2.49	1419.33		WITTRIG,1970

1 COMPONENTS	2	TYPE	NO. OF POINTS	TEMP. K	UNIQUAC PARAMETERS,K A12	A21	VARIANCE OF FIT	DATA REFERENCE
METHANOL	N-HEXANE	VL	26	322-336	-2.66	1636.05	49.26	RAAL,1972
	TOLUENE	VL	10	336-343	-43.91	951.82	39.58	BENEDICT,1945
	TOLUENE	VL	16	336-383	-27.12	872.31	262.95	BURKE,1964
	TRIETHYLAMINE	VL	18	337-356	-186.66	664.29	22.91	NAKANISHI,1967
	WATER	VL	16	373	-221.91	433.98	284.20	GRISWOLD,1952
	WATER	VL	21	339-368	-50.82	148.27	27.52	RAMALHO,1961
	WATER	VL	11	339-365	51.90	40.22	34.54	OTHMER,1945
N-BUTANOL	ACETIC ACID	VL	18	388-393	-296.30	546.68	10.54	RIUS,1959
	BENZENE	VL	9	318	-181.24	928.90	1.32	BROWN,1959
	BENZENE	VL	5	298	-135.92	754.21	19.75	ALLEN,1939
	CARBON TETRACHLORIDE	VL	21	308	-188.77	1248.17	1.22	PARASKEVOPOULOS,1962
	CARBON TETRACHLORIDE	VL	13	323	-201.94	1117.85	106.98	SMITH,1929
	CYCLOHEXANE	VL	14	298	-175.51	1299.00	13.51	SWIRNOVA,1969
	CYCLOHEXANE	VL	15	343	-217.82	1360.84	16.54	VONKA,1971
	CYCLOHEXANE	VL	12	318	-193.09	1591.50	9.83	SWIRNOVA,1959
	CYCLOHEXANE	VL	14	323	-196.90	1393.11	5.45	VONKA,1971
	N-DECANE	VL	19	373	-259.67	1430.77	44.79	LEE,1967
	N-HEPTANE	VL	10	387-401	-270.85	1326.41	171.69	VIJAYARAGHAVAN,1967
	N-HEPTANE	VL	8	367-382	-251.11	1291.98	44.46	KOGAN,1959
	N-HEXANE	VL	9	298	-159.24	1370.74	7.54	SWIRNOVA,1969
	N-OCTANE	VL	10	382-388	-236.21	1098.91	9.27	KOGAN,1959
	WATER	VL	13	365-384	20.94	1131.59	38.31	ELLIS,1960
	WATER	VL	22	365-384	30.64	1098.06	44.95	STOCKHARDT,1931
N-DECANE	ETHANOL	VL	16	352-384	38.69	1097.59	28.58	SMITH,1949
	I-PROPANOL	VL	13	352-433	1254.65	-127.48	65.76	ELLIS,1961
	N-BUTANOL	VL	7	363	1074.76	-207.27	4.44	RATCLIFF,1969
	N-PROPANOL	VL	19	373	1430.77	-259.67	4.79	LEE,1967
N-HEPTANE	ACETIC ACID	VL	13	293	1137.20	-201.82	6.37	RATCLIFF,1969
	ACETIC ACID	VL	13	303	311.32	8.24	93.50	MARKUZIN,1971
	ACETIC ACID	VL	15	313	342.57	-9.49	76.23	MARKUZIN,1971
	ACETIC ACID	VL	14	318	339.28	-1.74	80.87	MARKUZIN,1971
	ACETONITRILE	MS	1	318	545.79	23.71		PALMER,1972
	BENZENE	VL	15	333-347	31.35	19.07	1.77	PALMER,1972
	BENZENE	VL	9	233	87.50	-32.03	2.84	NIELSEN,1959
	2-BUTANONE	VL	12	350-369	58.52	-21.60	4.30	BROWN,1951
	CARBON TETRACHLORIDE	VL	17	323	242.53	-75.13	37.03	STEINHAUSER,1949
	ETHANE	VL	13	235-448	-59.13	88.30	24.20	SMITH,1929
	ETHANOL	VL	15	303	-112.61	264.89	4.81	KAY,1938
	ETHANOL	VL	15	333	1224.27	-117.60	170.57	SMITH,1929
	ETHANOL	VL	14	327-336	1356.90	-110.15	10.84	DIAZ PENA,1970B
	ETHANOL	VL	18	323	1190.60	-88.48	121.59	KATZ,1956
	ETHANOL	VL	12	313	1350.07	-115.90	111.15	SMITH,1929
	ETHANOL	VL	13	343	1435.59	-101.26	25.27	DIAZ PENA,1970B
	I-PROPANOL	VL	17	303	1192.47	-113.40	84.65	SMITH,1929
	I-PROPANOL	VL	18	303	1340.02	-151.85	10.02	VAN NESS,1967

1	COMPONENTS 2	TYPE	NO. OF POINTS	TEMP. K	UNIQUAC PARAMETERS A12	A21	VARIANCE OF FIT	DATA REFERENCE
N-HEPTANE	1-PROPANOL	VL	18	318	1295.60	-162.54	11.45	VAN NESS,1967
	1-PROPANOL	VL	18	333	1243.35	-173.79	15.73	VAN NESS,1967
	METHANOL	VL	7	332	1325.89	4.84	32.06	BENEDICT,1945
	METHANOL	MS	1	306	1419.33	2.49		WITTRIG,1970
	N-BUTANOL	VL	10	387-401	1326.41	-270.85	171.69	VIJAYARAGHAVAN,1967
	N-BUTANOL	VL	8	367-382	1291.98	-251.11	44.46	KOGAN,1959
	N-PROPANOL	VL	31	333	1306.22	-160.63	23.23	DIAZ PENA,1970C
	N-PROPANOL	VL	21	348	1083.59	-142.34	434.25	LEE,1967
	TOLUENE	VL	19	371-383	108.24	-72.96	12.34	STEINHAUSER,1949
	ACETONE	VL	14	318	261.51	-33.08	14.38	SCHAEFER,1958
N-HEXANE	ANILINE	MS	1	298	283.76	34.82		DARWENT,1943
	BENZENE	VL	12	341-350	62.19	-20.14	5.18	PRAHBU,19633
	BENZENE	VL	24	341-352	132.43	-77.13	2.87	RIDGWAY,1967
	BENZENE	VL	10	298	31.94	27.92	4.36	SMITH,1970
	BENZENE	VL	10	333	155.51	-87.13	38.30	BEYER,1965
	CYCLOHEXANE	VL	16	343-352	-145.56	172.73	3.22	RIDGWAY,1967
	ETHANOL	VL	16	331-349	1441.57	-108.93	16.15	SINOR,1960
	ETHANOL	VL	8	333	1376.23	-107.93	41.82	LINDBERG,1971
	METHYLCYCLOPENTANE	VL	9	333	-138.84	162.13	.51	BEYER,1965
	METHANOL	VL	26	322-335	1636.05	-2.65	49.26	RAAL,1972
	N-BUTANOL	VL	9	298	1370.74	-159.24	7.54	SMIRNOVA,1959
	N-PROPANOL	VL	20	323	1311.01	-152.85	31.76	DIAZ PENA,1970C
	2-NITROPROPANE	VL	5	318	1326.05	-144.11	10.03	BROWN,1969
	NITRO-ETHANE	VL	16	298	252.20	-32.95	37.14	SAUNDERS,1961
	TOLUENE	VL	12	318	230.64	-5.85	.71	EDWARDS,1962
	TOLUENE	VL	13	316-350	29.59	4.38	8.68	MYERS,1955
	TOLUENE	VL	25	344-375	220.93	-148.51	15.55	SIEG,1950
	TOLUENE	VL	12	344-374	218.17	-147.81	21.84	MYERS,1955
	TOLUENE	VL	13	298-329	34.27	4.30	8.86	MYERS,1955
N-OCTANE	ETHANOL	VL	17	318	1425.20	-103.04	2.18	BOUBLIKOVA,1969
	ETHANOL	VL	18	338	1388.59	-115.85	3.07	BOUBLIKOVA,1969
	ETHANOL	VL	19	328	1385.91	-109.09	2.71	BOUBLIKOVA,1969
	ETHANOL	VL	19	348	1354.92	-123.57	2.35	BOUBLIKOVA,1969
	I-PROPANOL	VL	19	338-370	1107.44	-166.18	118.81	PRABHU,1963A
	N-BUTANOL	VL	10	392-388	1098.91	-236.21	9.27	KOGAN,1959
	NITROETHANE	MS	1	308	308.14	-17.28		JOY,1969
N-PENTANE	NITRO-ETHANE	VL	11	308	333.48	-30.98	1.93	EDWARDS,1962
	PROPIONIC ACID	VL	13	394-412	556.12	-183.20	115.30	JOHNSON,1959
N-PROPANOL	ACETONE	VL	11	305-322	266.31	-22.83	202.14	LO,1962
	ACETIC ACID	VL	14	370-387	299.33	445.77	15.97	RIUS,1959
	BENZENE	VL	11	318	-155.10	928.50	14.47	BROWN,1959
	CARBON TETRACHLORIDE	VL	11	343	-194.57	1227.39	14.01	PAPOUSEK,1959
	CARBON TETRACHLORIDE	VL	20	308	-166.93	1336.03	4.14	PARASKEVOPOULOS,1962
	CARBON TETRACHLORIDE	VL	9	346-363	-186.06	1075.47	10.13	CARLEY,1949
	CYCLOHEXANE	VL	14	338	-173.42	1284.75	8.37	STRUBL,1970

* * UNIQUAC BINARY INTERACTION PARAMETERS * *

COMPONENTS 1	COMPONENTS 2	TYPE	NO. OF POINTS	TEMP. K	UNIQUAC PARAMETERS A_{12}	A_{21}	VARIANCE OF FIT	DATA REFERENCE
N-PROPANOL	CYCLOHEXANE	VL	10	298	-136.92	1328.49	12.30	SMIRNOVA,1969
	CYCLOHEXANE	VL	17	328	-168.39	1364.65	34.76	STRUBL,1970
	ETHYL ACETATE	VL	14	333	-190.31	539.64	21.17	MURTI,1958
	ETHANOL	VL	9	333	-67.70	210.95	7.57	UDOVENKO,1943
	N-DECANE	VL	11	363	-201.82	1137.23	6.37	RATCLIFF,1969
	N-HEPTANE	VL	31	333	-160.43	1306.22	23.73	DIAZ PENA,1970C
	N-HEPTANE	VL	21	348	-142.34	1083.59	434.25	LEE,1967
	N-HEXANE	VL	5	318	-144.11	1326.05	10.03	BROWN,1969
	N-HEXANE	VL	20	323	-152.85	1311.01	31.76	DIAZ PENA,1970C
	NITROETHANE	AZ	1	368	-94.30	574.22		HORSLEY,1962
	TOLUENE	VL	12	369-377	-195.40	818.34	13.67	LU,1957
	WATER	VL	13	333	78.37	583.03	11.17	SCHREIBER,1971
	WATER	VL	13	303	100.17	539.63	38.39	UDOVENKO,1972
	WATER	VL	18	333	23.26	587.18	62.84	MURTI,1958
1-NITROPROPANE	BENZENE	VL	8	298	535.16	-246.81	41.27	SAUNDERS,1961
	CARBON TETRACHLORIDE	VL	21	298	-95.68	267.13	13.28	SAUNDERS,1961
2-NITROPROPANE	BENZENE	VL	7	298	794.91	-344.15	387.65	SOUNDERS,1961
	CARBON TETRACHLORIDE	VL	13	298	-134.32	307.25	22.28	SAUNDERS,1961
	N-HEXANE	VL	16	298	-32.95	252.20	37.14	SAUNDERS,1961
NITRO-ETHANE	BENZENE	VL	9	298	16.61	46.05	11.52	SAUNDERS,1961
	CARBON TETRACHLORIDE	VL	17	298	73.06	295.06	12.97	SAUNDERS,1961
	N-HEXANE	VL	12	318	-5.86	230.64	.71	EDWARDS,1962
	N-OCTANE	VL	11	308	-30.98	333.48	1.93	EDWARDS,1962
	N-OCTANE	MS	1	308	-17.28	308.14		JOY,1969
	N-PROPANOL	AZ	1	368	574.22	-94.39		HORSLEY,1962
	224-TRIMETHYLPENTANE	VL	19	308	10.66	236.48	21.82	EDWARDS,1962
NITRO-METHANE	WATER	MS	1	298	920.08	139.44		MALONE,1967
	BENZENE	VL	10	298	29.71	134.43	11.30	SAUNDERS,1961
	BENZENE	VL	12	318	8.84	163.25	1.80	BROWN,1955B
	CARBON TETRACHLORIDE	VL	12	318	.59	398.40	18.55	BROWN,1955B
	CYCLOHEXANE	MS	1	298	105.01	517.19		WECK,1954
PROPIONIC ACID	METHYLISOBUTYLKETONE	VL	20	390-411	-78.49	136.46	17.07	WISNIAK,1976
	N-OCTANE	VL	13	394-412	-183.20	556.12	115.30	JOHNSON,1959
	WATER	VL	8	372-404	433.89	-164.67	78.46	ITO,1963
	WATER	VL	9	339-368	540.10	-188.72	42.87	ITO,1963
	WATER	VL	7	317-344	599.98	-190.19	82.66	ITO,1963
SEC-BUTANOL	BENZENE	VL	10	318	784.99	-168.83	2.77	BROWN,1969
TERT-BUTANOL	BENZENE	VL	11	318	744.89	-165.40	7.88	BROWN,1969
224-TRIMETHYLPENTANE	BENZENE	VL	11	353-368	91.65	-35.12	2.19	CHU,1956
	CYCLOHEXANE	VL	7	308	141.01	-112.65	.68	BATTINO,1966
	CYCLOHEXANE	VL	8	318	128.20	-105.65	.96	BATTINO,1966
	ETHANOL	VL	10	298	1443.34	-103.21	2.95	KRETSCHMER,1948
	ETHANOL	VL	13	323	1449.61	-120.42	3.26	KRETSCHMER,1949
	FURFURAL	VL	10	372-399	385.84	-63.19	172.99	THORTON,1951B
	FURFURAL	MS	1	298	410.08	-4.93		HENTY,1964

* * UNIQUAC BINARY INTERACTION PARAMETERS * *

COMPONENTS 1	2	TYPE	NO. OF POINTS	TEMP. K	UNIQUAC PARAMETERS,K A12	A21	VARIANCE OF FIT	DATA REFERENCE
224-TRIMETHYLPENTANE	1-PROPANOL	VL	12	350-366	994.00	-198.06	26.54	BURES,1959
	NITRO-ETHANE	VL	19	308	236.48	10.66	21.82	EDWARDS,1962
	TOLUENE	VL	11	372-382	-2.88	19.29	95.20	THORTON,1951B
TOLUENE	TOLUENE	VL	9	372-381	141.11	-94.60	2.21	PRENGLE,1957
	ACETIC ACID	VL	15	303	298.09	-67.91	25.58	MARKUZIN,1971
	BENZENE	VL	10	354-378	330.69	-220.57	22.04	ROSANOFF,1914
	2-BUTANONE	VL	23	352-383	123.57	-82.85	5.51	STEINHAUSFR,1949
	CARBON TETRACHLORIDE	VL	17	352-382	203.67	-168.53	5.47	PODGER,1969
	CYCLOHEXANE	VL	30	354-381	-44.04	83.67	3.72	MYERS,1956
	CYCLOHEXANE	VL	20	354-380	-25.90	53.31	9.97	SIEG,1950
	ETHYL ACETATE	VL	18	351-382	-214.26	309.41	5.55	CARR,1962
	ETHYL ACETATE	VL	33	314-341	-179.88	273.45	48.03	RIVENQ,1970
	ETHANOL	VL	10	328	1009.48	-141.16	2.20	KRETSCHMER,1949A
	ETHANOL	VL	10	308	1083.75	-132.12	1.10	KRETSCHMER,1949A
	ETHANOL	VL	11	358	805.83	-103.87	438.65	WRIGHT,1933
	FURFURAL	VL	11	384-426	74.87	244.12	268.89	THORTON,1951B
	METHYL CYCLOHEXANE	VL	15	373	-134.19	210.35	8.04	SCHNEIDER,1961
	METHYLCYCLOPENTANE	VL	27	345-381	-48.05	89.77	5.10	MYERS,1956
	METHANOL	VL	10	336-343	951.82?	-43.91	39.58	BENEDICT,1945
	METHANOL	VL	16	336-383	872.31	-27.12	262.95	BURKE,1964
	N-HEPTANE	VL	19	371-383	-72.96	108.24	12.34	STEINHAUSER,1949
	N-HEXANE	VL	13	298-329	4.30	34.27	8.86	MYERS,1955
	N-HEXANE	VL	13	316-350	4.38	29.59	8.68	MYERS,1955
	N-HEXANE	VL	12	344-374	-147.81	218.17	21.84	MYERS,1955
	N-HEXANE	VL	25	344-375	-148.51	220.93	15.55	SIEG,1950
	N-PROPANOL	VL	12	369-377	818.34	-195.40	13.67	LU,1957
	224-TRIMETHYLPENTANE	VL	11	372-382	19.29	-2.88	95.20	THORTON,1951B
	224-TRIMETHYLPENTANE	VL	9	372-381	-94.60	141.11	2.21	PRENGLE,1957
	WATER	MS	1	323	1371.36	305.71		STEPHEN,1964
TRIETHYLAMINE VINYL ACETATE	METHANOL	VL	18	337-356	664.29	-186.66	22.91	NAKANISHI,1967
	ACETONE	VL	8	330-342	110.60	-82.48	12.44	SERAFIMOV,1964
	ACETIC ACID	VL	14	348-386	360.56	-140.18	142.93	CAPKOVA,1963
	ACETALDEHYDE	VL	9	302-339	243.51	-117.74	96.59	FRIED,1963
WATER		MS	1	298	1541.72	130.05		SMITH,1941
	ACETONE	VL	12	309-339	-117.85	572.52	23.25	OTHMER,1945
	ACETONE	VL	10	330-362	-67.18	490.50	86.32	OTHMER,1945
	ACETONE	VL	10	295-321	-134.20	603.58	69.05	OTHMER,1945
	ACETONE	VL	13	331-368	-100.71	530.99	47.43	REINDERS,1947
	ACETONE	VL	21	329-365	-56.91	466.35	71.35	BRUNJFS,1943
	ACETONE	VL	13	318-344	-103.94	530.98	62.97	OTHMER,1945
	ACETONE	MS	1	298	-128.35	636.17	22.07	BEARE,1930
	ACRYLONTRILE	MS	1	298	141.17	575.58		VOLPICELLI,1968
	ACETIC ACID	VL	15	339-350	-291.10	602.74	14.56	ITO,1963
	ACETIC ACID	VL	14	317-326	-277.77	570.35	8.88	ITO,1963
	ACETIC ACID	VL	23	373-386	-296.05	518.98	68.65	SEBASTIANI,1967

* * UNIQUAC BINARY INTERACTION PARAMETERS * *

COMPONENTS 1	2	TYPE	NO. OF POINTS	TEMP. K	UNIQUAC PARAMETERS,K A12	A21	VARIANCE OF FIT	DATA REFERENCE
WATER	ACETIC ACID	VL	10	293	-258.51	555.01	1.18	LAZEEVA,1973
	ACETIC ACID	VL	16	373-389	-299.90	530.94	23.34	ITO,1963
	ACETONITRILE	VL	12	350-364	61.92	294.10	45.35	PRATT,1947
	ACETONITRILE	VL	13	349-363	83.42	301.93	74.17	MASLAN,1956
	ACETONITRILE	VL	11	349-359	23.81	388.46	412.28	BLACKFORD,1965
	BENZENE	MS	1	298	115.13	2057.42		STEPHEN,1964
	2-BUTANONE	VL	8	346-348	-23.29	593.57	18.46	OTHMER,1945
	2-BUTANONE	VL	8	333	-11.75	509.67	22.45	ALTSYBEEVA,1964
	2-BUTANONE	VL	7	346-363	-15.70	542.99	153.66	ALTSYBEEVA,1964
	2-BUTANONE	VL	27	346-370	-55.89	622.84	133.11	OTHMER,1952
	2-BUTANONE	VL	7	340-347	2.58	503.58	211.29	ALTSYBEEVA,1946
	2-BUTANONE	VL	16	346-371	-46.55	592.04	81.34	ELLIS,1960
	2-BUTANONE	VL	8	335-337	-8.88	547.14	25.60	OTHMER,1945
	2-BUTANONE	VL	7	329-337	-6.76	486.97	82.57	ALTSYBEEVA,1964
	2-BUTANONE	VL	5	325-330	-8.79	546.76	67.39	OTHMER,1945
	2-BUTANONE	VL	5	313	-6.36	519.97	104.54	OTHMER,1945
	2-BUTANONE	VL	5	313-325	-23.39	503.39	129.40	ALTSYBEEVA,1964
	DIOXANE	VL	18	360-372	-328.70	927.26	113.75	KATO,1970
	ETHYL ACETATE	MS	1	323	80.91	569.86		STEPHEN,1964
	ETHYL ACETATE	VL	8	343	-222.02	1081.40	8.26	MERTL,1972
	ETHYL ACETATE	VL	8	328	-174.08	1000.87	6.74	MERTL,1972
	ETHYL ACETATE	VL	8	313	-107.58	877.04	5.32	MERTL,1972
	ETHYL ACETATE	VL	8	343	437.92	-81.94	7.99	MERTL,1972
	ETHANOL	VL	13	313	284.81	-27.38	11.40	MERTL,1972
	ETHANOL	VL	34	351-372	387.38	-71.06	7.58	RIEDER,1949
	ETHANOL	VL	10	333	561.82	-129.66	154.75	UDOVENKO,19528
	ETHANOL	VL	13	328	380.68	-64.56	6.27	MERTL,1972
	FORMIC ACID	VL	12	374-380	-525.85	924.01	20.73	ITO,1963
	FORMIC ACID	VL	9	311-321	-483.39	1047.51	3.74	ITO,1963
	FORMIC ACID	VL	11	335-345	-508.85	1019.29	12.92	ITO,1963
	METHANOL	VL	11	339-365	40.22	51.90	34.54	OTHMER,1945
	METHANOL	VL	21	338-368	148.27	-50.82	27.52	PAMALHO,1961
	METHANOL	VL	16	373	433.98	-221.91	284.20	GRISWOLD,1952
	N-BUTANOL	VL	22	365-384	1098.06	30.64	44.95	STOCKHARDT,1931
	N-BUTANOL	VL	13	365-388	1131.59	20.94	38.31	ELLIS,1960
	N-BUTANOL	VL	16	365-384	1097.59	38.69	28.58	SMITH,1949
	N-PROPANOL	VL	18	333	587.18	23.26	62.84	MURTI,1958
	N-PROPANOL	VL	13	333	583.03	78.37	11.17	SCHREIBER,1971
	N-PROPANOL	VL	13	303	539.63	100.17	38.39	UDOVENKO,1972
	NITROETHANE	MS	9	298	138.44	920.08		MALONE,1967
	PROPIONIC ACID	VL	9	339-368	-188.72	540.10	42.87	ITO,1963
	PROPIONIC ACID	VL	8	372-404	-164.67	433.89	78.46	ITO,1963
	PROPIONIC ACID	VL	7	317-344	-190.19	599.98	82.66	ITO,1963
	TOLUENE	MS	1	323	305.71	1371.36		STEPHEN,1964
	VINYL ACETATE	MS	1	298	130.05	1541.72		SMITH,1941

202

DATA REFERENCES

ALLEN,B.B.,LINGO,S.P.,FELSING,W.A./J.PHYS.CHEM.43,425(1939).

ALTSYBEEVA,A.I.,MARACHEVSKII,A.G./RUSS.J.PHYS.CHEM.38, 849(1964).

AMER,H.H.,PAXTON,R.R.,VAN WINKLE,M./IND.ENG.CHEM.48, 142(1956).

BARKER,J.A.,BROWN,I.,SMITH,F./DISC.FARAD.SOC.15,142(1953).

BATTINO,R./J.PHYS.CHEM.70(11),3408(1966).

BEARE,W.G.,MCVICAR,G.A.,FERGUSSON,J.B./J.PHYS.CHEM. 34,1310(1930).

BENEDICT,M.,JOHNSON,C.A.,SOLOMON,E.,RUBIN,L.C./TRANS.AM. INST.CHEM.ENGRS.41,371(1945).

BEYER,W.,SCHUBERTH,H.,LEIBNITZ E./J.PRAKT.CHEM.27,276(1965).

BLACKFORD,D.S.,YORK,R./J.CHEM.ENG.DATA 10,313(1965).

BOUBLIKOVA,L.,LU,B.C.Y./J.APPL.CHEM.(LONDON) 19(3),89(1969).

BOUBLIK,T./COLL.CZECH.CHEM.COMMUN.28,1771(1963).

BROWN,I.,EWALD,A.H./AUSTR.J.SCI.RES.A3,306(1950).

BROWN,I.,EWALD,A.H./AUSTR.J.SCI.RES.A4,198(1951).

BROWN,I.,FOCK,W.,SMITH,F./AUSTR.J.CHEM.9,364(1956).

BROWN,I.,FOCK,W.,SMITH,F./J.CHEM.THERMODYN.1(3),273(1969).

BROWN,I.,SMITH,F./AUSTR.J.CHEM.7,264(1954A).

BROWN,I.,SMITH,F./AUSTR.J.CHEM.7,269(1954B).

BROWN,I.,SMITH,F./AUSTR.J.CHEM.8,62(1955A).

BROWN,I.,SMITH,F./AUSTR.J.CHEM.8,501(1955B).

BROWN,I.,SMITH,F./AUSTR.J.CHEM.10,423(1957).

BROWN,I.,SMITH,F./AUSTR.J.CHEM.12,407(1959).

BRUNJES,A.S.,BOGART,M.J.P./IND.ENG.CHEM.35,255(1943).

BURES,E.,CANO,C.,DE WIRTH,A./J.CHEM.ENG.DATA 4,199(1959).

BURKE,D.E.,WILLIAMS,G.C.,PLANK,C.A./J.CHEM.ENG.DATA 9,212(1964).

CAPKOVA,A.,FRIED,V./COLL.CZECH.CHEM.COMMUN.28,2237(1963).

CARLEY,J.F.,BERTELSEN,L.W./IND.ENG.CHEM.41,2806(1949).

CARR,A.D.,KROPHOLLER,H.W./J.CHEM.ENG.DATA 7,26(1962).

CHU,J.C./VAPOR—LIQUID EQUILIBRIUM DATA, ANN ARBOR, MICHIGAN (1956).

CONTI,J.J.,OTHMER,D.F.,GILMONT,R./J.CHEM.ENG.DATA, 5,301(1960).

DARWENT,B.,WINKLER,C./J.PHYS.CHEM.47,442(1943).

DIAZ PENA,M.,RODRIGUEZ CHEDA,D./AN.QUIM.66(9-10),721(1970A).

DIAZ PENA,M.,RODRIGUEZ CHEDA,D./AN.QUIM.66(9-10),737(1970B).

DIAZ PENA,M.,RODRIGUEZ CHEDA,D./AN.QUIM.66(9-10),747(1970C).

DONNELLY,J.K.,MALPANI,S.N.,MOORE,R.G./J.CHEM.ENG.DATA 20,171(1975).

EDWARDS,J.B./PH.D.DISSERTATION,GEORGIA INST.TECH.(1962).

ELLIS,S.R.M.,GARBETT,R.D./IND.ENG.CHEM.52(5),385(1960).

ELLIS,S.R.M.,SPURR,M.J./BRIT.CHEM.ENG.6,92(1961).

FRIED,A.,CAPKOVA,A.,SUSKA,J./COLL.CZECH.CHEM.COMMUN. 28,3171(1963).

GARNER,F.,ELLIS,S.,PEARCE,C./CHEM.ENG.SCI.3,48(1954).

GORDON,A.R.,HINES,W.G./CAN.J.RES.24B,254(1946).

GRISWOLD,J.,BUFORD,C.B./IND.ENG.CHEM.41,2347(1949).

GRISWOLD,J.,WONG,S.Y./CHEM.ENG.PROGR.SYMP.SER.48(3), 18(1952).

HEERTJES,P.M./CHEM.PROC.ENG.41,385(1960).

HENTY,C.J.,MCMANAMEY,W.J.,PRINCE,R.G.H./J.APPL.CHEM. 14,148(1964).

HILL,W.D.,VAN WINKLE,M./IND.ENG.CHEM.44(1),205(1952).

HORSLEY,L./AZEOTROPIC DATA, AD.CHEM.SER.,NO.35(1962).

HUDSON,J.W.,VAN WINKLE,M./J.CHEM.ENG.DATA 14(3),310(1969).

IINO,M.,SUDO,J.,HIRATA,M.,HIROSE,Y.,NAKAE,A./J.CHEM.ENG.DATA 15(3),446(1970).

ITO,T.,YOSHIDA,F./J.CHEM.ENG.DATA 8,315(1963).

JOHNSON,I.I.,FURTER,W.F.,BARRY,T.W./CAN.J.TECH. 32(5), 179(1959).

KATO,M.,KONISHI,H.,HIRATA,M./J.CHEM.ENG.DATA 15,501(1970).

KATZ,K.,NEWMAN,M./IND.ENG.CHEM.48,137(1956).

KAY,W.B./IND.ENG.CHEM.30,459(1938).

KIREEV,V.A.,SITNIKOV,I.P./ZH.FIZ.KHIM.15,492(1941).

KOGAN,V.B.,FRIDMAN,V.M.,ROMANOVA,T.G./RUSS.J.PHYS.CHEM. 33,34(1959).

KRAUS,J.,LINEK,J./COLL.CZECH.CHEM.COMMUN.36(7),2547(1971).

KRETSCHMER,C.B.,NOWAKOWSKA,J.,WIEBE,R./J.AM.CHEM.SOC. 70,1785(1948).

KRETSCHMER,C.B.,WIEBE,R./J.AM.CHEM.SOC.71,1793(1949A).

KRETSCHMER,C.B.,WIEBE,R./J.AM.CHEM.SOC.71,3176(1949B).

LAZEEVA,M.S.,MARKUZIN,N.P./J.APPL.CHEM.USSR.46,373(1973).

DATA REFERENCES

LEE,L.L.,SCHELLER,W.A./J.CHEM.ENG.DATA 12(4),497(1967).

LEE,S.C./J.PHYS,CHEM.35,3558(1931).

LINDBERG,G.W.,TASSIOS,D./J.CHEM.ENG.DATA 16(1),52(1971).

LINEK,J.,PROCHAZKA,K.,WICHTERLE,I./COLL.CZECH.CHEM.
COMMUN.37(9),3010(1972).

LO,T.C.,BIEBER,H.H.,KARR,A.E./J.CHEM.ENG.DATA
7(3),327(1962).

LU,B./CANN.J.TECH.34,468(1957).

MALONE,J.W.,VINING,R.W./J.CHEM.ENG.DATA 12,387(1967).

MARKUZIN,N.PAVLOVA,L./ZH.PRIK.KHIM.44,311(1971).

MASLAN,F.D.,STODDARD,E.A./J.PHYS.CHEM.60,1146(1956).

MC GLASHAN,M.L.,PRUE,J.E.,SAINSBURY,J.E.J./TRAN.FARAD.SOC.
50,1284(1954).

MERTL,I./COLL.CZECH.CHEM.COMMUN.37(2),366(1972).

MURTI,P.S.,VAN WINKLE,M./CHEM.ENG.DATA SERIES 3,72(1958).

MYERS,H.S./IND.ENG.CHEM.47,2215(1955).

MYERS,H.S./IND.ENG.CHEM.48,1104(1956).

MYLES,M.G.,WINGA-D,R.E./IND.ENG.CHEM.53,219(1961).

NAGAI,J.,ISII,N./J.SOC.CHEM.IND.JAPAN 38,8(1935).

NAGATA,I./J.CHEM.ENG.DATA 7,60(1962A).

NAGATA,I./J.CHEM.ENG.DATA 7,367(1962B).

NAGATA,I.,HAYASHIDA,H./J.CHEM.ENG.JAPAN 3,161(1970).

NAGATA,I.,OHTA,T.,TAKAHASHI,T./J.CHEM.ENG.JAPAN 5(3),
227(1972).

NAKANISHI,K.,SHIRAI,H.,MINAMIYAMA,T./J.CHEM.ENG.DATA
12(4),591(1967).

NAKANISHI,K.,SHIRAI,H.,NAKASATO,K./J.CHEM.ENG.DATA
13,189(1968).

NIELSEN,R.L.,WEBER,J.H./J.CHEM.ENG.DATA 4,145(1959).

OTHMER,D.F./IND.ENG.CHEM.20,743(1928).

OTHMER,D.F.,BENENATI,R.F./IND.ENG.CHEM.37,299(1945).

OTHMER,D.F.,CHUDGAR,M.M.,LEVY,S.L./IND.ENG.CHEM.44,
1872(1952).

PALMER,D.A.,SMITH,B.D./J.CHEM.ENG.DATA,17,71(1972).

PAPOUSEK,D.,PAPOUSKOVA,Z.,PAGO,L./Z.PHYS.CHEM.(LEIPZIG)
211,231(1959).

PARASKEVOPOULOS,G.C.,MISSEN,R.W./TRANS.FARADAY SOC.58,
869(1962).

PRABHU,P.S.,VAN WINKLE,M./J.CHEM.ENG.DATA 8,14(1963A).

DATA REFERENCES

PRAHBU,P.S.,VAN WINKLE,M./J.CHEM.ENG.DATA 8,210(1963B).

PRATT,H.R.C./TRANS.INST.CHEM.ENGRS.(LONDON) 25,43(1947).

PRENGLE,H.W.,PALM,G.F./IND.ENG.CHEM.49,1769(1957).

PROKHOROVA,V.V.,SERAFIMOV,L.V.,TAKHTAMYSHEVA,L.S./
RUSS.J.PHYS.CHEM.38,549(1964).

RAAL,J.D.,CODE,R.K.,BEST,D.A./J.CHEM.ENG.DATA
17(2),211(1972).

RAMALHO,R.S.,TILLER,F.M.,JAMES,W.J.,BUNCH,D.W./IND.ENG.CHEM.
53,895(1961).

RATCLIFF,G.A.,CHAO,K.C./CAN.J.CHEM.ENG.47,148(1969).

REINDERS,W.,DE MINJER,C.H./REC.TRAV.CHIM.59,369(1940).

REINDERS,W.,DE MINJER,C.H./REC.TRAV.CHIM.,66,573(1947).

RICHARDS,A.R.,HARGREAVES,E./IND.ENG.CHEM.36,805(1944).

RIDGWAY,K..BUTLER,P.A./J.CHEM.ENG.DATA 12(4),509(1967).

RIEDER,R.M.,THOMPSON,A.R./IND.ENG.CHEM.41,2905(1949).

RIUS,A.,OTERO,J.L.,MACARRON,A./CHEM.ENG.SCI.,10,105(1959).

RIVENQ,F./BULL.SOC.CHIM.(FRANCE) 11,3833(1970).

RODGER,A.H.,HSU,C.C.,FURTER,W.F./J.CHEM.ENG.DATA 14(3),
362(1969).

ROECK,H.,SCHROEDER,W./Z.PHYS.CHEM.(FRANKFURT)9,277(1956).

ROSANOFF,M.A.,BACON,C.W.,SCHULZE,J.F.W./J.AM.CHEM.SOC.
36,1999(1914).

SAUNDERS,D.F.,SPAULL,A.J.B./Z.PHYSIK.CHEM.(FRANKFURT)
28,332(1961).

SCATCHARD,G.,RAYMOND,C.L./J.AM.CHEM.SOC.60,1278(1938).

SCATCHARD,G.,SATKIEWICZ,F.G./J.AM.CHEM.SOC.86,130(1964).

SCATCHARD,G.,WOOD,S.E.,MOCHEL,J.M./J.AM.CHEM.SOC.61,
3206(1939).

SCATCHARD,G.,WOOD,S.E.,MOCHEL,J.M./J.AM.CHEM.SOC.
62,712(1940).

SCATCHARD,G.,WOOD,S.E.,MOCHEL,J.M./J.AM.CHEM.SOC.68,
1960(1946).

SCHAEFER,K.,RALL,W./ELEKTROCHEM.62(10),1090(1958).

SCHNEIDER,G./Z.PHYS.CHEM.(FRANKFURT)27(3-4),171(1961).

SCHREIBER,E.,SCHUTTAU,E./Z.PHUS.CHEM.(LEIPZIG) 247(1-2),
23(1971).

SEBASTIANI,E.,LACQUANITI,L./CHEM.ENG.SCI.22,1155(1967).

SERAFIMOV,L.A.,TIKHONOVA,N.K.,LVOV,S.V./RUSS.J.PHYS.CHEM.
38,1119(1964).

SEVERNS,W.H.,SESONSKE,A.,PERRY,R.H.,PIGFORD,R.L./AICHE.J.1,
401(1955).

SIEG,L./CHEM.ING.TECHN.22,322(1950).

SINOR,J.E.,WEBER,J.H./J.CHEM.ENG.DATA 5,243(1960).

SMIRNOVA,N.A.,KURTYNINA,L.M./RUSS.J.PHYS.CHEM.43(7),
1059(1969).

SMITH,C.P.,ENGEL,E.W./J.AM.CHEM.SOC.51,2660(1929).

SMITH,J.C./J.PHYS,CHEM.45,1301(1941).

SMITH,T.E.,BONNER,R.F./IND.ENG.CHEM.41,2867(1949).

SMITH,V.C.,ROBINSON,R.L./J.CHEM.ENG.DATA 15(3),391(1970).

SODAY,F.J.,BENNETT,G.W./J.CHEM.EDUC.7,1336(1930).

STEINHAUSER,H.H.,WHITE,R.R./IND.ENG.CHEM.41,2912(1949).

STEPHEN,H.,STEPHEN,T.EDS./SOLUBILITIES OF INORGANIC AND
ORGANIC COMPOUNDS,MACMILLAN CO.,NEW YORK(1964).

STOCKHARDT,J.S.,HU..,C.M./IND.ENG.CHEM.23(12),1438(1931).

STRUBL,K.,SVOBODA,V.,HOLUB,R./COLL.CZECH.CHEM.COMMUN.37(11),
3522(1972).

STRUBL,K.,SVOBODA,V.,HOLUB,R.,PICK,J./COLL.CZECH.CHEM.
COMMUN.35,3004(1970).

SUSAREV,M.P.,LYSLOVA,R.V./ZH.FIZ.FHIM.36,437(1962).

TEAGUE,P.C.,FELSING,W.A./J.AM.CHEM.SOC.,65,485(1943).

THORTON,J.D.,GARNER,F.H./J.APPL.CHEM.1,SUPPL.1,S61(1951A).

THORTON,J.D.,GARNER,F.H./J.APPL.CHEM.1,SUPPL.1,S74(1951B).

TYRER,D./J.CHEM.SOC.101,1104(1912).

UCHIDA,S.,OGAWA,S.,YAMAGUSHI,M./JAPAN SCI.REV.,ENG.SCI.
1(2),41(1950).

UDOVENKO,V.V.,FATKULINA,L.G./ZH.FIZ.KHIM.26,719(1952A).

UDOVENKO,V.V.,FATKULINA,L.G./ZH.FIZ.KHIM.26,1438(1952B).

UDOVENKO,V.,FRID,T./ZH.FIZ.KHIM.22,1135(1943).

UDOVENKO,V.V.,MAZANKO,T.F./IZV.VYSSH,UCHEB.ZAVED.KHIM.
KHIM.TEKHNOL.15,1654(1972).

VAN NESS,H.C.,SOCZEK,C.A.,PELOQUIN,G.L.,MACHADO,R.L./
J.CHEM.ENG.DATA 12,217(1967).

VIERK,A.L./Z.ANORG.CHEM.261,283(1950).

VIJAYARAGHAVAN,S.,DESHPANDE,P.,KULOOR,N./
J.CHEM.ENG.DATA 12,13(1967).

VOLPICELLI,G./J.CHEM.ENG.DATA 13,150(1968).

DATA REFERENCES

VONKA,P.,SVOBODA,V.,STRUBL,K.,HOLUB,R./COLL.CZECH.CHEM.COMMUN.36(1),18(1971).

WECK,H.I.,HUMT,H./IND.ENG.CHEM.46,2521(1954).

WEHE,A.H.,COATES,J./AICHE.J.1,241(1955).

WILLOCK,J.M.,VAN WINKLE,M./J.CHEM.ENG.DATA 15,281(1970).

WISNIAK,J.,TAMIR,A./J.CHEM.ENG.DATA 21, 88(1976).

WITTRIG,T.S./B.S. DEGREE THESIS,UNIVERSITY OF ILLINOIS,URBANA,ILL.(1977).

WRIGHT,W.A./J.PHYS.CHEM.37,233(1933).

YANG,C.P.,VAN WINKLE,M./IND.ENG.CHEM.47,293(1955).

YORK,R.,HOLMES,R.C./IND.ENG.CHEM.,34,345((1942).

ZAWIDZKI,J./Z.PHYS.CHEM.35,129(1900).

Appendix C-7

UNIQUAC Binary Parameters for Noncondensable
Components with Condensable Components.
Parameters Obtained from Vapor-Liquid Equilibrium
Data in the Dilute Region

Appendix C-7   Interaction Parameters for Noncondensables

Component 1	Component 2	Temperature Range, K	$\delta_{12}^{(0)}$	$\delta_{12}^{(1)}$, K	Data Reference
Hydrogen	Acetone	252-313	3.23	218	Stephen (1963)
	Methanol	294-413	3.94	218	Krichevskii (1937)
	n-Hexane	278-478	2.99	-4	Nichols (1957)
	m-Xylene	462-582	2.09	765	Sattler (1940)
	n-Octane	463-543	-0.50	1646	Connolly (1976)
	Benzene	433-533	1.97	810	Sattler (1940)
	Water	273-353	9.97	-880	Wiebe (1934a)
	Ammonia	273-373	0.68	1344	Wiebe (1934b)
Nitrogen	n-Butane	311-411	-1.67	814	Roberts (1961)
	n-Hexane	311-444	0.07	336	Poston (1966)
	Water	298-513	3.76	913	Saddington (1934)
	Ammonia	273-373	-3.35	2098	Wiebe (1937)
Carbon Monoxide	n-Octane	463-533	-3.38	1878	Connolly (1976)
	Benzene	433-533	-1.04	1062	Connolly (1976)
	Methanol	293-413	0.83	552	Krichevskii (1937)
	Acetone	273-313	0.48	408	Stephen (1963)
Methane	Propane	311-344	-2.01	604	Sage (1950)
	n-Butane	311-394	-2.05	731	Sage (1950)
	n-Pentane	311-411	-1.32	509	Sage (1950)
	n-Hexane	311-444	-0.06	94	Poston (1966)
	n-Decane	311-511	-0.24	150	Sage (1950)
	Benzene	293-333	-0.38	473	Stephen (1963)
	Acetone	273-313	-0.26	471	Stephen (1963)

Data Sources for Binary Systems with a Condensable Component and a Noncondensable Component

Connolly, J. F., Kandalic, G. A., Research and Development Dept., Amoco Oil Co., Naperville, Ill., personal communication (1976).

Krichevskii, I. R., Zhavoronkov, N. M., Tsiklis, D. S., Zh. Fiz. Khim., 9, 317 (1937).

Nichols, W. B., Reamer, H. H., Sage, B. H., A.I.Ch.E. J., 3, 262 (1957).

Poston, R. S., McKetta, J., J. Chem. Eng. Data, 11, 362 (1966a).

Poston, R. S., McKetta, J., J. Chem. Eng. Data, 11, 369 (1966b).

Roberts, L. K., McKetta, J., A.I.Ch.E. J., 7, 173 (1961).

Saddington, A. W., Krase, W. W., J. Am. Chem. Soc., 56, 353 (1934).

Sage, B. H., Lacey, W. N., "Thermodynamic Properties of Higher Paraffin Hydrocarbons and Nitrogen," Am. Petr. Inst., New York, N.Y. (1950).

Stephen, H., Stephen, T., "Solubilities of Inorganic and Organic Compounds," Pergamon Press, New York (1963).

Wiebe, R., Gaddy, V. L., J. Am. Chem. Soc., 56, 76 (1934a).

Wiebe, R., J. Am. Chem. Soc., 56, 2357 (1934b).

Wiebe, R., Gaddy, V. L., J. Am. Chem. Soc., 59, 1984 (1937).

Appendix D

BINARY PARAMETER ESTIMATION PROGRAM:
DESCRIPTION AND LISTING

VPLQFT is a computer program for correlating binary vapor-liquid equilibrium (VLE) data at low to moderate pressures. For such binary mixtures, the truncated virial equation of state is used to correct for vapor-phase nonidealities, except for mixtures containing organic acids where the "chemical" theory is used. The Hayden-O'Connell (1975) correlation gives either the second virial coefficients or the dimerization equilibrium constants, as required.

At the user's option, one of two methods can be used to calculate the liquid-phase reference fugacity: (1) An empirical equation is used for the vapor pressure as a function of temperature. This vapor pressure is corrected to give the pure-component liquid fugacity at zero pressure. (2) An empirical equation is used for the pure-component liquid fugacity directly as a function of temperature.

The user may also choose to use the vapor-pressure equation instead of the fugacity at zero pressure.

VLE data are correlated by any one of thirteen equations representing the excess Gibbs energy in the liquid phase. These equations contain from two to five adjustable binary parameters; these are estimated by a nonlinear regression method based on the maximum-likelihood principle (Anderson et al., 1978).

A number of correlation options are possible:
1. fit P-T-x-y data or P-T-x data, minimizing the errors in all experimental measurements;
2. fit P-T-x-y data but minimize the errors in only the P-y measurements;
3. fit P-T-x data and minimize the errors in only the P measurements.

An additional option allows the user to fit data for binary mixtures where one of the components is noncondensable. The mixture is treated as an ideal dilute solution. The solute

211

(i.e. the noncondensable) is designated component 1 and the solvent (i.e. the condensable) is component 2. The fugacity of the solvent in the liquid phase, f_2^L, is given as a function of liquid composition by

$$f_2^L = x_2 f_2^O$$

where f_2^O is the reference fugacity. For the solute, the fugacity in the liquid phase is given by

$$f_1^L = x_1 H_{1,2}$$

where $H_{1,2}$ is Henry's constant for component 1 in 2. For convenience, Henry's constant is separated into two contributions,

$$H_{1,2} = \gamma_{1,2}^{\infty} \, f_1^{OL}$$

where f_1^{OL} is a hypothetical, pure-component liquid reference fugacity for the noncondensable, and $\gamma_{1,2}^{\infty}$ is an activity coefficient which accounts for binary solute-solvent interactions. In reduced form, the hypothetical reference fugacity is given by the generalized equation

$$\ln \frac{f_1^{OL}}{P_c} = 7.224 - 7.534 \left(\frac{T}{T_c}\right)^{-1} - 2.598 \ln\left(\frac{T}{T_c}\right)$$

where P_c is the critical pressure and T_c is the critical temperature, both for the solute. The hypothetical reference fugacity accounts for the major effect of temperature on Henry's constant. For a specified temperature range, the correction term, $\gamma_{1,2}^{\infty}$, is given by

$$\ln \gamma_{1,2}^{\infty} = \delta_{12}^{(0)} + \frac{\delta_{12}^{(1)}}{T}$$

The two adjustable binary parameters, $\delta_{12}^{(0)}$ and $\delta_{12}^{(1)}$, are determined from binary vapor-liquid equilibrium data, using the computer program described here.

Detailed discussion of all input options are given in the Data Input section.

Equations for Liquid-Phase Nonidealities

A. Modified UNIQUAC Equation

212

$$\ln \gamma_1 = \ln \frac{\Phi_1}{x_1} + \frac{z}{2} q_1 \ln \frac{\theta_1}{\Phi_1} + \Phi_2 (\ell_1 - \frac{r_1}{r_2} \ell_2)$$

$$+ C q_1' \left[- \ln (\theta_1' + \theta_2' \tau_{21}) + \frac{\theta_2' \tau_{21}}{\theta_1' + \theta_2' \tau_{21}} - \frac{\theta_2' \tau_{12}}{\theta_2' + \theta_1' \tau_{12}} \right]$$

$$\ln \gamma_2 = \ln \frac{\Phi_2}{x_2} + \frac{z}{2} q_2 \ln \frac{\theta_2}{\Phi_2} + \Phi_1 (\ell_2 - \frac{r_2}{r_1} \ell_1)$$

$$+ C q_2' \left[- \ln (\theta_2' + \theta_1' \tau_{12}) + \frac{\theta_1' \tau_{12}}{\theta_2' + \theta_1' \tau_{12}} - \frac{\theta_1' \tau_{21}}{\theta_1' + \theta_2' \tau_{21}} \right]$$

where

$$\ell_1 = \frac{z}{2}(r_1 - q_1) - (r_1 - 1)$$

$$\ell_2 = \frac{z}{2}(r_2 - q_2) - (r_2 - 1)$$

$$\Phi_1 = \frac{x_1 r_1}{x_1 r_1 + x_2 r_2} \qquad\qquad \Phi_2 = \frac{x_2 r_2}{x_1 r_1 + x_2 r_2}$$

$$\theta_1 = \frac{x_1 q_1}{x_1 q_1 + x_2 q_2} \qquad\qquad \theta_2 = \frac{x_2 q_2}{x_1 q_1 + x_2 q_2}$$

$$\theta_1' = \frac{x_1 q_1'}{x_1 q_1' + x_2 q_2'} \qquad\qquad \theta_2' = \frac{x_2 q_2'}{x_1 q_1' + x_2 q_2'}$$

$$\tau_{12} = \exp\left(\frac{-\Delta u_{12}}{CRT}\right) \qquad\qquad \tau_{21} = \exp\left(\frac{-\Delta u_{21}}{CRT}\right)$$

and

z - coordination number, equal to 10

r_1, r_2 - structural size parameters

q_1, q_2 - structural area parameters

q_1', q_2' - modified structural area parameters

$\Delta u_{12}, \Delta u_{21}$ - adjustable energy parameters

C - adjustable binary parameter, usually set equal to unity

The parameters determined from binary VLE data are:

$$P_{(1)} = \Delta u_{12}$$
$$P_{(2)} = u_{21}$$
$$P_{(3)} = C$$

If only two parameters are estimated, C must be set to some value,

213

and only $P_{(1)}$ and $P_{(2)}$ are determined from the VLE data. If C = 1.0 and q' = q for both components, the original UNIQUAC equation is obtained (Abrams and Prausnitz, 1975).

B. Wilson Equation (Wilson, 1964; Renon and Prausnitz, 1969)

$$\ln \gamma_1 = C \left[-\ln(x_1 + \Lambda_{12}x_2) + \frac{\Lambda_{12}x_2}{x_1 + \Lambda_{12}x_2} - \frac{\Lambda_{21}x_2}{x_2 + \Lambda_{21}x_1} \right]$$

$$\ln \gamma_2 = C \left[-\ln(x_2 + {}_{21}x_1) + \frac{\Lambda_{21}x_1}{x_2 + \Lambda_{21}x_1} - \frac{\Lambda_{12}x_1}{x_1 + \Lambda_{12}x_2} \right]$$

where

$$\Lambda_{12} = \frac{v_2}{v_1} \exp\left(\frac{-\Delta\lambda_{12}}{RT}\right) \qquad \lambda_{21} = \frac{v_1}{v_2} \exp\left(\frac{-\Delta\lambda_{21}}{RT}\right)$$

and

v_1, v_2 - molar liquid volumes

$\Delta\lambda_{12}, \Delta\lambda_{21}$ - adjustable energy parameters

C   - adjustable binary parameter, usually set equal to unity

From binary VLE data, the parameters obtained are:

$$P_{(1)} = \Delta\lambda_{12}$$
$$P_{(2)} = \Delta\lambda_{21}$$
$$P_{(3)} = C$$

If only two parameters are fit, C must set to some arbitrary valu usually one, and only $P_{(1)}$ and $P_{(2)}$ are estimated from the VLE data.

C. NRTL Equation (Renon and Prausnitz, 1968)

$$\ln \gamma_1 = x_2^2 \left[\tau_{21}\left(\frac{G_{21}}{x_1 + x_2 G_{21}}\right)^2 + \frac{\tau_{12} G_{12}}{x_2 + x_1 G_{12}} \right]$$

$$\ln \gamma_2 = x_1^2 \left[\tau_{12}\left(\frac{G_{12}}{x_2 + x_1 G_{12}}\right)^2 + \frac{\tau_{21} G_{21}}{x_1 + x_2 G_{21}} \right]$$

where

$$G_{12} = \exp(-\alpha \tau_{12}) \qquad G_{21} = \exp(-\alpha \tau_{21})$$

$$\tau_{12} = \frac{\Delta g_{12}}{RT} \qquad \tau_{21} = \frac{\Delta g_{21}}{RT}$$

and

g_{12}, g_{21} - adjustable energy parameters

α - adjustable binary parameter related to the nonrandomness of the mixture

The three parameters determined from binary VLE data are:

$$P_{(1)} = g_{12}$$
$$P_{(2)} = g_{21}$$
$$P_{(3)} = \alpha$$

If only two parameters are fit, α must be set to some predetermined value, and only $P_{(1)}$ and $P_{(2)}$ are determined from the binary VLE data.

D. Van Laar Equation

$$\ln \gamma_1 = A_{12} \left[\frac{x_2 A_{21}}{x_1 A_{12} + x_2 A_{21}} \right]^2$$

$$\ln \gamma_2 = A_{21} \left[\frac{x_1 A_{12}}{x_1 A_{12} + x_2 A_{21}} \right]^2$$

The two adjustable parameters, A_{12} and A_{21}, are determined from binary VLE data.

$$P_{(1)} = A_{12}$$
$$P_{(2)} = A_{21}$$

E. Three-Suffix Margules Equation (2 parameters)

$$\ln \gamma_1 = x_2^2 [A_{12} + 2(A_{21}-A_{12})x_1]$$

$$\ln \gamma_2 = x_1^2 [A_{21} + 2(A_{12}-A_{21})x_2]$$

Parameters estimated from binary VLE data are

$$P_{(1)} = A_{12}$$
$$P_{(2)} = A_{21}$$

F. Four-Suffix Margules Equation (3 parameters)

$$\ln \gamma_1 = x_2^2 [A_{12} + 2(A_{21}-A_{12}-D)x_1 + 3Dx_1^2]$$

$$\ln \gamma_2 = x_1^2 [A_{21} + 2(A_{12}-A_{21}-D)x_2 + 3Dx_2^2]$$

Three parameters are estimated from binary VLE data and correspond to:

$$P_{(1)} = A_{12}$$
$$P_{(2)} = A_{21}$$
$$P_{(3)} = D$$

G. Five-Suffix Margules Equation (4 parameters)

$$\ln \gamma_1 = x_2^2[A_{12}+2(A_{21}-A_{12})x_1+2(B_{21}x_1+B_{12}x_2)x_1(x_1-x_2)-B_{21}x_1^2]$$

$$\ln \gamma_2 = x_1^2[A_{21}+2(A_{12}-A_{21})x_2+2(B_{12}x_2+B_{21}x_1)x_2(x_2-x_1)-B_{12}x_2^2]$$

Parameters determined from binary VLE data are:

$$P_{(1)} = A_{12}$$
$$P_{(2)} = A_{21}$$
$$P_{(3)} = B_{12}$$
$$P_{(4)} = B_{21}$$

H. Modified Five-Suffix Margules Equation
   (Abbott and Van Ness, 1975)

$$\ln \gamma_1 = x_2^2 \left[A_{12}+2(A_{21}-A_{12})x_1 - \frac{2C_{12}C_{21}x_1x_2}{(C_{12}x_1+C_{21}x_2+Dx_1x_2)} \right.$$
$$\left. + \frac{C_{12}C_{21}(C_{12}+Dx_2^2)x_1^2}{(C_{12}x_1+C_{21}x_2+Dx_1x_2)^2} \right]$$

$$\ln \gamma_2 = x_1^2 \left[A_{21}+2(A_{12}-A_{21})x_2 - \frac{2C_{12}C_{21}x_1x_2}{(C_{12}x_1+C_{21}x_2+Dx_1x_2)} \right.$$
$$\left. + \frac{C_{12}C_{21}(C_{21}+Dx_1^2)x_2^2}{(C_{12}x_1+C_{21}x_2+Dx_1x_2)^2} \right]$$

216

The five binary parameters determined from VLE data are:

$$P_{(1)} = A_{12}$$
$$P_{(2)} = A_{21}$$
$$P_{(3)} = C_{12}$$
$$P_{(4)} = C_{21}$$
$$P_{(5)} = D$$

If only four parameters are estimated, D is set equal to zero.

Main Program and Subroutine Descriptions

Program VPLQFT. VPLQFT is the driver program which reads or sets the control variables for the regression, inputs the pure-component and VLE data by calling subroutines PRDTA2 and VLDTA2, initializes parameters in the correlation for the vapor-phase nonidealities by calling subroutine BIJS2, and starts the parameter estimation by calling subroutine REGRES. Results of the non-linear regression are printed when subroutine OUTDAT is called.

Subroutine PRDTA2. This subroutine reads the pure-component and binary parameters required for the various correlations describing the liquid and vapor phases. All input parameters are printed for verification.

Subroutine VLDTA2. VLDTA2 loads the binary vapor-liquid equilibrium data to be correlated. If the data are in units other than those used internally, the correct conversions are made here. This subroutine also reads the estimated standard deviations for the measured variables and the initial parameter estimates. All input data are printed for verification.

Subroutine OUTDAT. OUTDAT prints the estimated parameters and other statistical results obtained during the regression. This subroutine also prints all the experimentally measured points, the estimated true values corresponding to each measured point, and the deviations between experimental and calculated points. Finally, root-mean-squared deviations are printed for the P-T-x-y measurements.

Subroutine REGRES. REGRES is the main subroutine responsible for performing the regression. It solves for the parameters in nonlinear models where all the measured variables are subject to error and are related by one or two constraints. It uses subroutines FUNC, FUNDR, SUMSQ, and SYMINV.

Subroutine SUMSQ. SUMSQ calculates the weighted sum of

217

squares for REGRES.

Subroutine SYMINV. This subroutine inverts a symmetric matrix.

Subroutine FUNC. This subroutine evaluates the constraint functions for all the experimental points. It uses subroutine EVAL.

Subroutine FUNDR. This subroutine calculates the required derivatives for REGRES by central difference, using EVAL to calculate the objective functions.

Subroutine EVAL. This subroutine calculates and returns to FUNC calculated values of the independent variables (total pressure and vapor composition as functions of temperature, liquid composition, and the most recent parameter estimates) for reduction of binary vapor-liquid equilibria data. This subroutine is also used by FUNDR in calculating the required derivatives by central differences. EVAL iteratively calculates the pressure and vapor composition using the secant method. Subroutine VPLQK is called to provide K-factors for given values of pressure, temperature, and composition at the most recent values of the parameters.

Subroutine VPLQK. VPLQK calculates K factors ($K_i = y_i/x_i$) for given values of pressure, temperature, liquid and vapor compositions, and the adjustable parameters. The K factors are calculated from the following relation (Prausnitz, 1969):

$$y_i \phi_i P = x_i \gamma_i f_i^{0L} \exp(P v_i^L / RT) \tag{1}$$

where y is the vapor-phase mole fraction, ϕ is the fugacity coefficient, and P is the pressure. On the right-hand side of Equation (1), x is the liquid-phase mole fraction, γ is the activity coefficient, and f^{0L} is the pure-liquid reference fugacity (reference pressure is zero). The exponential term is the Poynting correction; it corrects for pressure effects on the liquid reference fugacity. We have assumed that the liquid partial molar volume is approximately equal to the pure-component saturated liquid molar volume, v^L, and that it is constant with respect to pressure. (This is a good assumption at temperatures well away from the critical temperature.) Solving Equation (1) for the K factor yields

218

$$K_i = \frac{y_i}{x_i} = \frac{\gamma_i f_i^{0L} \exp(Pv_i/RT)}{\phi_i P} \qquad (2)$$

If the data are correlated assuming an ideal vapor, the reference fugacity is just the vapor pressure, P^S, the Poynting correction is neglected, and fugacity coefficient is assumed to be unity. Equation (2) then becomes

$$K_i = \frac{y_i}{x_i} = \frac{\gamma_i P_i^S}{P} \qquad (3)$$

Subroutine VPLQK uses subroutines MVOLM, ACTIV2, REFUG, and PHIS2.

Subroutine REFUG. This subroutine calculates the liquid reference fugacities. Three options are possible. First, an equation of the form

$$\ln f_i^{0L} = C_1 + C_2/(T+C_6) + C_3 T + C_4 \ln T + C_5 T^2 \qquad (4)^{\ddagger}$$

can be used to give the liquid reference fugacity at zero pressure directly as a function of temperature. Second, the vapor pressure can be calculated from an equation of the form

$$\ln P_i^S = C_1 + C_2/(T+C_6) + C_3 T + C_4 \ln T + C_5 T^2 \qquad (5)^{\ddagger}$$

The constants in Equation (5) are not the same as those in Equation (4). Using this saturation pressure, the pure-liquid reference fugacity at zero pressure is then calculated from the equation

$$f_i^{0L} = \phi_i^S P_i^S \exp(-P_i^S v_i^L/RT) \qquad (6)$$

where ϕ_i^S is the fugacity coefficient of pure i at its saturation pressure. The exponential term is the Poynting correlation which corrects the liquid fugacity to a zero-pressure reference, and v_i^L is the pure-component saturated-liquid molar volume.

Finally, at low pressures, the liquid fugacity can be calculated using Equation (5), i.e. we can assume that $\phi_i^S = 1$ and that the Poynting correction $= 1$.

Subroutine MVOLM. MVOLM calculates the liquid molar volume at a given temperature using the modified Rackett equation

---

‡In this work C_6 is always equal to zero.

(Spencer and Danner, 1972). This equation has been further
modified by O'Connell for reduced temperatures greater than 0.75.
The saturated-liquid molar volume is given by the equation

$$v_i^L = \frac{RT_{c_i} Z_{r_i}^\tau}{P_{c_i}} \qquad (7)$$

where

$$\tau = 1 + (1-T_r)^{0.286} \quad \text{for } T_r \le 0.75 \qquad (8a)$$

$$\tau = 1.6 + 0.00693/(T_r - 0.655) \quad \text{for } T_r > 0.75 \qquad (8b)$$

T_r is the reduced temperature, T_c is the critical temperature,
P_c is the critical pressure, and Z_r is the modified Rackett
parameter as given in the supplemental table for pure-component
properties.

Subroutine ACTIV2. This subroutine calculates the activity
coefficients using one of the liquid-phase equations discussed
in the previous section.

Subroutine PHIS2. PHIS2 calculates the vapor-phase fugacity
coefficients according to the method of Hayden and O'Connell
(1975). It calls subroutine BIJS2 which returns the second virial
coefficients. If organic acids or other strongly associating
compounds are present, the chemical theory is used to calculate
the fugacity coefficients; otherwise, the volume-explicit virial
equation of state is used. Subroutine MULLER is used by PHIS2
to calculate the equilibrium composition when the chemical theory
is used.

Subroutine MULLER. MULLER iteratively solves the equilibrium relations and computes the equilibrium vapor composition
when organic acids are present. These compositions are used by
subroutine PHIS2 to calculate fugacity coefficients by the
chemical theory.

Subroutine BIJS2. This subroutine calculates the pure-
component and cross second virial coefficients for binary mixtures
according to the method of Hayden and O'Connell (1975).

Data Input

A. The first four data cards contain control parameters which are
   read only once for a series of binary VLE data sets.

220

1. First card; FORMAT(20I2), control and execution codes for the regression program (all integer values and right adjusted).

ILIQ   cols 1&2   indicator for type of liquid-phase model to be used, default value is 1
> 1 - 2-parameter UNIQUAC equation
> 2 - 2-parameter NRTL equation
> 3 - 2-parameter Wilson equation
> 4 - 2-parameter van Laar equation
> 5 - 2-parameter Henry's constant
> 6 - 2-parameter 3-suffix Margules equation
> 7 - 3-parameter UNIQUAC equation
> 8 - 3-parameter NRTL equation
> 9 - 3-parameter Wilson equation
> 10 - 3-parameter 4-suffix Margules equation
> 11 - 4-parameter 5-suffix Margules equation
> 12 - 4-parameter modified Margules equation
> 13 - 5-parameter modified Margules equation

IVAP   cols 3&4   execution code for form of vapor-phase correction and liquid reference fugacity, default value is 1
> 1 - zero-pressure reference fugacity equation and Hayden-O'Connell correlation for the vapor phase
> 2 - vapor pressure equation and Hayden-O'Connell correlation for the vapor phase
> 3 - vapor pressure equation and ideal vapor phase

IST   cols 5&6   execution code for type of regression
> 0 - regression on P-T-x-y
> 1 - regression on P-T-x
> 2 - regression on P-y
> 3 - regression on P

IPRT   cols 7&8   print control, 0 to 6, default value is 2
> 0 - minimum printed information
> 6 - maximum printed information

ITMX cols 9&10   maximum number of iterations allowed, default
                  value is 15

LMV   cols 11&12 constraint control on the true values of the
                  independent variables; liquid composition and
                  temperature, default value is 1
                      1 – constrained to be positive
                      2 – no constraints

LMP   cols 13&14 constraint control on the parameters, default
                  value is 1
                      1 – maximum absolute change in the parameters
                          is limited to value of PRCG
                      2 – no constraints

2. Second card; FORMAT(8F10.2), control variables for the regression. This program uses a Newton-Raphson type iteration which is susceptible to convergence problems with
   poor initial parameter estimates. Therefore, several
   features are implemented which help control oscillations,
   prevent divergence, and determine when convergence has been
   achieved. These features are controlled by the parameters
   on this card. The default values are the result of con-
   siderable experience and are adequate for the majority of
   situations. However, convergence may be enhanced in some
   cases with user supplied values.

ALST cols 1-10   initial value of step-limiting parameter; nor-
                  mally set equal to 1.0, the default value.
                  Fractional values of 1.0 (e.g. 0.7 or 0.4) may
                  be desirable for new systems for which con-
                  ceivably no good parameter estimates are known.
                  If parameters are obtained with a value of ALST
                  < 1.0, the program should be rerun with ALST =
                  1.0 to ensure that the optimum parameters are
                  obtained.

BETA cols 11-20 oscillation control parameter; default value is
                  set equal to 0.25. To help prevent oscilla-
                  tions (thus slowing convergence) we not only
                  require that the sum of squares, SSQ, decreases

222

at each iteration, but we also require that the
decrease in the SSQ be some fraction, BETA,
of that predicted by the linearized equations.
For highly correlated parameters leading to a
highly oscillatory convergence, a value of BETA
closer to 1.0 (e.g. 0.65 to 0.85) may be desirable.

PRCG cols 21-30 the maximum allowable change in any of the parameters when LMP = 1, default value is 1000.
Limiting the change in the parameters prevents
totally unreasonable values from being attained
in the first several iterations when poor initial estimates are used. A value of PRCG equal
to the magnitude of that anticipated for the
parameters is usually appropriate.

RP cols 31-40 step-back parameter, default value is equal to
1.5. This value determines the amount the step-
size is reduced to satisfy the criteria of a
SSQ which decreases from one iteration to the
next. The amount of the decrease is equal to
the previous value of the step-limiting parameter divided by RP.

SSTL cols 41-50 regression convergences control, default value
is 5.E-4. Convergence is assumed to be achieved
when the relative change in the SSQ from one
iteration to the next is less than SSTL.

3. Third card; FORMAT(8F10.2), size of increments to be used
in central difference formula for calculating derivatives
with respect to the independent variables.

DX(1) cols 1-10 increment for derivative with respect to the
temperature, default value is 0.5
DX(2) cols 11-20 increment for derivative with respect to the
liquid mole fraction, default value is 0.001.

4. Fourth card; FORMAT(8F10.2), size of increments to be used
in central difference formula for calculating derivatives

223

with respect to the parameters. LL is the number of values read. It is determined by IKIQ, which specifies the type of model and sets the number of parameters, LL.

DP(1) cols 1-10   increment for parameter 1; if zero, values for all increments, DP(I), I=1, LL, are set internally; if greater than zero, then all the increments must be specified on this card. The magnitudes of these finite-difference increments are typically one per cent of the anticipated values for the corresponding parameters.

B. The next four data cards contain pure-component data for component one.

   1. First card; FORMAT(5A4, 2X, 2A4, 3X, F7.3, 4X, F6.2, 4X, F6.2, 5X, F5.1, 4X, 2A3).

NAM(I,J) cols 1-20   name
FOR(I,J) cols 23-30 molecular formula
MW(I)   cols 34-40 molecular weight
TC(I)   cols 45-50 critical temperature, K
PC(I)   cols 55-60 critical pressure, bars
VC(I)   cols 66-70 critical volume, cm^3/mole, not required in present correlations
NCD(I)   cols 75-80 six-character identification code

   2. Second card; FORMAT (5X, F5.1, 4X, F6.4, 5X, F5.3, 6X, F4.2, 5X, F2.5, 5X, F5.2, 5X, F5.2)

VSR(I)   cols 6-10   O'Connell v*, not required in present correlations
ZRA(I)   cols 15-20 Rackett parameter for saturated-liquid molar volume correlation
RD(I)   cols 26-30 mean radius of gyration, angstroms
DM(I)   cols 37-40 dipole moment, debyes
R(I)   cols 46-50 structural volume parameter for the UNIQUAC equation
Q(I)   cols 56-60 structural area parameter for the UNIQUAC equation
QP(I)   cols 66-70 modified structural area parameter for the UNIQUAC equation

224

3. Third and Fourth cards; FORMAT(3E20.14), constants for
   zero-pressure reference fugacity equation, vapor pressure
   equation, or hypothetical-liquid reference fugacity equation
   of the form

$$\left.\begin{array}{r} \ln f_i^{OL} \\[2mm] \text{or} \\[2mm] \ln P_i^s \end{array}\right\} = C_1 + \frac{C_2}{T + C_6} + C_3T + C_4\ln T + C_5T^2$$

where T is the temperature in Kelvin, and f_i^{OL} or P_i^s is the
reference fugacity or pressure in bars. Constants C_1, C_2.
and C_3 are read in on the third card, and constants C_4, C_5,
and C_6 are read in on the fourth card. Any constants not
used are left blank.

C. The next four data cards contain pure-component data for component two. The same format as used in part B is repeated
   here.

D. The next card supplies the solvation and association parameters,
   and the third parameter for either the UNIQUAC, NRTL, or Wilson
   equation, if this parameter is not being fitted.
   FORMAT(8F10.2)

ETA(1)   cols 1-10   association parameter for component 1
ETA(2)   cols 11-20 solvation parameter for the 1-2 binary
ETA(3)   cols 21-30 association parameter for component 2
CC   cols 31-40 third parameter for either the UNIQUAC, NRTL,
                      or Wilson equation, default value is 1.0.

E. The next data card contains information about the binary VLE
   data.   FORMAT(2A3, 4X, 2A3, 4X, I2, 18X, 5(I1,1X)).

NMD(1,J) cols 1-6   six-character identification code for com-
                     ponent one; this must match the code read in
                     with the pure-component data.
NMD(2,J) cols 11-16 six-character identification code for com-
                     ponent two; this must match the code read in
                     with pure-component data.
NN   cols 21-22 number of experimental VLE data points

IRF   cols 41   indicator for number of reference cards

                      0 - one reference card

                      1 - two reference cards

ITU   cols 43   execution code on the temperature units of
                      the VLE data

                      0 - centigrade

                      1 - Kelvin

                      2 - Fahrenheit

ICU   col 45   execution code on the composition units

                      0 - mole fraction

                      1 - weight fraction

IPU   col 47   execution code for the pressure units of the
                      VLE data

                      0 - mm Hg

                      1 - bars

                      2 - atm

IT   col 49   execution code for type of regression; allows
                      a particular data set to be correlated dif-
                      ferently from the other data sets in the same
                      run

                      0 - regression on P-T-x-y

                      1 - regression on P-T-x

                      2 - regression on P-y

                      3 - regression on P

F. The next card supplies the VLE data reference (2 cards if
   IRF=1). FORMAT(15A4).

REF(I,J) cols 1-60 documentation of VLE data source, 1 or 2
                    cards.

G. The next NN cards supply the VLE data. NN equals the number
   of experimental points. Each card has one set of data.
   FORMAT(8F10.2).

YM(I)   cols 1-10 measured pressure

XM(I,1)   cols 11-20 measured temperature

XM(I,2)   cols 21-30 measured liquid-phase composition of com-
                      ponent one (mole or weight fraction)

ZM(I)   cols 31-40 measured vapor-phase composition of component

H. The next cards provide estimates of the standard deviations
   of the experimental data. At least one card is needed with
   non-zero values. Units are the same as those of the VLE data.
   FORMAT(4F10.2,I2).

SDY(I)   cols 1-10   standard deviation of pressure measurement
SDX(I,1) cols 11-20   standard deviation of temperature measurement
SDX(I,2) cols 21-30   standard deviation of liquid composition
                       measurement
SDZ(I)   cols 31-40   standard deviation of vapor composition
                       measurement
ND   cols 41-42   number of times this set of standard devia-
                       tions is duplicated. If ND equals zero, the
                       standard deviations read on this card are as-
                       sumed to be the same for all data points. If
                       ND is nonzero, the first "ND" data points will
                       have standard deviations assigned correspond-
                       ing to those read on this card. Another card
                       is read with a new value for ND. The next
                       "ND" data points will have standard deviations
                       corresponding to those read on the second card.
                       Cards will continue to be read until standard
                       deviations have been assigned to all data
                       points. This allows the user to specify each
                       set of standard deviations for each data point
                       in the most general case or to assign a single
                       set of standard deviations which are applicable
                       to all data points.

I. The next card gives the initial parameter estimates.
   FORMAT(8F10.2)

P(I)   cols 1-50   initial parameters for I=1,LL; LL is deter-
                       mined by the liquid-phase model used.

     Multiple sets of binary VLE data may be correlated by con-
tinuing with another set of cards starting at part B. The last
set of cards must be followed with a blank card to end the program.

References

Abbott, M. M., Van Ness, H. C., A.I.Ch.E. J., 21, 62 (1975).

Abrams, D. S., Prausnitz, J. M., A.I.Ch.E. J., 21, 116 (1975).

Anderson, T. F., Abrams, D. S., Grens, E. A., A.I.Ch.E. J., 24, 20 (1978).

Hayden, J. G., O'Connell, J. P., Ind. Eng. Chem., Process Des. Dev., 14, 209 (1975).

Prausnitz, J. M., "Molecular Thermodynamics of Fluid-Phase Equilibria," Prentice Hall, Englewood Cliffs, N.J. (1969).

Renon, H., Prausnitz, J. M., A.I.Ch.E. J., 14, 135 (1968).

Renon, H., Prausnitz, J. M., A.I.Ch.E. J., 15, 785 (1969).

Spencer, C. F., Danner, R. P., J. Chem. Eng. Data, 17, 236 (1972).

Wilson, G. M., J. Am. Chem. Soc., 86, 127 (1964).

```
      PROGRAM VPLQFT (INPUT,OUTPUT,TAPE5=INPUT,TAPE6=OUTPUT)
C                                                                               C
C***************************************************************************C
C                                                                               C
C     PROGRAM VPLQFT (INPUT,OUTPUT,TAPES=INPUT,TAPE6=OUTPUT)                   C
C                                                                               C
C     PURPOSE                                                                   C
C        MAIN PROGRAM AND DRIVER FOR FITTING BINARY VLE DATA USING             C
C        METHOD BASED ON THE MAXIMUM LIKELIHOOD PRINCIPLE.  ONLY               C
C        CONTROL VARIABLES ARE READ IN THIS ROUTINE.                           C
C                                                                               C
C     PARAMETERS                                                                C
C        IRR    - CNTRL VARIABLE INDICATING AN ERROR IN INPUT OR IN            C
C                 THE REGRESSION                                                C
C        ISTP   - CNTRL FOR TERMINATING THE PROGRAM                            C
C        KK     - NUMBER OF INDEPENDENT VARIABLES, EQUAL TO 2                   C
C        LL     - NUMBER OF PARAMETERS, DETERMINED BY CHOICE OF LIQUID-        C
C                 PHASE MODEL                                                   C
C        ILIO   - INDICATOR FOR THE TYPE OF LIQUID-PHASE MODEL TO BE           C
C                 USED                                                          C
C                 1 - 2-PARAMETER UNIQUAC EQUATION                             C
C                 2 - 2-PARAMETER NRTL EQUATION                                C
C                 3 - 2-PARAMETER WILSON EQUATION                              C
C                 4 - 2-PARAMETER VAN LAAR EQUATION                            C
C                 5 - 2-PARAMETER HENRY,S CONSTANT                             C
C                 6 - 2-PARAMETER 3-SUFFIX MARGULES EQUATION                   C
C                 7 - 3-PARAMETER UNIQUAC EQUATION                             C
C                 8 - 3-PARAMETER NRTL EQUATION                                C
C                 9 - 3-PARAMETER WILSON EQUATION                              C
C                10 - 3-PARAMETER 4-SUFFIX MARGULES EQUATION                   C
C                11 - 4-PARAMETER 5-SUFFIX MARGULES EQUATION                   C
C                12 - 4-PARAMETER MODIFIED 5-SUFFIX MARGULES EQUATION          C
C                13 - 5-PARAMETER MODIFIED MARGULES EQUATION                   C
C                                                                               C
C        IVAP   - EXECUTION CODE FOR FORM OF VAPOR-PHASE CORRECTION            C
C                 AND LIQUID REFERENCE FUGACITY                                 C
C                 1 - ZERO PRESSURE REFERENCE FUGACITY EQUATION AND            C
C                     HAYDEN-O,CONNELL CORRELATION FOR VAPOR-PHASE             C
C                 2 - VAPOR PRESSURE EQUATION AND HAYDEN-O,CONNELL             C
C                     CORRELATION FOR VAPOR-PHASE                              C
C                 3 - VAPOR PRESSURE EQUATION AND IDEAL VAPOR PHASE            C
C        IST    - EXECUTION CODE FOR TYPE OF REGRESSION                        C
C                 0 - REGRESSION ON P - T - X - Y                              C
C                 1 - REGRESSION ON P - T - X                                  C
C                 2 - REGRESSION ON P - Y                                      C
C                 3 - REGRESSION ON P.                                         C
C        IPRT   - PRINT CONTROL                                                 C
C                 0 - MINIMUM PRINTED OUTPUT                                   C
C                 6 - MAXIMUM PRINTED OUTPUT                                   C
C        ITMX   - ITERATION LIMIT                                              C
C        LMV    - CONSTRAINT CONTROL ON VARIABLES                             C
C                 1 - CONSTRAINED TO BE POSITIVE                              C
C                 2 - NO CONSTRAINTS                                           C
C        LMP    - CONSTRAINT CONTROL ON PARAMETERS                           C
C                 1 - MAXIMUM ABSOLUTE CHANGE IS LIMITED TO VALUE             C
C                     OF PRCG                                                  C
C                 2 - NO CONSTRAINTS                                           C
C        ALST   - CONVERGENCE CNTRL, USUALLY = 1.0                            C
C        BETA   - CONVERGENCE CNTRL, USUALLY = 0.25                           C
C        PRCG   - MAXIMUM ABSOLUTE VALUE OF PARAMETER CHANGES WHEN            C
C                 LMP EQUAL 2                                                  C
C        RP     - CONVERGENCE CNTRL, USUALLY = 1.5                            C
C        SSTL   - CONVERGENCE TOLERANCE, DEFAULT VALUE EQUALS 5.E-4           C
```

229

```
C           DP      - VECTOR OF CENTRAL DIFFERENCE INCREMENTS FOR           C
C                     CALCULATING DERIVATIVES WRT THE PARAMETERS            C
C           DX      - VECTOR OF CENTRAL DIFFERENCE INCREMENTS FOR           C
C                     CALCULATING DERIVATIVES WRT THE INDEPENDENT VARIABLES C
C                                                                           C
C      SUBROUTINES USED                                                     C
C         PROTA2, VLDTA2, REGRES, OUTDAT                                    C
C                                                                           C
C      PROGRAMED BY T.F.ANDERSON JAN 78.                                    C
C                                                                           C
C***************************************************************************C
C                                                                           C
      COMMON /ALL/ NN, LL, KK, IPRT, IST
      COMMON /RGRS/ SSTL, BETA, ITMX, RP, LMV, LMP, ALST, PRCG
      COMMON /DIFF/ DP(5), DX(5)
      COMMON /THRM/ ILIQ, IVAP
      ISTP = 0
      KK = 2
C
C     READ CONTROL PARAMETERS
C
      READ (5,1001) ILIQ, IVAP, IST, IPRT, ITMX, LMV, LMP, LL
      IF (IPRT.EQ.0) IPRT = 2
      IF (ITMX.EQ.0) ITMX = 15
      IF (ILIQ.EQ.0) ILIQ = 1
      IF (IVAP.EQ.0) IVAP = 1
      IF (LMV.EQ.0) LMV = 1
      IF (LMP.EQ.0) LMP = 1
      ITO = IST
C
C     DETERMINE NUMBER OF PARAMETERS TO BE FIT
C
      IF (LL.GT.0) GO TO 5
      LL = 2
      IF (ILIQ.GT.6) LL = 3
      IF (ILIQ.GT.10) LL = 4
      IF (ILIQ.GT.12) LL = 5
    5 CONTINUE
C
C     READ REMAINING REGRESSION CONTROLS
C
      READ (5,1002) ALST, BETA, PRCG, RP, SSTL
      IF (ALST.LT.1.E-9) ALST = 1.0
      IF (BETA.LT.1.E-9) BETA = 0.25
      IF (PRCG.LT.1.E-9) PRCG = 1000.
      IF (RP.LT.1.E-9) RP = 1.5
      IF (SSTL.LT.1.E-9) SSTL = 5.E-4
C
C     SET STEP SIZES FOR CALCULATING DERIVATIVES W.R.T THE VARIABLES
C
      READ (5,1002) (DX(K),K=1,KK)
      IF (DX(1).LT.1.E-9) DX(1) = 0.5
      IF (DX(2).LT.1.E-9) DX(2) = 0.001
C
C     SET STEP SIZES FOR CALCULATING DERIVATIVES W.R.T THE PARAMETERS
C
      READ (5,1002) (DP(L),L=1,LL)
      IF (DP(1).GT.1.E-9) GO TO 15
      DO 10 L=1,LL
   10 DP(L) = 0.001
      IF (ILIQ.GT.8) GO TO 15
      IF (ILIQ.EQ.4.OR.ILIQ.EQ.5) GO TO 15
      DP(1) = 1.0
      DP(2) = 1.0
```

```
      15 CONTINUE
C
C      DOCUMENT CONTROL PARAMETERS
C
       WRITE (6,2001)
       WRITE (6,2002) KK, LL, ILIQ, IVAP, IST, IPRT, ITMX, LMV, LMP
       WRITE (6,2003) ALST, BETA, PRCG, RP, SSTL
       WRITE (6,2004) (K,DX(K),K=1,KK)
       WRITE (6,2005) (L,DP(L),L=1,LL)
      20 CONTINUE
       IRR = 0
       IST = ITO
C
C      LOAD PURE COMPONENT DATA
C
       CALL PRDTA2 (ISTP)
       IF (ISTP.GT.0) GO TO 60
C
C      LOAD VLE DATA, ESTIMATED STANDARD DEVIATIONS, AND INITIAL
C      PARAMETER ESTIMATES
C
       CALL VLDTA2 (IRR)
       IF (IRR.GT.0) GO TO 60
C
C      INITIALIZE PARAMETERS IN BIJS2
C
       IF (IVAP.LT.3) CALL BIJS2 (-1,300.)
C
C      START REGRESSION
C
       CALL REGRES (IRR)
C
C      PRINT RESULTS
C
       CALL OUTDAT (IRR)
       GO TO 20
      60 STOP
    1001 FORMAT (20I2)
    1002 FORMAT (8F10.2)
    2001 FORMAT (1H1,57HMAXIMUM LIKELIHOOD ESTIMATION OF PARAMETERS FROM VL
      1E DATA//1X,40HCONTROL PARAMETERS WERE SET AS FOLLOWS -/)
    2002 FORMAT (1X,5HKK  =,I2/1X,5HLL   =,I2/1X,5HILIQ=,
      1        I2/1X,5HIVAP=,I2/1X,5HIST =,I2/1X,5HIPRT=,I2/1X,5HITMX=,
      2        I2/1X,5HLMV =,I2/1X,5HLMP =,I2)
    2003 FORMAT (1X,5HALST=,F7.2/1X,5HBETA=,F7.2/1X,5HPRCG=,F7.1/1X,
      1        5HRP  =,F7.2/1X,5HSSTL=,F7.5)
    2004 FORMAT (/1X,42HFINITE-DIFFERENCE INCREMENTS FOR VARIABLES//
      1        (1X,3HDX(,I1,2H)=,F6.4))
    2005 FORMAT (/1X,43HFINITE-DIFFERENCE INCREMENTS FOR PARAMETERS//
      1        (1X,3HDP(,I1,2H)=,F6.4))
       END
```

```
      SUBROUTINE PRDTA2 (ISTP)
C                                                                          C
C*****************************************************************************C
C                                                                          C
C     SUBROUTINE PRDTA2 (ISTP)                                             C
C                                                                          C
C     PURPOSE                                                              C
C        LOADS PURE COMPONENT AND BINARY DATA FOR USE IN THE VARIOUS      C
C        CORRELATIONS FOR LIQUID AND VAPOR PHASE NONIDEALITIES, THEN      C
C        DOCUMENTS THE INPUT DATA.                                        C
C                                                                          C
C     PARAMETERS                                                          C
C        ISTP   - STOP CONTROL                                            C
C        NCMP   - NUMBER OF COMPONENTS, EQUALS 2                          C
C        NAM    - COMPONENT NAME                                          C
C        NCD    - NAME CODE                                               C
C        FOR    - MOLECULAR FORMULA                                       C
C        MW     - MOLECULAR WEIGHT                                        C
C        TC     - CRITICAL TEMPERATURE, KELVIN                            C
C        PC     - CRITICAL PRESSURE, BARS                                 C
C        VSR    - O,CONNELL V-STAR CC/GR-MOLE                             C
C        VC     - CRITICAL VOLUME, CC/GR-MOLE                             C
C        ZRA    - RACKET PARAMETER FOR SATURATED LIQUID MOLAR VOLUME      C
C                 CORRELATION                                             C
C        RD     - MEAN RADIUS OF GYRATION, ANGSTROMS                      C
C        DM     - DIPOLE MOMENT, DEBYES                                   C
C        R      - STRUCTURAL VOLUME PARAMETER OF UNIQUAC EQUATION         C
C        Q      - STRUCTURAL AREA PARAMETER FOR UNIQUAC EQUATION          C
C        QP     - STRUCTURAL AREA PARAMETER FOR MODIFIED UNIQUAC          C
C                 EQUATION                                                C
C        ETA    - ASSOCIATION AND SOLVATION PARAMETERS                    C
C        FO     - CONSTANTS FOR ZERO PRESSURE REFERENCE FUGACITY          C
C                 EQUATION IF IVAP.EQ.1 OR CONSTANTS FOR VAPOR            C
C                 PRESSURE EQUATION IF IVAP.GT.1                          C
C        CC     - THRID PARAMETER FOR EITHER THE UNIQUAC, NRTL, OR        C
C                 WILSON EQUATION                                         C
C                                                                          C
C     PROGRAMED BY T.F.ANDERSON JAN 78.                                   C
C                                                                          C
C*****************************************************************************C
C                                                                          C
      REAL MW
      DIMENSION EL(2)
      COMMON /SIZE/ R(2), Q(2), QP(2), CC
      COMMON /FUG/ FO(2,6)
      COMMON /SYSTM/ NAM(2,5), NCD(2,2), FOR(2,2)
      COMMON /PRM/ MW(2), TC(2), PC(2), VC(2), VSR(2), ZRA(2), RD(2),
     1             DM(2), ETA(3)
      COMMON /THRM/ ILIQ, IVAP
      NCMP = 2
C
C     READ PURE COMPONENT DATA
C
      DO 10 I=1,NCMP
      READ (5,1001) (NAM(I,J),J=1,5), (FOR(I,J),J=1,2), MW(I),
     1              TC(I), PC(I), VC(I), (NCD(I,J),J=1,2)
      IF (TC(I).GT.1.E-5) GO TO 5
      ISTP = 1
      GO TO 200
    5 CONTINUE
      READ (5,1002) VSR(I), ZRA(I), RD(I), DM(I), R(I), Q(I), QP(I)
      IF (QP(I).LT.1.E-6) QP(I) = Q(I)
      READ (5,1003) (FO(I,J),J=1,3)
      READ (5,1003) (FO(I,J),J=4,6)
```

```
   10 CONTINUE
C
C     READ SOLVATION AND ASSOCIATION PARAMETERS
C
      II = NCMP*(NCMP+1)/2
      READ (5,1004) (ETA(I),I=1,II), CC
      IF (ABS(CC).LT.1.E-9) CC = 1.0
C
C     PRINT SUMMARY OF PURE COMPONENT DATA
C
      WRITE (6,2001) NCMP, (I, (NCD(I,J),J=1,2), (FOR(I,J),J=1,2),
     1               (NAM(I,J),J=1,5), I=1,NCMP)
      WRITE (6,2002)
      Z = 10.0
      DO 85 I=1,NCMP
      EL(I) = (Z/2.0)*(P(I)-Q(I)) - (R(I)-1.0)
   85 WRITE (6,2003)  I, MW(I), TC(I), PC(I), VC(I), VSR(I), ZPA(I),
     1               RD(I), DM(I), EL(I), R(I), Q(I), QP(I)
      IF (IVAP.EQ.1) WRITE (6,2004)
      IF (IVAP.GT.1) WRITE (6,2006)
      WRITE (6,2005) (I, (FO(I,J),J=1,5), I=1,NCMP)
      WRITE (6,2008)
      DO 90 I=1,NCMP
      II = (I+1)*I/2
   90 WRITE (6,2009) (NCD(I,J),J=1,2), ETA(II)
      WRITE (6,2010)
      IF (NCMP.LE.1) GO TO 110
      DO 100 I=2,NCMP
      II = I-1
      DO 100 J=1,II
      IJ = (I-1)*I/2 + J
  100 WRITE (6,2011) (NCD(I,K),K=1,2), (NCD(J,K),K=1,2), ETA(IJ)
  110 CONTINUE
      WRITE (6,2012) CC
  200 RETURN
 1001 FORMAT (5A4,2X,2A4,3X,F7.3,4X,F6.2,4X,F6.2,5X,F5.1,4X,2A3)
 1002 FORMAT (5X,F5.1,4X,F6.4,5X,F5.3,6X,F4.2,5X,F5.2,5X,F5.2,5X,F5.2)
 1003 FORMAT (3E20.14)
 1004 FORMAT (8F10.2)
 2001 FORMAT (1H1,I2,1X,16HCOMPONENT SYSTEM//(5X,1H(,I1,1H),3X,2A3,
     1         3X,2A4,3X,5A4/))
 2002 FORMAT (1H0,32X,25HPURE COMPONENT PROPERTIES//2X,2HNO,4X,6HMOL WT,
     1         5X,2HTC,7X,2HPC,7X,2HVC,5X,4HVSTR,4X,3HZRA,6X,2HRD,5X,
     2         2HDM,6X,1HL,7X,1HR,7X,1HQ,6X,2HQP//)
 2003 FORMAT (1X,1H(,I1,1H),3X,F7.3,3X,F6.2,3X,F6.2,3X,F5.1,3X,F5.1,3X,
     1  F5.4,3X,F5.3,3X,F4.2,3X,F5.2,3X,F5.2,3X,F5.2,3X,F5.2)
 2004 FORMAT (56H0CONSTANTS FOR ZERO PRESSURE REFERENCE FUGACITY EQUATIO
     1N//)
 2005 FORMAT (1X,1H(,I1,1H),3E23.14/4X,2E23.14)
 2006 FORMAT (38H0CONSTANTS FOR VAPOR PRESSURE EQUATION//)
 2007 FORMAT (1X,1H(,I1,1H),4E16.7)
 2008 FORMAT (23H0ASSOCIATION PARAMETERS//)
 2009 FORMAT (5X,2A3,3X,F4.2)
 2010 FORMAT (21H0SOLVATION PARAMETERS//)
 2011 FORMAT (5X,2A3,1H/,2A3,3X,F4.2)
 2012 FORMAT (1H0,53HUNIQUAC, WILSON, OR NRTL THIRD PARAMETER SET EQUAL
     1TO,F5.2)
      END
```

```
      SUBROUTINE VLDTA2 (IRP)
C                                                                              C
C******************************************************************************C
C                                                                              C
C      SUBROUTINE VLDTA2 (IRP)                                                 C
C                                                                              C
C      PURPOSE                                                                 C
C         LOADS BINARY VAPOR-LIQUID EQUILIBRIA DATA TO BE CORRELATED           C
C                                                                              C
C      PARAMETERS                                                              C
C                                                                              C
C         CONTROL VARIABLES                                                    C
C                                                                              C
C         NN      - NUMBERS OF DATA POINTS.                                    C
C         IPU     - EXECUTION CODE ON PRESS UNITS OF VLEQ DATA AND             C
C                   ESTIMATED ERROR VARRIANCES                                 C
C                   0 - MM HG                                                  C
C                   1 - BARS                                                   C
C                   2 - ATM.                                                   C
C         ITU     - EXECUTION CODE ON TEMP UNITS OF VLEQ DATA AND              C
C                   ESTIMATED ERROR VARRIANCES                                 C
C                   0 - CENTIGRADE                                             C
C                   1 - KELVIN                                                 C
C                   2 - FARENHEIGHT.                                           C
C         ICU     - EXECUTION CODE ON COMPOSITION UNITS                        C
C                   0 - MOLE FRACTION                                          C
C                   1 - WEIGHT FRACTION                                        C
C         IRF     - INDICATOR FOR NUMBER OF REFERENCE LINES, COLS 0-60         C
C                   0 - ONE REFERENCE LINE                                     C
C                   1 - TWO REFERENCE LINES                                    C
C         IST     - EXECUTION CODE FOR TYPE OF REGRESSION                      C
C                   0 - REGRESSION ON P - T - X - Y                            C
C                   1 - REGRESSION ON P - T - X                                C
C                   2 - REGRESSION ON P - Y                                    C
C                   3 - REGRESSION ON P.                                       C
C                                                                              C
C         PURE COMPONENT PARAMETERS                                            C
C                                                                              C
C         NAM     - COMPONENT NAME                                             C
C         NCD     - NAME CODE                                                  C
C         FOR     - MOLECULAR FORMULA                                          C
C         MW      - MOLECULAR WEIGHT                                           C
C                                                                              C
C         VAPOR-LIQUID EQUILIBRIUM DATA                                        C
C                                                                              C
C         YM(I)     - MEASURED PRESSURE.                                       C
C         XM(I,1)   - MEASURED TEMP.                                           C
C         XM(I,2)   - MEASURED LIQUID-PHASE COMPOSITION.                       C
C         ZM(I)     - MEASURED VAPOR-PHASE COMPOSITION.                        C
C         EVY(I)    - ERROR VARRIANCE OF PRESS MEASUREMENT.                    C
C         EVX(I,1)  - ERROR VARRIANCE OF TEMP MEASUREMENT.                     C
C         EVX(I,2)  - ERROR VARRIANCE OF THE LIQUID-PHASE COMPOSITION          C
C                     MEASUREMENT.                                             C
C         EVZ(I)    - ERROR VARRIANCE OF THE VAPOR-PHASE COMPOSITION           C
C                     MEASUREMENT.                                             C
C         SDY(I)    - STD DEV OF PRESS MEASUREMENT.                            C
C         SDX(I,1)  - STD DEV OF TEMPERATURE MEASUREMENT.                      C
C         SDX(I,2)  - STD DEV OF THE LIQUID-PHASE COMPOSITION                  C
C                     MEASUREMENT.                                             C
```

234

```
C            SDZ(I)    - STD DEV OF THE VAPOR-PHASE COMPOSITION              C
C                        MEASUREMENT.                                        C
C            P(I)      - INITIAL ESTIMATES OF PARAMETERS                     C
C                                                                           C
C       PROGRAMED BY T.F.ANDERSON JAN 78.                                   C
C                                                                           C
C*************************************************************************C
C                                                                           C
        REAL MW
        DIMENSION REF(15,2), NMD(2,2)
        DIMENSION SDX(50,5), SDY(50), SDZ(50)
        COMMON /DATA/ X(50,5), XM(50,5), EVX(50,5), DFX(50,5), DGX(50,5),
       1              P(5), DFP(50,5), DGP(50,5), Y(50), YM(50), EVY(50),
       2              ZM(50), Z(50), EVZ(50)
        COMMON /PRM/ MW(2), TC(2), PC(2), VC(2), VSR(2), ZRA(2), RD(2),
       1             DM(2), ETA(3)
        COMMON /ALL/ NN, LL, KK, IPRT, IST
        COMMON /SYSTM/ NAM(2,5), NCD(2,2), FOR(2,2)
        COMMON /INOT/ PA
C
C       READ REFERENCE AND VAPOR-LIQUID EQUILIBRIUM DATA
C
        READ (5,1001) ((NMD(I,J),J=1,2),  I=1,2),NN,IRF,ITU,ICU,IPU,IT
        IF (IST.EQ.0) IST = IT
C
C       READ COMPONENT NAMES, NUMBER OF DATA POINTS, AND EXECUTION
C       CODES
C
C
C       CHECK FOR MISMATCH BETWEEN PURE COMPONENT DATA AND VLE DATA
C
        IF (NMD(1,1).EQ.NCD(1,1).AND.NMD(2,1).EQ.NCD(2,1)) GO TO 20
        WRITE (6,3001) ((NMD(I,J),J=1,2),(NCD(I,J),J=1,2),I=1,2)
        IRR = 2
     20 CONTINUE
C
C       READ REFERENCE LINES FOR VLE DATA
C
        IRF = IRF + 1
        READ (5,1002) ((REF(I,J),I=1,15), J=1,IRF)
C
C       READ IN VLE DATA
C
        PA = 0.0
        DO 30 I=1,NN
        READ (5,1003) YM(I), XM(I,1), XM(I,2), ZM(I)
        PA = PA + YM(I)
        IF (ZM(I).GT.1.E-6) GO TO 25
        ZM(I) = XM(I,2)
        IF (IST.EQ.0.OR.IST.EQ.2) IST = 1
     25 CONTINUE
     30 CONTINUE
        PA = PA/FLOAT(NN)
C
C       READ STANDARD DEVIATIONS OF VLE MEASUREMENTS.  ND IS THE NUMBER
C       OF STANDARD DEVIATIONS WHICH ARE DUPLICATED (1.LE.ND.LE.NN).
C       THE SUM OF ALL THE ND,S MUST BE EQUAL TO NN.
C
        M1 = 1
        M2 = 0
     40 READ (5,1004) SDY(M1), SDX(M1,1), SDX(M1,2), SDZ(M1), ND
        IF (ND.EQ.0) ND = NN
        M2 = M2 + ND
        DO 50 I=M1,M2
```

```
      SDX(I,1) = SDX(M1,1)
      SDX(I,2) = SDX(M1,2)
      SDY(I) = SDY(M1)
      SDZ(I) = SDZ(M1)
   50 CONTINUE
      M1 = M1 + ND
      IF (M2.LT.NN) GO TO 40
C
C     CORRECT DATA TO INTERNAL UNITS
C
      IF (ITU.EQ.0) GO TO 100
      IF (ITU.EQ.1) GO TO 120
      DO 90 I=1,NN
      SDX(I,1) = SDX(I,1)/1.8
   90 XM(I,1) = (XM(I,1) + 459.69)/1.8
      GO TO 120
  100 CONTINUE
      DO 110 I=1,NN
  110 XM(I,1) = XM(I,1) + 273.15
  120 CONTINUE
      IF (ICU.EQ.0) GO TO 140
      DO 130 I=1,NN
      X1 = XM(I,2)/MW(1)
      X2 = (1.-XM(I,2))/MW(2)
      XM(I,2) = X1/(X1+X2)
      Y1 = ZM(I)/MW(1)
      Y2 = (1.-ZM(I))/MW(2)
      ZM(I) = Y1/(Y1+Y2)
  130 CONTINUE
  140 CONTINUE
      IF (IPU.EQ.0) GO TO 180
      IF (IPU.EQ.1) GO TO 200
      IF (IPU.EQ.2) GO TO 160
      DO 150 I=1,NN
      YM(I) = YM(I)/14.508
      PA = PA/14.508
  150 SDY(I) = SDY(I)/14.508
      GO TO 200
  160 CONTINUE
      DO 170 I=1,NN
      PA = PA/0.98692
      YM(I) = YM(I)/0.98692
  170 SDY(I) = SDY(I)/0.98692
      GO TO 200
  180 CONTINUE
      DO 190 I=1,NN
      PA = PA/750.06
      YM(I) = YM(I)/750.06
  190 SDY(I) = SDY(I)/750.06
  200 CONTINUE
C
C     DOCUMENT THE INPUT VAPOR-LIQUID EQUILIBRIUM DATA
C
      WRITE (6,2001)     (NCD(1,J),J=1,2),     (NCD(2,J),J=1,2)
      WRITE (6,2002) ((REF(I,J),I=1,15), J=1,IRF)
      IF (PA.LE.5.0) WRITE (6,2003)
      IF (PA.GT.5.0) WRITE (6,2008)
      DO 210 I=1,NN
      TEMP = XM(I,1) - 273.15
      PRES = YM(I)
      IF (PA.LE.5.0) PRES = PRES*750.06
      X2 = 1.-XM(I,2)
      Y2 = 1.-ZM(I)
      WRITE (6,2004) I, PRES, TEMP, XM(I,2), X2, ZM(I), Y2
```

236

```
  210 CONTINUE
      WRITE (6,2005)
      DO 215 I=1,NN
      SDP = SDY(I)
      IF (PA.LE.5.0) SDP = SDP*750.06
  215 WRITE (6,2006) I, SDP, SDX(I,1), SDX(I,2), SDZ(I)
C
C     COMPUTE ERROR VARIANCES
C
      DO 220 I=1,NN
      EVX(I,1) = SDX(I,1)**2
      EVX(I,2) = SDX(I,2)**2
      EVY(I) = SDY(I)**2
      EVZ(I) = SDZ(I)**2
  220 CONTINUE
C
C     READ IN AND DOCUMENT INITIAL PARAMETER ESTIMATES.
C
      READ (5,1003) (P(L),L=1,LL)
      WRITE (6,2007) (L,P(L),L=1,LL)
      RETURN
 1001 FORMAT (2A3,4X,2A3,4X,I2,18X,5(I1,1X))
 1002 FORMAT (15A4)
 1003 FORMAT (8F10.2)
 1004 FORMAT (4F10.2,I2)
 2001 FORMAT (37H1BINARY VAPOR-LIQUID EQUILIBRIUM DATA,5X,1H1,2X,2A3,
     1        3X,1H2,2X,2A3)
 2002 FORMAT (1H0,4X,3HREF,3X,15A4/11X,15A4)
 2003 FORMAT (1H0,2HNO,4X,6HP (MM),4X,5HT (C),5X,2HX1,7X,2HX2,7X,
     1        2HY1,7X,2HY2//)
 2004 FORMAT (1X,I2,3X,F7.2,3X,F6.2,4(3X,F6.4))
 2005 FORMAT (20HOSTANDARD DEVIATIONS//)
 2006 FORMAT (1X,I2,3X,F7.2,3X,F6.2,3X,F6.4,12X,F6.4)
 2007 FORMAT (19HOINITIAL PARAMETERS//(5X,1HP,I1,2H =,F10.3))
 2008 FORMAT (1H0,2HNO,4X,6HP(BAR),4X,5HT (C),5X,2HX1,7X,2HX2,7X,
     1        2HY1,7X,2HY2//)
 3001 FORMAT (//1X,41HVLE AND PURE COMPONENT NAMES DO NOT MATCH/1X,
     1        4(2X,2A3)/)
      END
```

```
       SUBROUTINE OUTDAT (IRR)
C
C******************************************************************************
C
C      SUBROUTINE OUTDAT (IRR)
C
C      PURPOSE
C          PRINTS THE RESULTS OF THE REGRESSION-OPTIMUM PARAMETERS,
C          VARRIANCE-COVARRIANCE MATRIX, CORRELATION COEFFICIENT MATRIX,
C          AND THE VARRIANCE OF THE FIT.  ALSO, PRINTED ARE THE MEASURED
C          DATA, THEIR ESTIMATED TRUE VALUES, AND THE DEVIATIONS FOR
C          ALL NN DATA POINTS.  FINALLY, THE ROOT-MEAN-SQUARED
C          DEVIATIONS ARE GIVEN FOR EACH OF THE MEASUREMENTS.
C
C      PROGRAMED BY T.F.ANDERSON JAN 73.
C
C******************************************************************************
C
       COMMON /ALL/ NN, LL, KK, IPRT, IST
       COMMON /STAT/ COV(5,5), RHO(5,5), SIGMA(5), SSQN, ESTSIG, IT
       COMMON /DATA/ X(50,5), XM(50,5), EVX(50,5), DFX(50,5), DGX(50,5),
      1             DFP(50,5), DGP(50,5), Y(50), YM(50), EVY(50),
      2             ZM(50), Z(50), EVZ(50)
       COMMON /INOT/ PA
C
C      PRINT STATISTICAL RESULTS
C
       IF (IRR.GT.0) GO TO 40
       WRITE (6,2001)
       WRITE (6,2002) IT
       WRITE (6,2003) SSQN, ESTSIG
       WRITE (6,2004) (L,P(L),SIGMA(L),L=1,LL)
       WRITE (6,2005)
       DO 20 I=1,LL
   20  WRITE (6,2006) (COV(I,J),J=1,LL)
       WRITE (6,2007)
       DO 30 I=1,LL
   30  WRITE (6,2006) (RHO(I,J),J=1,LL)
C
C      CONVERT TO OUTPUT UNITS AND PRINT SUMMARY OF RESULTS
C
   40  CONTINUE
       WRITE (6,2008)
       PS = 0.0
       TS = 0.0
       X1S = 0.0
       Y1S = 0.0
       DO 50 I=1,NN
       PM = YM(I)
       PC = Y(I)
       IF (PA.GT.5.0) GO TO 45
       PM = PM*750.06
       PC = PC*750.06
   45  CONTINUE
       PD = PM - PC
       PS = PS + PD**2
       TM = XM(I,1) - 273.15
       TC = X(I,1) - 273.15
       TD = TM - TC
       TS = TS + TD**2
       X1M = XM(I,2)
       X1C = X(I,2)
       X1D = X1M - X1C
       X1S = X1S + X1D**2
```

238

```
      Y1M = ZM(I)
      Y1C = Z(I)
      Y1D = Y1M - Y1C
      Y1S = Y1S + Y1D**2
      WRITE (6,2009) I,PM,PC,PD,TM,TC,TD,X1M,X1C,X1D,Y1M,Y1C,Y1D
   50 CONTINUE
C
C     COMPUTE ROOT MEAN SQUARED DEVIATIONS
C
      TPT = FLOAT(NN)
      PS = SQRT(PS/TPT)
      TS = SQRT(TS/TPT)
      X1S = SQRT(X1S/TPT)
      Y1S = SQRT(Y1S/TPT)
      WRITE (6,2010)
      WRITE (6,2011) PS,TS,X1S,Y1S
      RETURN
 2001 FORMAT (1H1,5X,18HREGRESSION SUMMARY)
 2002 FORMAT (1H0,I2,20H ITERATIONS REQUIRED)
 2003 FORMAT (1H0,51HTHE SUM OF THE SQUARES OF THE WEIGHTED RESIDUALS IS
     1,G15.5//1X,40HAND THE ESTIMATED VARIANCE OF THE FIT IS,G15.5)
 2004 FORMAT (1H0,5X,53HPARAMETERS AND ESTIMATES OF THEIR STANDARD DEVIA
     1TIONS/(/1X,1HP,I1,3H = ,F9.2,4H +/-,F7.2))
 2005 FORMAT (1H0,5X,17HCOVARIANCE MATRIX)
 2006 FORMAT (1H0,5G15.5)
 2007 FORMAT (1H0,5X,30HCORRELATION COEFFICIENT MATRIX)
 2008 FORMAT (1H1,5X,61HMEASURED VALUES, CALCULATED VALUES, AND RESIDUAL
     1S (MEAS-CALC)//1X,2HNO,6X,2HPM,7X,2HPC,7X,2HPD,7X,2HTM,6X,2HTC,
     2 6X,2HTD,6X,3HX1M,5X,3HX1C,5X,3HX1D,6X,3HY1M,5X,3HY1C,5X,3HY1D//)
 2009 FORMAT (1X,I2,1X,3(2X,F7.2),1X,2(2X,F6.2),2X,F5.2,1X,3(2X,F6.4),
     1          1X,3(2X,F6.4))
 2010 FORMAT (1H0,28HROOT MEAN SQUARED DEVIATIONS/)
 2011 FORMAT (25X,F6.2,19X,F5.2,19X,F6.4,19X,F6.4)
      END
```

```
      SUBROUTINE REGRES (IRR)
C                                                                                C
C*****************************************************************************C
C                                                                                C
C     SUBROUTINE REGRES (IRR)                                                     C
C                                                                                C
C     PURPOSE                                                                     C
C         SOLVES FOR THE PARAMETERS IN NON-LINEAR MODELS WHERE ALL THE            C
C         MEASURED VARIABLES ARE SUBJECT TO ERROR AND ARE RELATED BY              C
C         ONE OR TWO CONSTRAINTS.                                                 C
C                                                                                C
C     PARAMETERS                                                                  C
C                                                                                C
C         CONTROL VARIABLES                                                       C
C         NN     - NUMBER OF DATA POINTS                                          C
C         LL     - NUMBER OF PARAMETERS                                           C
C         KK     - NUMBER OF INDEPENDENT VARIABLES                                C
C         IPRT   - PRINT CONTROL                                                  C
C                     0 - MINIMUM PRINTED OUTPUT                                  C
C                     6 - MAXIMUM PRINTED OUTPUT                                  C
C         IST    - EXECUTION CODE FOR TYPE OF REGRESSION                          C
C                     0 - TWO CONSTRAINTS, MINIMIZE ERROR IN ALL VARIABLES        C
C                     1 - ONE CONSRAINT, MINIMIZE ERROR IN ALL VARIABLES          C
C                     2 - TWO CONSTRAINTS, MINIMIZE ERROR IN DEPENDENT            C
C                         VARIABLES ONLY                                          C
C                     3 - ONE CONSTRAINT, MINIMIZE ERROR IN SINGLE                C
C                         DEPENDENT VARIABLE                                      C
C         LMV    - CONSTRAINT CONTROL ON VARIABLES                                C
C                     1 - CONSTRAINED TO BE POSITIVE                              C
C                     2 - NO CONSTRAINTS                                          C
C         LMP    - CONSTRAINT CONTROL ON PARAMETERS                               C
C                     1 - MAXIMUM ABSOLUTE CHANGE IS LIMITED TO VALUE             C
C                         OF PRCG                                                 C
C                     2 - NO CONSTRAINTS                                          C
C         PRCG   - MAXIMUM ABSOLUTE VALUE OF PARAMETER CHANGES WHEN               C
C                     LMP EQUAL 2                                                 C
C         ALST   - CONVERGENCE CNTRL, USUALLY = 1.0                               C
C         SSTL   - CONVERGENCE TOLERANCE                                          C
C         ITMX   - ITERATION LIMIT                                                C
C         BETA   - CONVERGENCE CNTRL, USUALLY = 0.25                              C
C         RP     - CONVERGENCE CNTRL, USUALLY = 1.5                               C
C                                                                                C
C         PROGRAM VARIABLES TO BE SUPPLIED                                        C
C         XM     - ARRAY OF MEASURED INDEP VARIABLES                              C
C         EVX    - ARRAY OF ERROR VARRIANCES OF INDEP VARIABLES                   C
C         YM     - VECTOR OF FIRST MEASURED DEPENDENT VARIABLES                   C
C         EVY    - VECTOR OF ERROR VARRIANCES OF FIRST MEASURED                   C
C                  DEPENDENT VARIABLES                                            C
C         ZM     - VECTOR OF SECOND MEASURED DEPENDENT VARIABLES                  C
C         EVZ    - VECTOR OF ERROR VARRIANCES OF SECOND MEASURED                  C
C                  DEPENDENT VARIABLES                                            C
C         DFX    - ARRAY OF DERIVATIVES OF FIRST CONST FUNC WRT INDEP             C
C                  VARIABLES                                                      C
C         DFP    - ARRAY OF DERIVATIVES OF FIRST CONST FUNC WRT                   C
C                  PARAMETERS                                                     C
C         DGX    - ARRAY OF DERIVATIVES OF SECOND CONST FUNC WRT INDEP            C
C                  VARIABLES                                                      C
C         DGP    - ARRAY OF DERIVATIVES OF SECOND CONST FUNC WRT                  C
C                  PARAMETERS                                                     C
C         P      - VECTOR OF INITIAL PARAMETER ESTIMATES                          C
C                                                                                C
C         CALCULATED VARIABLES                                                    C
C         X      - ARRAY OF TRUE VALUES OF INDEP VARIABLES                        C
C         Y      - VECTOR OF TRUE VALUES OF FIRST DEPENDENT VARIABLE             C
```

```
C          Z      - VECTOR OF TRUE VALUES OF SECOND DEPENDENT VARIALBE     C
C          P      - VECTOR OF FINAL PARAMETER ESTIMATES                    C
C                                                                          C
C      SUBROUTINES USED                                                    C
C          SYMINV                                                          C
C          SUMSQ                                                           C
C                                                                          C
C      METHOD                                                              C
C          PARAMETER ESTIMATION BASED CN PRINCIPLE OF MAXIMUM LIKELIHOOD   C
C          WHILE TAKING INTO ACCOUNT ERROR IN ALL MEASURED VARIABLES.      C
C                                                                          C
C      REMARKS                                                             C
C          USES CONVERGENCE TECHNIQUES DESCRIBED BY LAW AND BAILEY         C
C                                                                          C
C      REFERENCES                                                          C
C          ABRAMS,D.S., PH.D. THESIS, UNIV OF CALIF, BERKELEY (1974).      C
C          BRITT,H.I., AND R.A.LUECKE, TECHNCMETRICS, 15, 233 (1973).      C
C          BRYSON,A.E., AND Y.B.HO, *APPLIED OPTIMAL CCNTROL,              C
C              OPTIMIZATION, ESTIMATION, AND CONTROL,* BLAISDELL PUBL,     C
C              WALTHAM, MASS., 1969.                                       C
C          CLIFFORD,A.A., *MULTIVARIATE ERROR ANALYSIS,* HALSTED PRESS,    C
C              N.Y., 1973.                                                 C
C          DEMMING,W.E., *STATISTICAL ADJUSTMENT CF DATA,* WILEY,          C
C              NEW YORK, (1943).                                           C
C          DRAPPER,N.R., AND H.SMITH, *APPLIED REGRESSION ANALYSIS,*       C
C              WILEY, N.Y., 1966.                                          C
C          FABRIES,J., AND H.RENCN, AICHE J., 21, 735 (1975).              C
C          LAW,V.J., AND R.V.BAILEY, CHEM ENG SCI, 18, 189 (1963).        C
C                                                                          C
C      PROGRAMED BY T.F.ANDERSON DEC 75.                                   C
C                                                                          C
C***************************************************************************C
C                                                                          C
      DIMENSION DXO(50,5),DZO(50),DYO(50),DELX(50,5),DELP(5),C(5),U(5),
     1          CC(5,5),D(50,5,5),S(50,5,5),DUM(50,5,5),Q(50,5),T(5,5),
     2          FP(50),FX(50),GP(50),GX(50)
      COMMON /STAT/ COV(5,5), RHO(5,5), SIGMA(5), SSQN, ESTSIG, IT
      COMMON /RGRS/ SSTL, BETA, ITMX, RP, LMV, LMP, ALST, PRCG
      COMMON /DATA/ X(50,5), XM(50,5), EVX(50,5), DFX(50,5), DGX(50,5),
     1              P(5), DFP(50,5), DGP(50,5), Y(50), YM(50), EVY(50),
     2              ZM(50), Z(50), EVZ(50)
      COMMON /ALL/ NN, LL, KK, IPRT, IST
C
C      INITIALIZE INTERNAL PARAMETERS.
C
      IT = 0
      ILP = 0
C
C      INITIAL ESTIMATES OF TRUE VALUES OF THE VARIABLES ARE SET EQUAL
C      TO THE MEASURED VALUES.
C
      DO 5 I=1,NN
      Y(I) = YM(I)
      Z(I) = ZM(I)
      DO 3 J=1,KK
      DELX(I,J) = 0.0
      DFX(I,J) = 0.0
      DGX(I,J) = 0.0
      X(I,J) = XM(I,J)
    3 CONTINUE
      DO 4 L=1,LL
      DFP(I,L) = 0.0
      DGP(I,L) = 0.0
```

241

```
    4 CONTINUE
    5 CONTINUE
C
C      BEGIN ITERATIONS FOR REGRESSICN SCLUTION.
C
   10 CONTINUE
      AL = ALST
C
C      CALL SUMSQ TO OBTAIN NEW VALUES CF DEPENDENT VARIABLES,
C      DERIVATIVES, AND SSQ.
C
      CALL FUNC (IRR)
      IF (IRR.GT.0) GO TO 570
      CALL FUNDR (IRR)
      IF (IRR.GT.0) GO TO 570
      CALL SUMSQ (SSQ)
      IF (IPRT.LT.1) GO TO 15
      WRITE (6,2001) IT,SSQ,(P(I),I=1,LL)
 2001 FORMAT (12HOITERATION =,I2,5X,22HSUM OF THE SQUARES IS ,G15.6//5X,
     1        16HPARAMETERS ARE -/(1X,G15.5))
      IF (IPRT.LT.4) GO TO 15
      WRITE (6,2008)
 2008 FORMAT (35HOESTIMATED TRUE VALUES OF VARIABLES//)
      DO 11 I=1,NN
   11 WRITE (6,2009) I, (X(I,J),J=1,KK), Y(I), Z(I)
 2009 FORMAT (1X,I2,2X,7G11.4)
      IF (IPRT.LT.7) GO TO 15
      WRITE (6,2010)
 2010 FORMAT (42H1DERIVATIVES WITH RESPECT TO THE VARIABLES/)
      DO 12 I=1,NN
      WRITE (6,2011) I, (DFX(I,K),K=1,KK)
   12 WRITE (6,2012) (DGX(I,K),K=1,KK)
 2011 FORMAT (/1X,I2,5X,5G15.5)
 2012 FORMAT (8X,8G15.5)
      WRITE (6,2013)
 2013 FORMAT (43H1DERIVATIVES WITH RESPECT TO THE PARAMETERS/)
      DO 13 I=1,NN
      WRITE (6,2011) I, (DFP(I,L),L=1,LL)
   13 WRITE (6,2012) (DGP(I,L),L=1,LL)
   15 CONTINUE
      IT = IT + 1
C
C      CALC DEVIATIONS OF VARIABLES.
C
      DO 17 I=1,NN
      DYO(I) = Y(I) - YM(I)
      DZO(I) = Z(I) - ZM(I)
      IF (IST.GE.2) GO TO 17
      DO 16 J=1,KK
   16 DXO(I,J) = X(I,J) - XM(I,J)
   17 CONTINUE
C
C      CALC MATRIX D INVERSE.
C
      IF (IST.GE.2) GO TO 31
      DO 25 I=1,NN
      DO 18 J=1,KK
      DO 18 K=1,KK
      T(J,K) = DFX(I,K)*DFX(I,J)/EVY(I) + DGX(I,K)*DGX(I,J)/EVZ(I)
      IF (J.EQ.K) T(J,K) = T(J,K) + 1./EVX(I,J)
   18 CONTINUE
      CALL SYMINV (T,KK,IRR)
      IF (IRR.GT.0) GO TO 520
      DO 22 J=1,KK
```

```
      DO 22 K=1,KK
   22 D(I,J,K) = T(J,K)
   25 CONTINUE
C
C     CALC VECTOR Q.
C
      DO 30 I=1,NN
      DO 28 K=1,KK
      Q(I,K) = DXO(I,K)/EVX(I,K) + DFX(I,K)*DYO(I)/EVY(I) +
     1          DGX(I,K)*DZO(I)/EVZ(I)
C
C     CALC VECTOR S.
C
      DO 28 L=1,LL
      S(I,K,L) = DFX(I,K)*DFP(I,L)/EVY(I) + DGX(I,K)*DGP(I,L)/EVZ(I)
   28 CONTINUE
   30 CONTINUE
   31 CONTINUE
C
C     CALC VECTOR U.
C
      DO 35 L=1,LL
      U(L) = 0.0
      DO 35 I=1,NN
   35 U(L) = U(L) + DFP(I,L)*DYO(I)/EVY(I) + DGP(I,L)*DZO(I)/EVZ(I)
C
C     CALC T MATRIX.
C
      DO 40 II=1,LL
      DO 40 JJ=1,LL
      T(II,JJ) = 0.0
      DO 38 I=1,NN
   38 T(II,JJ) = T(II,JJ) + DFP(I,II)*DFP(I,JJ)/EVY(I) +
     1          DGP(I,II)*DGP(I,JJ)/EVZ(I)
   40 CONTINUE
      IF (IST.GE.2) GO TO 52
C
C     CALC DUMMY MATRIX.
C
      DO 50 I=1,NN
      DO 48 II=1,LL
      DO 48 JJ=1,KK
      DUM(I,II,JJ) = 0.0
      DO 46 K=1,KK
   46 DUM(I,II,JJ) = DUM(I,II,JJ) + S(I,K,II)*D(I,K,JJ)
   48 CONTINUE
   50 CONTINUE
   52 CONTINUE
C
C     CALC CC MATRIX.
C
      DO 55 II=1,LL
      DO 55 JJ=1,LL
      IF (IST.LE.1) GO TO 53
      CC(II,JJ) = T(II,JJ)
      GO TO 55
   53 CC(II,JJ) = 0.0
      DO 54 I=1,NN
      DO 54 K=1,KK
   54 CC(II,JJ) = CC(II,JJ) + DUM(I,II,K)*S(I,K,JJ)
      CC(II,JJ) = T(II,JJ) - CC(II,JJ)
   55 CONTINUE
C
C     CALC C VECTOR.
```

```
C
      DO 60 L=1,LL
      IF (IST.LE.1) GO TO 57
      C(L) = -U(L)
      GO TO 60
   57 C(L) = 0.0
      DO 58 I=1,NN
      DO 58 K=1,KK
   58 C(L) = C(L) + DUM(I,L,K)*Q(I,K)
      C(L) = C(L) - U(L)
   60 CONTINUE
C
C     CALC CC INVERSE.
C
      CALL SYMINV (CC,LL,IRR)
      IF (IRR.GT.0) GO TO 510
C
C     CALC DELP.
C
      DO 65 II=1,LL
      DELP(II) = 0.0
      DO 65 JJ=1,LL
   65 DELP(II) = DELP(II) + CC(II,JJ)*C(JJ)
C
C     CALC FP AND GP.
C
      DO 70 I=1,NN
      FP(I) = 0.0
      GP(I) = 0.0
      DO 68 L=1,LL
      FP(I) = FP(I) + DFP(I,L)*DELP(L)
   68 GP(I) = GP(I) + DGP(I,L)*DELP(L)
   70 CONTINUE
C
C     CALC VECTOR Q PRIME.
C
      IF (IST.GE.2) GO TO 86
      DO 75 I=1,NN
      DO 73 K=1,KK
      DO 73 L=1,LL
   73 Q(I,K) = Q(I,K) + S(I,K,L)*DELP(L)
   75 CONTINUE
C
C     CALC DELX.
C
      DO 80 I=1,NN
      DO 78 K=1,KK
      DELX(I,K) = 0.0
      DO 78 J=1,KK
   78 DELX(I,K) = DELX(I,K) - D(I,K,J)*Q(I,J)
   80 CONTINUE
C
C     CALC FX AND GX.
C
      DO 85 I=1,NN
      FX(I) = 0.0
      GX(I) = 0.0
      DO 83 K=1,KK
      FX(I) = FX(I) + DFX(I,K)*DELX(I,K)
   83 GX(I) = GX(I) + DGX(I,K)*DELX(I,K)
   85 CONTINUE
   86 CONTINUE
C
```

```
C         COMPUTE CHANGE IN LINEARIZED SUM OF SQUARES (DSL).
C
          DSL = 0.0
          DO 95 I=1,NN
          IF (IST.LE.1) GO TO 89
          DSL = DSL + DYO(I)*FP(I)/EVY(I) + DZO(I)*GP(I)/EVZ(I)
          GO TO 95
       89 DO 90 J=1,KK
       90 DSL = DSL + DXO(I,J)*DELX(I,J)/EVX(I,J)
          DSL = DSL + DYO(I)*(FX(I) + FP(I))/EVY(I) +
         1            DZO(I)*(GX(I) + GP(I))/EVZ(I)
       95 CONTINUE
          WRITE (6,2003) DSL
     2003 FORMAT (34HODELTA LINEARIZED SUM OF SQUARES =,G15.5)
          WRITE (6,2004) (DELP(L),L=1,LL)
     2004 FORMAT (22HOCHANGES IN PARAMETERS/(5X,G15.5))
          IF (IPRT.LT.3) GO TO 97
          WRITE (6,2005)
     2005 FORMAT (33HOCHANGES IN INDEPENDENT VARIABLES//)
          DO 96 I=1,NN
       96 WRITE (6,2006) (DELX(I,J),J=1,KK)
     2006 FORMAT (5X,5G15.5)
       97 CONTINUE
C
C         IF DSL POSITIVE, CHANGE DIRECTION OF STEPS TO INSURE LOCAL
C         CONVERGENCE.
C
          IF (DSL.LT.0.0) GO TO 110
          DSL = -DSL
          IF (IST.GE.2) GO TO 102
          DO 100 I=1,NN
          DO 100 J=1,KK
      100 DELX(I,J) = -DELX(I,J)
      102 DO 105 L=1,LL
      105 DELP(L) = -DELP(L)
C
C         IF PARAMETERS OR VARIABLES ARE LIMITED, CHECK TO MAKE SURE
C         LIMITS ARE NOT EXCEEDED.
C
      110 IF (IST.GE.2) GO TO 113
          IF (LMV.EQ.2) GO TO 113
      111 IF (AL.LT.1.E-4) GO TO 540
          DO 112 I=1,NN
          DO 112 J=1,KK
          IF (X(I,J) + AL*DELX(I,J).GE.0.0) GO TO 112
          AL = AL/RP
          GO TO 111
      112 CONTINUE
      113 IF (LMP.EQ.2) GO TO 115
     1111 IF (AL.LT.1.E-5) GO TO 560
          DO 114 L=1,LL
          IF (PRCG-ABS(DELP(L)*AL).GT.0.0) GO TO 114
          AL = AL/RP
          GO TO 1111
      114 CONTINUE
C
C         COMPUTE NEW PARAMETERS AND TRUE VALUES OF THE VARIABLES.
C
      115 CONTINUE
          DO 116 I=1,NN
          DO 116 J=1,KK
      116 X(I,J) = X(I,J) + DELX(I,J)*AL
      118 DO 120 L=1,LL
      120 P(L) = P(L) + DELP(L)*AL
```

```
      APPLY STEP LIMITING TECHNIQUE TO PREVENT OSCILLATION AND SLOW
      CONVERGENCE.

      ILP = 0
  130 CONTINUE
      ILP = ILP + 1
      IF (AL.LT.1.E-5) GO TO 530
      CALL FUNC (IRR)
      IF (IRR.GT.0) GO TO 133
      CALL SUMSQ (SSQN)
      GO TO 134
  133 SSQN = 1.E+20
      IRR = 0
  134 CONTINUE
      IF (IPRT.LT.2) GO TO 135
      IF (ILP.EQ.1) WRITE (6,2007)
 2007 FORMAT (1H0,5X,2HAL,15X,4HSSQN//)
      WRITE (6,2002) AL,SSQN
 2002 FORMAT (1X,2G15.5)
  135 CONTINUE
      DS = SSQN - SSQ
      IF (ILP.GT.1) GO TO 138
      IF (ABS(DS)/SSQN.LT.SSTL) GO TO 200
  138 CONTINUE
      IF (DS.GT.0.0) GO TO 140
      IF ((DS - BETA*(2.*AL - AL*AL)*DSL).LT.0.0) GO TO 180
  140 AL = AL/RP
      IF (IST.GE.2) GO TO 148
      DO 145 I=1,NN
      DO 145 J=1,KK
  145 X(I,J) = X(I,J) - DELX(I,J)*(RP - 1.)*AL
  148 DO 150 L=1,LL
  150 P(L) = P(L) - DELP(L)*(RP - 1.)*AL
      GO TO 130

      CHECK FOR CONVERGENCE, IF NOT RETURN TO 10.

  180 DIFF = (SSQ - SSQN)/SSQN
      IF (ABS(DIFF).LT.SSTL) GO TO 200
      IF (IT.LT.ITMX) GO TO 10
      GO TO 500

      CALCULATE ESTIMATED VARRIANCE OF FIT.

  200 CONTINUE
      ESTSIG = SSQN/FLOAT(NN - LL)

      CALC COVARRIANCE MATRIX AND CORRELATION COEFFICIENT MATRIX.

      DO 210 I=1,LL
      DO 210 J=1,LL
  210 COV(I,J) = ESTSIG*CC(I,J)
      DO 215 L=1,LL
  215 SIGMA(L) = SQRT(COV(L,L))
      DO 220 I=1,LL
      DO 220 J=1,LL
  220 RHO(I,J) = COV(I,J)/(SIGMA(I)*SIGMA(J))
      RETURN
```

```
  500 IRR = 1
      WRITE (6,1001) IT
 1001 FORMAT (24H0MAX ITER EXCEEDED    IT=,I2)
      RETURN
  510 WRITE (6,1002)
 1002 FORMAT (29H0ILL-DEFINED PARAMETER MATRIX)
      RETURN
  520 WRITE (6,1003)
 1003 FORMAT (23H0ILL-DEFINED VARIABLE MATRIX)
      RETURN
  530 IRR = 4
      WRITE (6,1004) IT,AL
 1004 FORMAT (68H0LINEARIZED FUNCTION DOES NOT SUFFICIENTLY APPROXIMATE
     1TRUE FUNCTION,5X,3HIT=,I2,5X,3HAL=,G15.5)
      RETURN
  540 IRR = 5
      WRITE (6,1005) IT,AL
 1005 FORMAT (12H0DELX TO BIG,5X,3HIT=,I2,5X,3HAL=,G15.5)
      RETURN
  560 IRR = 6
      WRITE (6,1006) IT,AL
 1006 FORMAT (13H0DELP TO BIG,5X,3HIT=,I2,5X,3HAL=,G15.5)
      RETURN
  570 WRITE (6,1007) IRR
 1007 FORMAT (39H0ERROR IN FUNCTION EVALUATION    IRR = ,I2)
      RETURN
      END
```

```
      SUBROUTINE SYMINV (A,N,IRR)

***************************************************************************

      SUBROUTINE SYMINV (A,N,IRR)

      PURPOSE
          INVERT SYMETRIC MATRIX.

      PARAMETERS
          N     - SIZE OF MATRIX.
          A     - INITIALLY CONTAINS THE MATRIX TO BE INVERTED WHICH
                  IS DESTROYED AND ULTIMATELY CONTAINS THE INVERSE.
          IRR   - FAILURE PARAMETER WHEN NO INVERSE CAN BE FOUND.

      REFERENCE
          RUTISHAUSER,H., ALGORITHM 150 SYMETRIC MATIRX INVERSION,
              COMM. A.C.M., 6, 67 (1963).

      PROGRAMED BY D.S.ABRAMS FALL 74.

***************************************************************************

      DIMENSION A(5,5), P(5), Q(5), R(5)
      INTEGER R
      IRR = 0
      DO 10 I=1,N
  10  R(I) = 1
      DO 150 I=1,N
      BIG=0.
      DO 40 J=1,N
      TEST=ABS(A(J,J))
      IF (TEST-BIG) 40,40,20
  20  IF (R(J)) 160,40,30
  30  BIG=TEST
      K=J
  40  CONTINUE
      IF (BIG.LT.1.E-19) GO TO 160
      R(K)=0
      Q(K)=1./A(K,K)
      P(K)=1.
      A(K,K)=0.0
      KP1=K+1
      KM1=K-1
      IF (KM1) 160,80,50
  50  DO 70 J=1,KM1
      P(J)=A(J,K)
      Q(J)=A(J,K)*Q(K)
      IF (R(J)) 160,70,60
  60  Q(J)=-Q(J)
  70  A(J,K)=0.
  80  IF (K-N) 90,130,160
  90  DO 120 J=KP1,N
      P(J)=A(K,J)
      IF (R(J)) 160,100,110
 100  P(J)=-P(J)
 110  Q(J)=-A(K,J)*Q(K)
 120  A(K,J)=0.0
 130  DO 140 J=1,N
      DO 140 K=J,N
      A(J,K)=A(J,K)+P(J)*Q(K)
```

```
140  A(K,J)=A(J,K)
150  CONTINUE
     RETURN
160  IRR = 1
     RETURN
     END
```

```
      SUBROUTINE SUMSQ (SSQ)

***********************************************************************

      SUBROUTINE SUMSQ (SSQ)

      PURPOSE
          CALCULATES THE SUM OF THE SQUARES OF THE DEVIATIONS OF ALL
          MEASURED VARIABLES FROM THEIR TRUE VALUES FOR REGRES.

      PARAMETERS
          SSQ   - SUM OF RESIDUALS SQUARED.
          IRR   - ERROR FLAG RETRUNED FROM FUNC.
          IST   - EXECUTION CODE FOR TYPE OF REGRESSION
                  0 - TWO CONSTRAINTS, MINIMIZE ERROR IN ALL VARIABLES
                  1 - ONE CONSRAINT, MINIMIZE ERROR IN ALL VARIABLES
                  2 - TWO CONSTRAINTS, MINIMIZE ERROR IN DEPENDENT
                      VARIABLES ONLY
                  3 - ONE CONSTRAINT, MINIMIZE ERRCR IN SINGLE
                      DEPENDENT VARIABLE

      PROGRAMED BY T.F.ANDERSON NOV 75.

***********************************************************************

      COMMON /ALL/ NN, LL, KK, IPRT, IST
      COMMON /DATA/ X(50,5), XM(50,5), EVX(50,5), DFX(50,5), DGX(50,5),
     1              P(5), DFP(50,5), DGP(50,5), Y(50), YM(50), EVY(50),
     2              ZM(50), Z(50), EVZ(50)
      SSQ = 0.
      DO 10 I=1,NN
      IF (IST.GT.1) GO TO 8
      DO 5 J=1,KK
    5 SSQ = SSQ + (X(I,J)-XM(I,J))**2/EVX(I,J)
    8 SSQ = SSQ + (Y(I)-YM(I))**2/EVY(I)
      IF (IST.EQ.1.OR.IST.EQ.3) GO TC 10
      SSQ = SSQ + (Z(I) - ZM(I))**2/EVZ(I)
   10 CCNTINUE
      RETURN
      END
```

```
      SUBROUTINE FUNC (IRR)
C                                                                         C
C*************************************************************************C
C                                                                         C
C     SUBROUTINE FUNC (IRR)                                               C
C                                                                         C
C     PURPOSE                                                             C
C        INTERMEDIATE ROUTINE FOR CALCULATING ALL THE CONSTRAINT         C
C        FUNCTICNS FOR NN EXPERIMENTAL PCINTS.                           C
C                                                                         C
C     PARAMETERS                                                          C
C                                                                         C
C        CONTROL VARIABLES                                               C
C                                                                         C
C        NN     - NUMBER OF EXPERIMENTAL POINTS.                         C
C        IRR    - ERROR FLAG RETURNED FROM EVAL INDICATING FAILURE OF    C
C                 PROGRAM TO CALCULATE THE DESIRED PARAMETERS.           C
C                                                                         C
C        CALCULATED VARIABLES                                           C
C                                                                         C
C        X      - ARRAY OF TRUE VALUES CORRESPONDING TO THE MEASURED     C
C                 VALUES IN THE XM ARRAY                                 C
C        Y      - VECTOR OF TRUE VALUES CORRESPONDING TO THE MEASURED    C
C                 VALUES IN THE YM VECTOR                                C
C        Z      - VECTOR OF TRUE VALUES CORRESPONDING TO THE MEASURED    C
C                 VALUES IN TH ZM VECTOR                                 C
C        XV     - TRANSFER VECTOR FOR THE INDEPENDENT VARIABLES          C
C        P      - VECTOR OF MOST RECENT ESTIMATES OF THE PARAMETERS      C
C                                                                         C
C     SUBROUTINES USED                                                    C
C        EVAL                                                             C
C                                                                         C
C     PROGRAMED BY T.F.ANDERSON JAN 78.                                  C
C                                                                         C
C*************************************************************************C
C                                                                         C
      DIMENSION XV(5)
      COMMCN /ALL/ NN, LL, KK, IPRT, IST
      COMMON /DATA/ X(50,5), XM(50,5), EVX(50,5), DFX(50,5), DGX(50,5),
     1              P(5), DFP(50,5), DGP(50,5), Y(50), YM(50), EVY(50),
     2              ZM(50), Z(50), EVZ(50)
      DO 20 I=1,NN
      DO 15 J=1,KK
   15 XV(J) = X(I,J)
      YI = Y(I)
      ZI = Z(I)
      CALL EVAL (P,XV,YI,ZI,YO,ZC,IRR)
      Y(I) = YO
      Z(I) = ZO
      IF (IRR.GT.0) RETURN
   20 CONTINUE
      RETURN
      END
```

```
      SUBROUTINE FUNDR (IRR)
C                                                                        C
C************************************************************************C
C                                                                        C
C     SUBROUTINE FUNDR (IRR)                                             C
C                                                                        C
C     PURPOSE                                                            C
C        CALCULATES THE REQUIRED DERIVATIVES FOR REGRES BY CENTRAL       C
C        DIFFERENCE UTILIZING EVAL TO CALCULATE THE CONSTRAINTS.         C
C                                                                        C
C     PARAMETERS                                                         C
C                                                                        C
C        CONTROL VARIABLES                                               C
C                                                                        C
C        NN   - NUMBER OF EXPERIMENTAL POINTS                            C
C        IRR  - ERROR FLAG RETURNED FROM EVAL INDICATING FAILURE OF      C
C               PROGRAM TO CALCULATE THE DESIRED PARAMETERS.             C
C                                                                        C
C        CALCULATED VARIABLES                                            C
C                                                                        C
C        X    - ARRAY OF TRUE VALUES CORRESPONDING TO THE MEASURED       C
C               VALUES IN THE XM ARRAY                                   C
C        Y    - VECTOR OF TRUE VALUES CORRESPONDING TO THE MEASURED      C
C               VALUES IN THE YM VECTOR                                  C
C        Z    - VECTOR OF TRUE VALUES CORRESPONDING TO THE MEASURED      C
C               VALUES IN TH ZM VECTOR                                   C
C        XV   - TRANSFER VECTOR FOR THE INDEPENDENT VARIABLES            C
C        P    - VECTOR OF MOST RECENT ESTIMATES OF THE PARAMETERS        C
C        PR   - TRANSFER A                                               C
C        PR   - TRANSFER VECTOR FOR PARAMETERS                           C
C        DP   - VECTOR OF CENTRAL DIFFERENCE INCREMENTS FOR              C
C               CALCULATING DERIVATIVES WRT THE PARAMETERS               C
C        DX   - VECTOR OF CENTRAL DIFFERENCE INCREMENTS FOR              C
C               CALCULATING DERIVATIVES WRT THE INDEPENDENT VARIABLES C
C        DFP  - ARRAY OF DERIVATIVES OF CONSTRAINT 1 WRT THE             C
C               PARAMETERS                                               C
C        DGP  - ARRAY OF DERIVATIVES OF CONSTRAINT 2 WRT THE             C
C               PARAMETERS                                               C
C        DFX  - ARRAY OF DERIVATIVES OF CONSTRAINT 1 WRT THE             C
C               INDEPENDENT VARIABLES                                    C
C        DGX  - ARRAY OF DERIVATIVES OF CONSTRAINT 2 WRT THE             C
C               INDEPENDENT VARIABLES                                    C
C                                                                        C
C     SUBROUTINES USED                                                   C
C        EVAL                                                            C
C                                                                        C
C     PROGRAMED BY T.F.ANDERSON JAN 78.                                  C
C                                                                        C
C************************************************************************C
C                                                                        C
      DIMENSION XV(5),PR(5)
      COMMON /ALL/ NN, LL, KK, IPRT, IST
      COMMON /DATA/ X(50,5), XM(50,5), EVX(50,5), DFX(50,5), DGX(50,5),
     1              P(5), DFP(50,5), DGP(50,5), Y(50), YM(50), EVY(50),
     2              ZM(50), Z(50), EVZ(50)
C     COMMON /DIFF/ DP(5), DX(5)
C
C     CALC DERIVATIVES FOR REGRES BY CENTRAL DIFFERENCE.
C
      DO 10 L=1,LL
   10 PR(L) = P(L)
      DO 50 I=1,NN
      DO 20 J=1,KK
   20 XV(J) = X(I,J)
```

252

```
C
C      CALC DERIVATIVES W.R.T. THE PARAMETERS.
C
       DO 30 L=1,LL
       PP(L) = P(L) + DP(L)
       CALL EVAL (PP,XV,Y(I),Z(I),YYP,ZZP,IRR)
       PR(L) = P(L) - DP(L)
       CALL EVAL (PP,XV,Y(I),Z(I),YYM,ZZM,IRR)
       PR(L) = P(L)
       DFP(I,L) = (YYP-YYM)/(2.*DP(L))
       IF (IST.EQ.1.OR.IST.EQ.3) GO TO 30
       DGP(I,L) = (ZZP-ZZM)/(2.*DP(L))
   30 CONTINUE
       IF (IST.GT.1) GO TO 50
C
C      CALC DERIVATIVES W.R.T. THE VARIABLES.
C
       DO 40 J=1,KK
       XV(J) = X(I,J) + DX(J)
       CALL EVAL (PP,XV,Y(I),Z(I),YYP,ZZP,IRR)
       XV(J) = X(I,J) - DX(J)
       CALL EVAL (PP,XV,Y(I),Z(I),YYM,ZZM,IRR)
       XV(J) = X(I,J)
       DFX(I,J) = (YYP-YYM)/(2.*DX(J))
       IF (IST.GT.0) GO TO 40
       DGX(I,J) = (ZZP-ZZM)/(2.*DX(J))
   40 CONTINUE
   50 CONTINUE
       RETURN
       END
```

```
      SUBROUTINE EVAL (PR,XV,A1,A2,B1,B2,IRR)
C                                                                          C
C**************************************************************************C
C                                                                          C
C     SUBROUTINE EVAL (PR,XV,A1,A2,B1,B2,IRR)                              C
C                                                                          C
C     PURPOSE                                                              C
C        CALCULATES THE CONSTRAINT FUNCTIONS FOR BINARY VAPOR-LIQUID       C
C        EQUILIBRIUM DATA REDUCTION.  THE CONSTRAINT FUNCTIONS             C
C        RELATING THE TRUE VALUES OF THE MEASURED VARIABLES ARE            C
C        (1) PRESS = F(LIQ COMP,TEMP,PARAMETERS)                           C
C        (2) VAPOR COMP = G(LIQ COMP,TEMP,PARAMETERS)                      C
C        THIS REPRESENTS ESSENTIALLY A BUBBLE POINT-PRESSURE               C
C        CALCULATION USING K-FACTORS CALCULATED IN SUBROUTINE VPLQK        C
C                                                                          C
C     SUBROUTINES REQUIRED                                                 C
C        VPLQK                                                             C
C                                                                          C
C     PROGRAMED BY T.F.ANDERSON JAN 78.                                    C
C                                                                          C
C**************************************************************************C
C                                                                          C
      REAL K(2)
      DIMENSION X(2), Y(2)
      DIMENSION PR(5),XV(5)
      COMMON /ALL/ NN, LL, KK, IPRT, IST
      COMMON /LPRM/ PAR(5)
C
C     TRANSFER CURRENT PARAMETERS TO SUBROUTINE ACTIV
C
      DO 10 L=1,LL
   10 PAR(L) = PR(L)
C
C     INITIALIZE PARAMETERS FOR PRESSURE-VAPOR COMPOSITION CALCULATION
C
      P = A1
      T = XV(1)
      X(1) = XV(2)
      X(2) = 1.0 - X(1)
      Y(1) = A2
      Y(2) = 1.0 - Y(1)
C
C     THIS SECTION CALCULATES THE VAPOR COMPOSITION AND PRESSURE
C     CORRESPONDING TO A GIVEN LIQUID COMPOSITION AND TEMPERATURE.
C
C     INITIALIZE VARIABLES
C
      KEY = 1
      IT = 0
      EPS = 1.E-8
C
C     START ITERATIVE PROCEDURE.
C
   25 CONTINUE
      IT = IT + 1
      CALL VPLQK (KEY,P,T,X,Y,K,IRR)
      IF (IRR.GT.0) GO TO 50
      KEY = 3
      S = 0.0
      DO 30 I=1,2
      Y(I) = X(I)*K(I)
   30 S = S + Y(I)
C
C     NORMALIZE NEW X OR Y.
```

254

```
C
      DO 35 I=1,2
   35 Y(I) = Y(I)/S
C
C     CALCULATE NEW PRESSURE
C
      IF (S.LT.5.0.AND.S.GT.0.2) GO TO 45
      IF (S.LT.1.0) GO TO 40
      S = 5.0
      GO TO 45
   40 S = 0.2
   45 CONTINUE
      P = P*S
C
C     CHECK FOR CONVERGENCE
C
      IF (ABS(ALOG(S)).LT.EPS) GO TO 100
      IF (P.LT.0.0.OR.P.GT.2000.) GO TO 60
      IF (IT.GT.15) GO TO 55
      GO TO 25
   50 WRITE (6,3001)
 3001 FORMAT (/1X,*ERROR IN VPLQK*)
      GO TO 60
   55 WRITE (6,3002)
 3002 FORMAT (/1X,*MAX ITERATIONS EXCEEDED IN EVAL *)
      IRR = 2
   60 CONTINUE
      WRITE (6,3003) P, T, X(1), Y(1)
 3003 FORMAT (1H0,5HPRES=,F8.4,3X,5HTEMP=,F8.2,3X,5HX(1)=,F8.5,3X,
     1        5HY(1)=,F8.5)
      B1 = A1
      B2 = A2
      RETURN
  100 CONTINUE
      B1 = P
      B2 = Y(1)
      RETURN
      END
```

255

```
      SUBROUTINE VPLQK (KEY,P,T,X,Y,K,IRR)
C                                                                             C
C****************************************************************************C
C                                                                             C
C     SUBROUTINE VPLQK (KEY,P,T,X,Y,K,IRR)                                   C
C                                                                             C
C     PURPOSE                                                                C
C        CALCULATES K-FACTORS GIVEN P-T-X-Y                                  C
C                                                                             C
C     PARAMETERS                                                             C
C        KEY     - CNTRL VARIABLE.  IF (KEY.EQ.1) TEMP-DEPENDENT            C
C                  QUANTITIES ARE EVALUATED.  IF (KEY.GT.1) TEMP IS         C
C                  UNCHANGED AND THESE QUANTITIES ARE NOT RE-EVALUATED.     C
C        P       - PRESS, BARS                                              C
C        T       - TEMP, KELVIN                                             C
C        X       - LIQ PHASE MOLE FRACTION                                  C
C        Y       - VAPOR-PHASE MOLE FRACTION                                C
C        PHI     - FUGACITY COEFFICIENT                                     C
C        VLIQ    - SATURATED LIQUID MOLAR VOLUME, CC/GR-MOLE                C
C        FR      - ZERO PRESS REF FUGACITY IF IVAP.LE.2 OR VAPOR PRESS      C
C                  IF IVAP.EQ.3                                             C
C        K       - K-FACTOR                                                 C
C                                                                             C
C     SUBROUTINES REQUIRED                                                  C
C        ACTIV2, MVOLM, REFUG, PHIS2                                        C
C                                                                             C
C     PROGRAMED BY T.F.ANDERSON JAN 78.                                     C
C                                                                             C
C****************************************************************************C
C                                                                             C
      REAL K(2)
      DIMENSION PHI(2), X(2), Y(2)
      COMMON /MVLM/ VLIQ(2)
      COMMON /ATCF/ GAM(2)
      COMMON /REST/ FR(2), RT
      COMMON /THRM/ ILIQ, IVAP
      IF (KEY.GT.1) GO TO 20
      RT = T*83.1473
      CALL MVOLM (T)
      CALL ACTIV2 (T,X,IRR)
      IF (IRR.GT.0) RETURN
      CALL REFUG (T)
   20 CONTINUE
      IF (IVAP.EQ.3) GO TO 40
      CALL PHIS2 (KEY,Y,P,T,PHI,IRR)
      DO 30 I=1,2
   30 K(I) = GAM(I)*FR(I)*EXP(P*VLIQ(I)/RT)/(PHI(I)*P)
      RETURN
   40 CONTINUE
      DO 50 I=1,2
   50 K(I) = GAM(I)*FR(I)/P
      RETURN
      END
```

```
      SUBROUTINE REFUG (T)
C                                                                          C
C**************************************************************************C
C     SUBROUTINE REFUG (T)                                                 C
C                                                                          C
C     PURPOSE                                                              C
C        CALCULATE PURE COMPONENT LIQUID FUGACITY AT SPECIFIED TEMP        C
C        AND ZERO PRESSURE IF IVAP.LE.2 OR PURE COMPONENT VAPOR            C
C        PRESSURE IF IVAP.EQ.3                                             C
C                                                                          C
C     PARAMETERS                                                           C
C        FO      - CONSTANTS FOR ZERO PRESSURE REFERENCE FUGACITY          C
C                  EQUATION IF IVAP.EQ.1 OR CONSTANTS FOR VAPOR            C
C                  PRESSURE EQUATION IF IVAP.GT.1                          C
C        VLIQ    - SATURATED LIQUID MOLAR VOLUME, CC/GR-MOLE               C
C        FR      - ZERO PRESS REF FUGACITY IF IVAP.LE.2 OR VAPOR PRESS     C
C                  IF IVAP.EQ.3                                            C
C        PHS - FUGACITY COEFFICIENT AT SATURATION TEMP AND PRESS           C
C                                                                          C
C     SUBROUTINES REQUIRED                                                 C
C        PHIS2                                                             C
C                                                                          C
C     PROGRAMED BY T.F.ANDERSON JAN 78.                                    C
C                                                                          C
C**************************************************************************C
C                                                                          C
      DIMENSION Y(2), PHS(2)
      COMMON /MVLM/ VLIQ(2)
      COMMON /FUG/ FO(2,6)
      COMMON /REST/ FR(2), RT
      COMMON /THRM/ ILIQ, IVAP
      DO 10 I=1,2
      FR(I) = EXP(FO(I,1) + FO(I,2)/(T+FO(I,6)) + FO(I,3)*T +
     1         FO(I,4)*ALOG(T) + FO(I,5)*T**2)
      IF (IVAP.EQ.1.OR.IVAP.EQ.3) GO TO 10
      PS = FR(I)
      Y(1) = 1.0
      Y(2) = 0.0
      IF (I.EQ.1) GO TO 5
      Y(1) = 0.0
      Y(2) = 1.0
    5 CALL PHIS2 (1,Y,PS,T,PHS,IRP)
      FR(I) = PHS(I)*PS*EXP(-PS*VLIQ(I)/RT)
   10 CONTINUE
      RETURN
      END
```

```
      SUBROUTINE MVOLM (T)
C                                                                          C
C************************************************************************C
C                                                                          C
C     SUBROUTINE MVOLM (T)                                                 C
C                                                                          C
C     PURPOSE                                                              C
C        CALCULATE LIQUID MOLAR VOLUMES.                                   C
C                                                                          C
C     PARAMETERS                                                           C
C        PC    - PURE COMPONENT CRITICAL PRESS (BARS).                     C
C        T     - TEMP (KELVIN).                                            C
C        TC    - PURE COMPONENT CRITICAL TEMP. (KELVIN).                   C
C        VLIQ  - VECTOR OF LIQUID MOLAR VOLUMES (CC/G-MOLE).               C
C        ZRA   - RACKETT PARAMETER.                                        C
C        ALL OTHER CONSTANTS AND VARIABLES INTERNAL.                       C
C                                                                          C
C     METHOD                                                               C
C        MODIFIED RACKETT EQUATION FURTHER MODIFIED EMPIRICALLY FOR        C
C        T/TC .GT. 0.75 .                                                  C
C                                                                          C
C     REFERENCE                                                            C
C        SPENCER,C.F. AND P.P.DANNER, J CHEM ENG DATA, VOL 17, NO 2,       C
C        P236 (1972).                                                      C
C        COMMUNICATION PROF J.P.OCONNELL, DEPT CHEM ENG, UNIV FLORIDA.     C
C                                                                          C
C     PROGRAMED BY T.F.ANDERSON AUG 75.                                    C
C                                                                          C
C************************************************************************C
C                                                                          C
C                                                                          C
      COMMON /MVLM/ VLIQ(2)
      COMMON /PRM/ MW(2), TC(2), PC(2), VC(2), VSR(2), ZRA(2), RD(2),
     1             DM(2), ETA(3)
      DATA R,E/83.1473,0.28571429/
      DO 5 I=1,2
         TR = T/TC(I)
         IF (TR.GT.0.75) GO TO 3
         TAU = 1.0 + (1.0 - TR)**E
         GO TO 4
    3    TAU = 1.6 + 6.93026E-3/(TR - 0.555)
    4    VLIQ(I) = R*TC(I)*ZRA(I)**TAU/PC(I)
    5 CONTINUE
      RETURN
      END
```

```
      SUBROUTINE ACTIV2 (T,X,IRR)
C                                                                           C
C*********************************************************************C
C                                                                           C
C     SUBROUTINE ACTIV2 (T,X,IRR)                                           C
C                                                                           C
C     PURPOSE                                                               C
C         CALCULATES ACTIVITY COEFFICIENTS FOR A BINARY MIXTURE USING       C
C         1 OF 12 POSSIBLE EQUATIONS AS DETERMINED BY ILIQ                   C
C                                                                           C
C     PARAMETERS                                                            C
C         T      - TEMP , KELVIN                                            C
C         X      - LIQUID-PHASE MOLE FRACTION                               C
C         GAM    - ACTIVITY COEFFICIENTS                                    C
C         ILIQ   - INDICATOR FOR THE TYPE OF LIQUID-PHASE MODEL TO BE       C
C                  USED                                                     C
C                   1 - 2-PARAMETER UNIQUAC EQUATION                        C
C                   2 - 2-PARAMETER NRTL EQUATION                           C
C                   3 - 2-PARAMETER WILSON EQUATION                         C
C                   4 - 2-PARAMETER VAN LAAR EQUATION                       C
C                   5 - 2-PARAMETER HENRY,S CONSTANT                        C
C                   6 - 2-PARAMETER 3-SUFFIX MARGULES EQUATION              C
C                   7 - 3-PARAMETER UNIQUAC EQUATION                        C
C                   8 - 3-PARAMETER NRTL EQUATION                           C
C                   9 - 3-PARAMETER WILSON EQUATION                         C
C                  10 - 3-PARAMETER 4-SUFFIX MARGULES EQUATION              C
C                  11 - 4-PARAMETER 5-SUFFIX MARGULES EQUATION              C
C                  12 - 4-PARAMETER MODIFIED 5-SUFFIX MARGULES EQUATION     C
C                  13 - 5-PARAMETER MODIFIED MARGULES EQUATION              C
C                                                                           C
C     PROGRAMED BY T.F.ANDERSON JAN 78.                                     C
C                                                                           C
C*********************************************************************C
C                                                                           C
      REAL LG1, LG2
      COMMON /MVLM/ VLIQ(2)
      COMMON /THRM/ ILIQ, IVAP
      COMMON /LQPRM/ PAR(5)
      COMMON /SIZE/ R(2), Q(2), QP(2), CC
      COMMON /ATCF/ GAM(2)
      DIMENSION X(2)
      RT = 1.987*T
      IF (ILIQ.EQ.1.OR.ILIQ.EQ.6) GO TO 20
      IF (ILIQ.EQ.2.OR.ILIQ.EQ.8) GO TO 40
      IF (ILIQ.EQ.3.OR.ILIQ.EQ.9) GO TO 60
      IF (ILIQ.EQ.4) GO TO 80
      IF (ILIQ.EQ.5) GO TO 5
      IF (ILIQ.LE.13) GO TO 100
      WRITE (6,3001) ILIQ
 3001 FORMAT (1H0,50HLIQUID PHASE MODEL NOT AVAILABLE FOR ILIQ EQUAL TO,
     1        I3 )
      IRR = 1
      RETURN
C
C     FIT HENRY,S CONSTANT DATA
    5 CONTINUE
      GAM(1) = EXP(PAR(1) + PAR(2)/T)
      GAM(2) = 1.0
      RETURN
C
C     USE 2 OR 3 PARAMETER UNIQUAC EQUATION
C
   20 CONTINUE
      T12 = EXP(-PAR(1)/T)
```

```
          T21 = EXP(-PAR(2)/T)
          IF (ILIQ.EQ.7) CC = PAR(3)
          EL1 = 5.*(R(1)-Q(1)) - (R(1)-1.)
          EL2 = 5.*(R(2)-Q(2)) - (R(2)-1.)
          PHS = X(1)*R(1) + X(2)*R(2)
          PH1 = X(1)*R(1)/PHS
          PH2 = 1. - PH1
          THS = X(1)*Q(1) + X(2)*Q(2)
          THP1 = X(1)*QP(1)/(X(1)*QP(1) + X(2)*QP(2))
          THP2 = 1. - THP1
          S1 = THP2 + THP1*T12
          S2 = THP1 + THP2*T21
          LG1 = ALOG(R(1)/PHS) + 5.*Q(1)*ALOG(Q(1)*PHS/R(1)/THS) +
         1        PH2*(EL1 - EL2*R(1)/R(2))
          LG2 = ALOG(R(2)/PHS) + 5.*Q(2)*ALOG(Q(2)*PHS/R(2)/THS) +
         1        PH1*(EL2 - EL1*R(2)/R(1))
          LG1 = LG1 + CC*QP(1)*(-ALOG(S2) + THP2*(T21/S2 - T12/S1))
          LG2 = LG2 + CC*QP(2)*(-ALOG(S1) + THP1*(T12/S1 - T21/S2))
          GO TO 200
C
C     USE 2 OR 3 PARAMETER NRTL EQUATION
C
   40 CONTINUE
          T12 = PAR(1)/RT
          T21 = PAR(2)/RT
          IF (ILIQ.EQ.8) CC = PAR(3)
          G12 = EXP(-CC*T12)
          G21 = EXP(-CC*T21)
          GS12 = (X(2)+X(1)*G12)**2
          GS21 = (X(1)+X(2)*G21)**2
          LG1 = X(2)**2*(T21*G21**2/GS21 + T12*G12/GS12)
          LG2 = X(1)**2*(T12*G12**2/GS12 + T21*G21/GS21)
          GO TO 200
C
C     USE 2 OR 3 PARAMETER WILSON EQUATION
C
   60 CONTINUE
          T12 = VLIQ(2)/VLIQ(1)*EXP(-PAR(1)/RT)
          T21 = VLIQ(1)/VLIQ(2)*EXP(-PAR(2)/RT)
          IF (ILIQ.EQ.9) CC = PAR(3)
          TS12 = X(1) + X(2)*T12
          TS21 = X(2) + X(1)*T21
          LG1 = CC*(-ALOG(TS12) + X(2)*(T12/TS12 - T21/TS21))
          LG2 = CC*(-ALOG(TS21) + X(1)*(T21/TS21 - T12/TS12))
          GO TO 200
C
C     USE VAN LAAR EQUATION
C
   80 CONTINUE
          A12 = PAR(1)
          A21 = PAR(2)
          SUM = A12*X(1) + A21*X(2)
          LG1 = A12*(A21*X(2)/SUM)**2
          LG2 = A21*(A12*X(1)/SUM)**2
          GO TO 200
C
C     USE ONE OF SEVERAL FORMS OF THE MARGULES EQUATION
C
  100 CONTINUE
          A12 = PAR(1)
          A21 = PAR(2)
          LG1 = A12 + 2.*(A12-A21)*X(1)
          LG2 = A21 + 2.*(A12-A21)*X(2)
          IF (ILIQ.EQ.6) GO TO 130
```

```
      IF (ILIQ.GT.10) GO TO 120
      LG1 = LG1 - 2.*PAR(3)*X(1)
      LG2 = LG2 - 2.*PAR(3)*X(2)
      GO TO 180
120   CONTINUE
      B12 = PAR(3)
      B21 = PAR(4)
      IF (ILIQ.GT.11) GO TO 140
      CK = B12*X(1) + B21*X(2)
      LG1 = LG1 + 2.*CK*X(1)*(X(1)-X(2)) - B21*X(1)**2
      LG2 = LG2 + 2.*CK*X(2)*(X(2)-X(1)) - B12*X(2)**2
      GO TO 180
140   CONTINUE
      D = 0.0
      IF (ILIQ.EQ.13) D = PAR(5)
      E = B12*B21
      F = B12*X(1) + B21*X(2) + D*X(1)*X(2)
      LG1 = LG1 + E*X(1)/F*(-2.*X(2) + X(1)*(B12+D*X(2)**2)/F)
      LG2 = LG2 + E*X(2)/F*(-2.*X(1) + X(2)*(B21+D*X(1)**2)/F)
180   CONTINUE
      LG1 = LG1*X(2)**2
      LG2 = LG2*X(1)**2
200   CONTINUE
      GAM(1) = EXP(LG1)
      GAM(2) = EXP(LG2)
      RETURN
      END
```

```
      SUBROUTINE BIJS2 (KEY,T)
C                                                                            C
C****************************************************************************C
C                                                                            C
C     SUBROUTINE BIJS2 (KEY,T)                                               C
C                                                                            C
C     PURPOSE                                                                C
C        CALCULATE PURE COMPONENT AND CROSS SECOND VIRIAL COEFFICIENTS       C
C        FOR A BINARY MIXTURE AS A FUNCTION OF TEMPERATURE.                  C
C                                                                            C
C     NECESSARY INPUT PARAMETERS                                             C
C        DM(I)     -  DIPOLE MOMENT OF COMPONENT I (DEBYES).                 C
C        ETA(IJ)   -  ASSOCIATION OR SOLVATION PARAMETER FOR THE (I,J)       C
C                     PAIR                                                    C
C        KEY       -  CONTROL VARIABLE                                       C
C                     0 - ALL PARAMETERS ARE EVALUATED.                      C
C                     1 - TEMPERATURE DEPENDENT PARAMETERS ONLY ARE          C
C                         EVALUATED.                                         C
C                    -1 - TEMPERATURE INDEPENDENT PARAMETERS ONLY ARE        C
C                         CALCULATED.                                        C
C        PC(I)     -  CRITICAL PRESSURE OF COMPONENT I (BARS).               C
C        RD(I)        - MEAN RADIUS OF GYRATION OF COMPONENT I (A).          C
C        TC(I)     -  CRITICAL TEMPERATURE OF COMPONENT I (DEGREES K).       C
C        T         -  TEMPERATURE OF MIXTURE (DEGREES K).                    C
C                                                                            C
C     CALCULATED PARAMETERS                                                  C
C        A(IJ)     -  O.CONNELL A.                                           C
C        BD(IJ)    -  METASTABLE, BOUND, AND CHEMICAL CONTRIBUTIONS TO       C
C                     THE SECOND VIRIAL COEFFICIENTS, WHEN ICHM=1.           C
C        B(IJ)     -  TOTAL SECOND VIRIAL COEFFICIENT IF ICHM=0.             C
C                     FREE CONTRIBUTION ONLY TO THE SECOND VIRIAL            C
C                     COEFFICIENT WHEN ICHM=1.                               C
C        BO(IJ)    -  O.CONNELL BO.                                          C
C        DH(IJ)    -  O.CONNELL DELTA-H.                                     C
C        E(IJ)     -  PARAMETER USED TO CALCULATE PART OF CHEMICAL           C
C                     CONTRIBUTION TO THE SECOND VIRIAL COEFFICIENT.         C
C                     CALCULATED ONE OF TWO WAYS DEPENDING ON THE            C
C                     VALUE OF ETA(IJ).                                      C
C        EOK(IJ)   -  ENERGY PARAMETER DIVIDED BY BOLTXMAN CONSTANT.         C
C        ICHM      -  CONTROL PARAMETER NORMALLY ZERO WHICH IS SET           C
C                     EQUAL TO 1 WHEN ORGANIC ACIDS ARE PRESENT              C
C                     (ANY ETA(IJ).GE.4.5).                                  C
C        RDP(IJ)   -  REDUCED DIPOLE MOMENT PRIME.  ALSO, REDUCED            C
C                     DIPOLE MOMENT IN EARLIER PART OF THE PROGRAM.          C
C        W(IJ)     -  EFFECTIVE NON-POLAR ACENTRIC FACTOR.                   C
C                                                                            C
C     INTERNAL PARAMETERS                                                    C
C        TS        -  O.CONNELL T-STAR.                                      C
C        TSP       -  INVERSE OF O.CONNELL T-STAR PRIME.                     C
C        XI        -  O.CONNELL XI.                                          C
C        SGM(IJ)   -  SIGMA CUBED FOR AN IJ PAIR.                           C
C                                                                            C
C     METHOD                                                                 C
C        THE PREDICTIVE METHOD OF HAYDEN AND O.CONNELL IS USED.             C
C                                                                            C
C     REMARKS                                                                C
C        INITIAL CALL TO THIS SUBROUTINE MUST BE MADE WITH KEY.EQ.0 OR       C
C        -1 FOR A GIVEN MIXTURE.  REMAINDER OF CALLS WITH KEY=1 SINCE        C
C        ONLY THE TEMPERATURE DEPENDENT PARAMETERS NEED BE                   C
C        RECALCULATED.                                                       C
C                                                                            C
C     REFERENCES                                                             C
C        HAYDEN,J.G.,O.CONNELL,J.P./IND.ENG.CHEM.,PROC.DES.DEV.,             C
C        14,209(1975).                                                       C
```

```
C            NOTHNAGEL,K.H.,ABRAMS,D.S.,PRAUSNITZ,J.M./IND.ENG.CHEM.,     C
C         PROC.DES.DEV.,12,25(1973).                                      C
C            PRAUSNITZ,J.M./MOLECULAR THERMODYNAMICS OF FLUID PHASE       C
C         EQUILIBRIA, PRENTICE-HALL, ENGLEWOOD CLIFFS, N.J.(1969).        C
C                                                                         C
C         PROGRAMED BY T.F.ANDERSON FEB 77.                              C
C                                                                         C
C*************************************************************************C
C                                                                         C
      REAL MW
      DIMENSION SGM(3)
      COMMON /VIRIAL/ ICHM, B(3), BD(3)
      COMMON /PRM/ MW(2), TC(2), PC(2), VC(2), VSR(2), ZRA(2), RD(2),
     1             DM(2), ETA(3)
      COMMON /BSTR/ W(3), EOK(3), RDP(3), A(3), DH(3), E(3), BO(3)
C
C     CHECK CONTROL (KEY). IF -1 OR 0 CALCULATE TEMPERATURE INDEPENDENT
C     PARAMETERS.
C
      IF (KEY.GT.0) GO TO 45
      ICHM = 0
C
C     CALCULATE TEMP INDEPENDENT PURE COMP PARAMETERS.
C
      DO 15 I=1,2
         J = (I + 1)*I/2
         W(J) = RD(I)*(RD(I)*(-0.001366*RD(I) + 0.02096) + 0.006026)
         EOK(J) = TC(I)*(0.748 + 0.91*W(J))
         IF (ETA(J).GT.1.E-19) EOK(J) = EOK(J) - TC(I)*0.4*ETA(J)/
     1                                 (2. + 20.*W(J))
         SGM(J) = (2.4507 - W(J))**3*TC(I)/PC(I)
         IF (DM(I).GT.1.E-19) GO TO 5
         RDP(J) = 0.
         GO TO 15
    5    IF (DM(I).LT.1.45) GO TO 10
C
C     ANGLE AVERAGING IS USED TO ACCOUNT FOR THE EFFECT OF POLARITY
C     DUE TO A LARGE DIPOLE ON THE ENERGY AND SIZE PARAMETERS.
C
         PN = 16. + 400.*W(J)
         P1 = PN/(PN - 6.)
         P2 = 3./(PN - 6.)
         CONT = 2.882 - 1.882*W(J)/(0.03 + W(J))
         XI = 1.7941E07*DM(I)**4/(CONT*EOK(J)*SGM(J)**2*TC(I))
         EOK(J) = EOK(J)*(1. - XI*P1 + P1*(1. + P1)*XI*XI/2.)
         SGM(J) = SGM(J)*(1. + XI*P2)
   10    RDP(J) = 7243.8*DM(I)**2/(EOK(J)*SGM(J))
   15 CONTINUE
C
C     CALCULATE TEMP INDEP CROSS PARAMETERS.
C
      IJ = 2
      I = 1
      J = 3
      W(IJ) = 0.5*(W(I) + W(J))
      EOK(IJ) = 0.7*SQRT(EOK(I)*EOK(J)) + 0.6/(1./EOK(I) + 1./EOK(J))
      SGM(IJ) = SQRT(SGM(I)*SGM(J))
C
C     ANGLE AVERAGING IS USED TO ACCOUNT FOR AN INDUCED DIPOLE IN
C     POLAR-NONPOLAR PAIR.
C
      IF (DM(1).GT.1.E-19) GO TO 17
      RDP(IJ) = 0.
      IF (DM(2).LE.2.0) GO TO 25
```

263

```
          XI = DM(2)**2*(EOK(I)**2*SGM(I))**(1./3.)/(EOK(IJ)*SGM(J))
          GO TO 20
   17     IF (DM(2).GT.1.E-19) GO TO 23
          RDP(IJ) = 0.
          IF (DM(1).LE.2.0) GO TO 25
          XI = DM(1)**2*(EOK(J)**2*SGM(J))**(1./3.)/(EOK(IJ)*SGM(I))
   20     PN = 16. + 400.*W(IJ)
          P1 = PN/(PN - 6.)
          P2 = 3./(PN - 6.)
          SGM(IJ) = SGM(IJ)*(1. - XI*P2)
          EOK(IJ) = EOK(IJ)*(1. + XI*P1)
          GO TO 25
   23     RDP(IJ) = 7243.8*DM(1)*DM(2)/(EOK(IJ)*SGM(IJ))
   25 CONTINUE
C
C     CALCULATE REMAINDER OF TEMPERATURE INDEPENDENT PARAMETERS.
C
      DO 40 I=1,3
         BO(I) = 1.2618*SGM(I)
         A(I) = -0.3 - 0.05*RDP(I)
        DH(I) = 1.99 + 0.2*RDP(I)**2
C
C     CALCULATE THE MODIFIED REDUCED DIPOLE TO BE USED IN CALCULATING
C     THE FREE-POLAR CONTRIBUTION TO THE VIRIAL COEFFICIENT.
C
         IF (RDP(I).LT.0.25) GO TO 26
         RDP(I) = RDP(I) - 0.25
         GO TO 30
   26    IF (RDP(I).LT.0.04) GO TO 30
         RDP(I) = 0.
   30    CONTINUE
C
C     CALCULATE E(IJ) VALUES, DEPENDING ON ETA(IJ).   ICHM IS SET EQUAL
C     TO 1 IF ANY ETA(IJ).GE.4.5.
C
         E(I) = 0.0
         IF (ETA(I).LT.1.E-19) GO TO 40
         IF (ETA(I).GE.4.4999) GO TO 35
         E(I) = ETA(I)*(650./(EOK(I)+300.) - 4.27)
         GO TO 40
   35    E(I) = ETA(I)*(42800./(EOK(I)+22400.) - 4.27)
         ICHM = 1
   40 CONTINUE
      IF (KEY.LT.0) RETURN
C
C     CALCULATE THE TEMPERATURE DEPENDENT SECOND VIRIAL COEFFICIENTS.
C
   45 DO 60 I=1,3
         TS = T/EOK(I)
         TSP = 1./TS - 1.6*W(I)
C
C     CALCULATE THE FREE CONTRIBUTION TO THE SECOND VIRIAL COEFFICIENT.
C
         B(I) = 0.94 - 1.47*TSP - 0.85*TSP**2 + 1.015*TSP**3
         IF (RDP(I).LT.1.E-19) GO TO 50
         B(I) = B(I) - RDP(I)*(0.75 - 3.0*TSP + 2.1*TSP**2 +
     1         2.1*TSP**3)
   50    B(I) = BO(I)*B(I)
C
C     CALCULATE METASTABLE, BOUND, AND CHEMICAL CONTRIBUTIONS.
C
         BD(I) = A(I)*EXP(DH(I)/TS)
         IF (ETA(I).LT.1.E-19) GO TO 55
         BD(I) = BD(I) + EXP(E(I)) - EXP(1500.*ETA(I)/T + E(I))
```

```
55      BD(I) = BD(I)*BD(I)
        IF (ICHM.GT.0) GO TO 60
        B(I) = B(I) + BD(I)
60 CONTINUE
   RETURN
   END
```

```
      SUBROUTINE PHIS2 (KEY,Y,P,T,PHI,IRR)
C                                                                       C
C**********************************************************************C
C                                                                       C
C     SUBROUTINE PHIS2 (KEY,Y,P,T,PHI,IRR)                              C
C                                                                       C
C     PURPOSE                                                           C
C        CALCULATES VAPOR PHASE FUGACITY COEFFICIENTS FOR PURE          C
C        AND BINARY MIXTURES.                                           C
C                                                                       C
C     PARAMETERS                                                        C
C        B      - TOTAL SECOND VIRIAL COEFFICIENT, IF ICHM=0.           C
C                 FREE CONTRIBUTION ONLY TO SECOND VIRIAL COEFFICIENTS  C
C                 WHEN ICHM=1.                                          C
C        BD     - METASTABLE, BOUND, AND CHEM CONTRIBUTIONS TO          C
C                 SECOND VIRIAL COEFFICIENTS WHEN ICHM=1.               C
C        C      - EQUILIBRIUM CONSTANTS FOR THE CHEMICAL THEORY.        C
C        ICHM   - CNTRL VARIABLE SET BY BIJS2                           C
C                 0 - NO ORGANIC ACIDS                                  C
C                 1 - ONE OR BOTH COMP ARE ORGANIC ACIDS.              C
C        IRR    - ERROR FLAG INDICATING FAILURE OF SUBROUTINE MULLER    C
C                 TO CONVERGE (USED ONLY WHEN ORGANIC ACIDS PRESENT).   C
C        KEY    - CNTRL VARIABLE                                        C
C                 0 - BIJS2 CALLED AND ALL PARAMETERS CALC.             C
C                 1 - BIJS2 CALLED AND ONLY TEMP DEPENDENT PARAMETERS   C
C                     CALCULATED.                                       C
C                 2 - TEMP UNCHANGED, BIJS2 NOT CALLED, SECOND VIRIAL   C
C                     COEFFICIENTS ARE UNCHANGED.                       C
C        PHI    - OUTPUT OF CALCULATED FUGACITY COEFFICIENTS.           C
C        P      - PRESS (BARS).                                         C
C        T      - TEMP (K).                                             C
C        Y      - VAPOR-PHASE MOLE FRACTION (STOICHIOMETRIC).           C
C        Z      - TRUE VAPOR-PHASE MOLE FRACTION.  CALCULATED WHEN      C
C                 THE CHEMICAL THEORY IS USED.                          C
C                                                                       C
C     SUBROUTINES USED                                                  C
C        MULLER, BIJS2                                                  C
C                                                                       C
C     METHOD                                                            C
C        IF BINARY SYSTEM CONTAINS NO ORGANIC ACIDS, THE SECOND         C
C        VIRIAL COEFFICIENTS ARE USED IN A VOLUME EXPLICIT EQUATION     C
C        OF STATE TO CALCULATE THE FUGACITY COEFFICIENTS.  FOR          C
C        ORGANIC ACIDS FUGACITY COEFFICIENTS ARE PREDICTED FROM THE     C
C        CHEMICAL THEORY FOR NON-IDEALITY WITH EQUILIBRIUM CONSTANTS    C
C        OBTAINED FROM METASTABLE, BOUND, AND CHEMICAL CONTRIBUTIONS    C
C        TO THE SECOND VIRIAL COEFFICIENTS.                             C
C                                                                       C
C     REFERENCES                                                        C
C          HAYDEN,J.G.,O,CONNELL,J.P./IND.ENG.CHEM.,PROC.DES.DEV.,      C
C     14,209(1975).                                                     C
C          NOTHNAGEL,K.H.,ABRAMS,D.S.,PRAUSNITZ,J.M./IND.ENG.CHEM.,     C
C     PROC.DES.DEV.,12,25(1973).                                        C
C          PRAUSNITZ,J.M./MOLECULAR THERMODYNAMICS OF FLUID PHASE       C
C     EQUILIBRIA, PRENTICE-HALL, ENGLEWOOD CLIFFS, N.J.(1969).          C
C                                                                       C
C     PROGRAMED BY T.F.ANDERSON FEB 77. LOGIC DEVELOPED BY E.A.GRENS.   C
C                                                                       C
C**********************************************************************C
C                                                                       C
      DIMENSION PHI(2),Y(2)
      COMMON /VIRIAL/ ICHM, B(3), BD(3)
      DATA R/83.1473/
      PORT = P/R/T
      Y1 = Y(1)
```

```
      Y2 = Y(2)
C
C     CALCULATE SECOND VIRIAL COEFFICIENTS UNLESS TEMPERATURE IS
C     UNCHANGED.
C
      IF (KEY.GT.1) GO TO 5
      CALL BIJS2 (KEY,T)
    5 CONTINUE
C
C     IF THE CHEMICAL THEORY IS TO BE USED, PROCEED TO STATEMENT 30,
C     OTHERWISE USE THE VOLUME EXPLICIT EQUATION OF STATE
C
      IF (ICHM.GT.0) GO TO 30
C
C     CALCULATE FUGACITY COEFFICIENTS FOR NON-ASSOCIATING COMPONENTS.
C
      BM = Y1*Y1*B(1) + Y2*Y2*B(3) + 2.*Y1*Y2*B(2)
      PHI(1) = EXP((2.*(Y1*B(1) + Y2*B(2)) - BM)*PORT)
      PHI(2) = EXP((2.*(Y2*B(3) + Y1*B(2)) - BM)*PORT)
      RETURN
C
C     CALCULATE FUGACITY COEFFICIENTS FOR ASSOCIATING COMPONENTS WITH
C     CHEMICAL THEORY.  FIRST CALCULATE THE EQUILIBRIUM CONSTANTS.
C
   30 CONTINUE
      C1 = -BD(1)*PORT*EXP(PORT*B(1))
      C2 = -2.*BD(2)*PORT*EXP(PORT*(B(1) + B(3) - B(2)))
      C3 = -BD(3)*PORT*EXP(PORT*B(3))
C
C     CHECK FOR Y1 OR Y2 EQUAL TO ZERO.
C
      IF (Y1.GT.1.E-8) GO TO 35
      Z2 = (SQRT(1. + 4.*C3) - 1.)/(2.*C3)
      ZOY = (1.+C3*Z2**2)/(1.+C2*Z2)
      PHI(1) = EXP(B(1)*PORT)*ZOY
      PHI(2) = Z2*EXP(PORT*B(3))
      RETURN
   35 IF (Y2.GT.1.E-8) GO TO 40
      Z1 = (SQRT(1. + 4.*C1) - 1.)/(2.*C1)
      ZOY = (1.+C1*Z1**2)/(1.+C2*Z1)
      PHI(1) = Z1*EXP(PORT*B(1))
      PHI(2) = EXP(B(3)*PORT)*ZOY
      RETURN
   40 CONTINUE
      IF (C1.LT.0.0.OR.C2.LT.0.0.OR.C3.LT.0.0) GO TO 70
C
C     USE MULLER.S METHOD TO SOLVE NONLINEAR EQUATIONS FOR THE TRUE
C     VAPOR-PHASE COMPOSITION.
C
      CALL MULLER (Y2,C1,C2,C3,Z1,Z2,IR)
      IF (IR.GT.0) GO TO 75
C
C     CALCULATE FUGACITY COEFFICIENTS
C
      PHI(1) = Z1/Y1*EXP(PORT*B(1))
      PHI(2) = Z2/Y2*EXP(PORT*B(3))
      RETURN
C
C     SET UP ERROR RETURN
C
   70 WRITE (6,2001) C1,C2,C3
      PHI(1) = 1.0
      PHI(2) = 1.0
      IRR = 7
```

```
      RETURN
   75 WRITE (6,2002) IR
      PHI(1) = 1.0
      PHI(2) = 1.0
      IRR = 5
      RETURN
 2001 FORMAT (/1X,*EQULIBRIUM CONSTANTS ARE NEGATIVE*/1X,3G15.5)
 2002 FORMAT(/1X,*MULLER DID NOT CONVERGE - IT =*,I3)
      END
```

```
      SUBROUTINE MULLER (Y2,C1,C2,C3,Z1,Z2,IR)                            C
C                                                                         C
C*************************************************************************C
C                                                                         C
C     SUBROUTINE MULLER (Y2,C1,C2,C3,Z1,Z2,IR)                            C
C                                                                         C
C     PURPOSE                                                             C
C         ITERATIVELY SOLVES THE EQUILIBRIUM RELATIONS AND COMPUTES       C
C         THE EQUILIBRIUM VAPOR COMPOSISTION WHEN ORGANIC ACIDS ARE       C
C         PRESENT.  THESE COMPOSISTICNS ARE USED BY SUBROUTINE PHIS2 TO   C
C         CALCULATE FUGACITY COEFFICIENTS BY THE CHEMICAL THEORY.         C
C                                                                         C
C     PARAMETERS                                                          C
C         R      - RATIO OF VAPOR MOLE FRAC PRIOR TO ASSOCIATION.         C
C         Y1     - MOLE FRAC OF COMPONENT CNE.                            C
C         Y2     - MOLE FRAC CF COMPONENT TWO.                            C
C                  (Y2/Y1).                                               C
C         C1     - CONSTANT FOR DIMERIZATION OF COMP 1.                   C
C         C2     - CCNSTANT FOR ASSOCIATION OF COMP 1 WITH COMP 2.        C
C         C3     - CONSTANT FOR DIMERIZATION CF COMP 2.                   C
C         Z1     - CALCULATED EQUILIBRIUM MOLE FRAC COMP 1.               C
C         Z2     - CALCULATED EQUILIBRIUM MOLE FRAC COMP 2.               C
C         IRR    - ERROR FLAG INDICATING FAILURE TO CONVERGE.             C
C                                                                         C
C     METHOD                                                              C
C         MULLERS METHOD IS USED.                                         C
C                                                                         C
C     REFERENCES                                                          C
C         YOUNG,D.M., AND R.T.GREGORY, A SURVEY OF NUMERICAL MATHAMTICS,  C
C             VOL I, ADDISON-WESLEY PUBL CO, READING, MASS., 1972.        C
C         MULLER,D.E., MATH OF COMP, 10, 208 (1956).                      C
C                                                                         C
C     PROGRAMED BY T.F.ANDERSON OCT 75.                                   C
C                                                                         C
C*************************************************************************C
C                                                                         C
      SOLVE (AT,BT,CT) = (-BT + SQRT(BT**2 - 4.*AT*CT))/(2.*AT)
      FINAL(U,V) = U + V + C1*U*U + C3*V*V + C2*U*V - 1.0
      IR = 0
      IT = 0
      Y1 = 1. - Y2
      R = Y1/Y2
      Z11 = 2.*Y1
      IF (Z11.GT.1.0) Z11 = 1.0
      A = 2.*R*C3
      B = R + R*C2*Z11 - C2*Z11
      C = -Z11*(1. + 2.*C1*Z11)
      Z21 = SOLVE (A,B,C)
      IF (Z21.GT.(2.*Y2)) Z21 = 2.*Y2
      IF (Z21.GT.1.0) Z21 = 1.0
      R = 1./R
      A = 2.*R*C1
      B = R + R*C2*Z21 - C2*Z21
      C = -Z21*(1. + 2.*C3*Z21)
      Z11 = SOLVE (A,B,C)
      F1 = FINAL(Z11,Z21)
      Z22 = 0.0
      Z12 = 0.0
      F2 = -1.0
      Z2 = Z21/2.0
   50 IT = IT + 1
      A = 2.*R*C1
      B = R + R*C2*Z2 - C2*Z2
      C = -Z2*(1. + 2.*C3*Z2)
      Z1 = SOLVE (A,B,C)
```

269

```
      F = FINAL(Z1,Z2)
      IF (ABS(F).LT.1.E-6) RETURN
      IF (IT.GT.8) GO TO 70
      AA = Z21 - Z2
      BB = Z22 - Z2
      CC = (Z21 - Z22)*AA*BB
      FA = F1 - F
      FB = F2 - F
      AHT = (BB*FA - AA*FB)/CC
      BHT = (AA*AA*FB - BB*BB*FA)/CC
      CHT = F
      CONT = SQRT(BHT*BHT - 4.*AHT*CHT)
      D1 = (-BHT + CONT)/(2.*AHT)
      D2 = (-BHT - CONT)/(2.*AHT)
      ZNEW = D1 + Z2
      IF (ABS(D2).LT.ABS(D1)) ZNEW = D2 + Z2
      IF (F.LT.0.0) GO TO 60
      Z21 = Z2
      F1 = F
      GO TO 65
   60 Z22 = Z2
      F2 = F
   65 Z2 = ZNEW
      GO TO 50
   70 WRITE (6,2001) Z11,Z21,F1,Z12,Z22,F2,Z1,Z2,F
      IR = IT
 2001 FORMAT (/1X,9G13.5)
      RETURN
      END
```

```
         1         2         3         4         5         6         7         8
12345678901234567890123456789012345678901234567890123456789012345678901234567890

                         B L A N K
                         B L A N K
                         B L A N K
                         B L A N K
WATER    .2380    H2O   .615    18.020   647.37   221.20          56.0   WATER
      46.4                        1.83      .92      1.40         1.00   WATER
     .57041587918252E+02-.70048416152596E+04 .35888444369749E-02        FO1 WATER
    -.66689387843423E+01-.85054287344029E-060.                          FO2 WATER
ACTACD   .2240    C2H4O2  2.595   60.050   594.80   57.85         171.0  ACTACD
      158.0                        1.74      2.23     2.04                ACTACD
     .38697292580096E+03-.15090637989661E+05 .16773947362187E+00        FO1 ACTACD
    -.67641965668654E+02-.72738490059886E-040.                          FO2 ACTACD
1.70    2.50    4.50
WATER    ACTACD   16    P=760
ITO,T.,YOSHIDA,F./J.CHEM.ENG.DATA,8,315(1963).
760.0   116.5   0.022   0.058             WATER    FRMACD
760.0   114.6   0.054   0.123             WATER    FRMACD
760.0   113.4   0.086   0.168             WATER    FRMACD
760.0   113.5   0.099   0.183             WATER    FRMACD
760.0   113.1   0.101   0.188             WATER    FRMACD
760.0   110.6   0.189   0.298             WATER    FRMACD
760.0   107.8   0.303   0.433             WATER    FRMACD
760.0   106.1   0.413   0.545             WATER    FRMACD
760.0   104.4   0.522   0.649             WATER    FRMACD
760.0   103.1   0.624   0.735             WATER    FRMACD
760.0   102.3   0.696   0.792             WATER    FRMACD
760.0   101.6   0.778   0.851             WATER    FRMACD
760.0   100.8   0.876   0.914             WATER    FRMACD
760.0   100.5   0.923   0.944             WATER    FRMACD
760.0   100.4   0.945   0.960             WATER    FRMACD
760.0   100.1   0.985   0.989             WATER    FRMACD
1.0     0.05    0.001   0.003
-400.   800.

                         B L A N K
```

Example Calculation: Output

MAXIMUM LIKELIHOOD ESTIMATION OF PARAMETERS FROM VLE DATA

CONTROL PARAMETERS WERE SET AS FOLLOWS -

```
KK   = 2
LL   = 2
ILIQ= 1
IVAP= 1
IST =-0
IPRT= 2
ITMX=15
LMV = 1
LMP = 1
ALST=    1.00
BETA=     .25
PRCG= 1000.0
RP  =    1.50
SSTL= .00050
```

FINITE-DIFFERENCE INCREMENTS FOR VARIABLES

```
DX(1)= .5000
DX(2)= .0010
```

FINITE-DIFFERENCE INCREMENTS FOR PARAMETERS

```
DP(1)=1.0000
DP(2)=1.0000
```

Example Calculation: Output

2 COMPONENT SYSTEM

(1) WATER   H2O   WATER

(2) ACTACD C2H4O2 ACETIC ACID

PURE COMPONENT PROPERTIES

NO	MOL WT	TC	PC	VC	VSTP	ZRA	PD	DM	L	R	Q	QP
(1)	18.020	647.37	221.20	56.0	46.4	.2380	.615	1.83	-2.30	.92	1.40	1.00
(2)	60.050	594.80	57.85	171.0	158.0	.2240	2.595	1.74	-.28	2.23	2.04	2.04

CONSTANTS FOR ZERO PRESSURE REFERENCE FUGACITY EQUATION

(1) $.5704158791825E+02$ $-.7004841615259E+04$ $.35884443697409E-02$
  $-.6658938784343E+01$ $-.850542873440299E-06$

(2) $.3869772925809E+03$ $-.1509063789661E+05$ $.1677394736218E+00$
  $-.6764196566865E+02$ $-.727384990598E-04$

ASSOCIATION PARAMETERS

WATER   1.70
ACTACD 4.50

SOLVATION PARAMETERS

ACTACD/WATER   2.50

UNIQUAC, WILSON, OR NRTL THIRD PARAMETER SET EQUAL TO 1.00

BINARY VAPOR—LIQUID EQUILIBRIUM DATA   1   WATER   2   ACTACD

   REF   ITO,T.,YOSHIDA,F./J.CHEM.ENG.DATA,8,315(1963).

NO	P (MM)	T (C)	X1	X2	Y1	Y2
1	760.00	116.50	.0220	.9780	.0580	.9420
2	760.00	114.60	.0540	.9460	.1230	.8770
3	760.00	113.40	.0860	.9140	.1680	.8320
4	760.00	113.50	.0990	.9010	.1830	.8170
5	760.00	113.10	.1010	.8990	.1880	.8120
6	760.00	110.60	.1890	.8110	.2980	.7020
7	760.00	107.80	.3030	.6970	.4330	.5670
8	760.00	106.10	.4130	.5870	.5450	.4550
9	760.00	104.40	.5220	.4780	.6490	.3510
10	760.00	103.10	.6240	.3760	.7350	.2650
11	760.00	102.30	.6960	.3040	.7920	.2080
12	760.00	101.60	.7780	.2220	.8510	.1490
13	760.00	100.80	.8760	.1240	.9140	.0860
14	760.00	100.50	.9230	.0770	.9440	.0560
15	760.00	100.40	.9450	.0550	.9600	.0400
16	760.00	100.10	.9850	.0150	.9890	.0110

STANDARD DEVIATIONS

1	1.00	.05	.0010	.0030
2	1.00	.05	.0010	.0030
3	1.00	.05	.0010	.0030
4	1.00	.05	.0010	.0030
5	1.00	.05	.0010	.0030
6	1.00	.05	.0010	.0030
7	1.00	.05	.0010	.0030
8	1.00	.05	.0010	.0030
9	1.00	.05	.0010	.0030
10	1.00	.05	.0010	.0030
11	1.00	.05	.0010	.0030
12	1.00	.05	.0010	.0030
13	1.00	.05	.0010	.0030
14	1.00	.05	.0010	.0030
15	1.00	.05	.0010	.0030
16	1.00	.05	.0010	.0030

INITIAL PARAMETERS

   P1 =   100.000
   P2 =   200.000

ITERATION = 0   SUM OF THE SQUARES IS   .542482E+06

   PARAMETERS ARE —
   100.00000
   200.00000

Example Calculation: Output

REGRESSION SUMMARY

7 ITERATIONS REQUIRED

THE SUM OF THE SQUARES OF THE WEIGHTED RESIDUALS IS   326.86411

AND THE ESTIMATED VARIANCE OF THE FIT IS   23.34744

PARAMETERS AND ESTIMATES OF THEIR STANDARD DEVIATIONS

P1 =   -299.81 +/-   13.99

P2 =   530.66 +/-   41.25

COVARIANCE MATRIX

195.30714   -568.22587

-568.22587   1701.37321

CORRELATION COEFFICIENT MATRIX

1.00000   -.98574

-.98574   1.00000

Example Calculation: Output

MEASURED VALUES, CALCULATED VALUES, AND RESIDUALS (MEAS-CALC)

NO	PM	PC	PD	TM	TC	TD	X1M	X1C	X1D	Y1M	Y1C	Y1D
1	760.00	758.39	1.61	116.50	116.60	-.10	.0220	.0250	-.0030	.0580	.0543	.0037
2	760.00	756.87	3.13	114.60	114.79	-.19	.0540	.0584	-.0044	.1230	.1189	.0041
3	760.00	757.85	2.15	113.40	113.53	-.13	.0860	.0878	-.0018	.1660	.1698	-.0018
4	760.00	762.28	-2.28	113.50	113.36	.14	.0990	.0966	.0024	.1830	.1842	-.0012
5	760.00	759.91	.09	113.10	113.11	-.01	.1010	.1006	.0004	.1880	.1906	-.0026
6	760.00	762.36	-2.36	110.60	110.45	.15	.1890	.1854	.0036	.2980	.3117	-.0137
7	760.00	760.88	-.88	107.80	107.74	.06	.3030	.3009	.0021	.4330	.4466	-.0136
8	760.00	762.40	-2.40	106.10	105.94	.16	.4130	.4109	.0021	.5450	.5560	-.0110
9	760.00	760.23	-.23	104.40	104.38	.02	.5220	.5215	.0005	.6490	.6536	-.0046
10	760.00	758.51	1.49	103.10	103.20	-.10	.6240	.6243	-.0003	.7350	.7357	-.0007
11	760.00	757.35	2.65	102.30	102.48	-.18	.6960	.6968	-.0008	.7920	.7897	.0023
12	760.00	757.21	2.79	101.60	101.79	-.19	.7780	.7788	-.0008	.8510	.8480	.0030
13	760.00	756.77	3.23	100.80	101.02	-.22	.8760	.8765	-.0005	.9140	.9162	-.0022
14	760.00	757.44	2.56	100.50	100.67	-.17	.9230	.9232	-.0002	.9440	.9491	-.0051
15	760.00	758.27	1.73	100.40	100.52	-.12	.9450	.9451	-.0001	.9600	.9644	-.0044
16	760.00	759.16	.84	100.10	100.16	-.06	.9850	.9851	-.0001	.9890	.9912	-.0022

ROOT MEAN SQUARED DEVIATIONS

		2.12				.14			.0020			.0062

Example Calculation: Input

```
              1         2         3         4         5         6         7         8
     1234567890123456789012345678901234567890123456789012345678901234567890123456789012345678901234567890

3 2

                                  B L A N K
                                  B L A N K
                                  B L A N K

ACETONE                 C3H60     58.900    509.10    47.60    211.0    ACETON
     200.0       .2470  2.740     2.86      2.57      2.34              ACETON
  -.16019809744826E+02-.34511875839852E+04-.13784672778720E-01                 VP1 ACETON
  +.53569233060743E+01                                                           VP2 ACETON
BENZENE                 C6H6      78.110    562.16    48.98    260.0    BENZEN
     255.0       .2696  3.004     0.        3.19      2.40              BENZEN
  +.10620674022835E+03-.72779401204870E+04+.16575564030299E-01                 VP1 BENZEN
  -.15586107092867E+02                                                           VP2 BENZEN
0.90      0.50      0.0
ACETON    BENZEN    11        T=45
BROWN,I.,SMITH,F./AUSTR.J.CHEM.10,423(1957).
250.73    45.00     .0470     .1444              ACETON    BENZEN
275.02    45.00     .0963     .2574              ACETON    BENZEN
324.25    45.00     .2207     .4417              ACETON    BENZEN
348.40    45.00     .2936     .5204              ACETON    BENZEN
379.88    45.00     .4011     .6139              ACETON    BENZEN
399.73    45.00     .4759     .6697              ACETON    BENZEN
432.95    45.00     .6125     .7614              ACETON    BENZEN
453.99    45.00     .7045     .8201              ACETON    BENZEN
475.39    45.00     .8081     .8805              ACETON    BENZEN
495.32    45.00     .9084     .9418              ACETON    BENZEN
503.96    45.00     .9519     .9699              ACETON    BENZEN

1.0       0.05      0.001     0.003
300.      -50.

                                  B L A N K
```

277

Example Calculation: Output

```
MAXIMUM LIKELIHOOD ESTIMATION OF PARAMETERS FROM VLE DATA

CONTROL PARAMETERS WERE SET AS FOLLOWS -

KK  = 2
LL  = 2
ILIQ= 3
IVAP= 2
IST =-0
IPRT= 2
ITMX=15
LMV = 1
LMP = 1
ALST=    1.00
BETA=     .25
PRCG= 1000.0
RP  =    1.50
SSTL= .00050

FINITE-DIFFERENCE INCREMENTS FOR VARIABLES

DX(1)= .5000
DX(2)= .0010

FINITE-DIFFERENCE INCREMENTS FOR PARAMETERS

DP(1)=1.0000
DP(2)=1.0000
```

Example Calculation: Output

2 COMPONENT SYSTEM

    (1)   ACETON   C3H6O   ACETONE

    (2)   BENZEN   C6H6   BENZENE

PURE COMPONENT PROPERTIES

NO	MOL WT	TC	PC	VC	VSTR	ZRA	RD	DM	L	R	Q	QP
(1)	58.800	509.10	47.60	211.0	200.0	.2470	2.740	2.86	-.42	2.57	2.34	2.34
(2)	78.110	562.16	48.98	260.0	255.0	.2695	3.004	0.	1.76	3.19	2.40	2.40

CONSTANTS FOR VAPOR PRESSURE EQUATION

(1) -.16019809744826E+02   -.34511875838852E+04   -.13784672778720E-01
      .53569233060743E+01   -0.

(2)   .10620674022835E+03   -.72779401204870E+04   .16575564030299E-01
     -.15586107092867E+02   -0.

ASSOCIATION PARAMETERS

    ACETON   .90
    BENZEN   0.

SOLVATION PARAMETERS

    BENZEN/ACETON   .50

UNIQUAC, WILSON, OR NRTL THIRD PARAMETER SET EQUAL TO 1.00

279

Example Calculation: Output

BINARY VAPOR-LIQUID EQUILIBRIUM DATA   1 ACETON   2 BENZEN

REF   BROWN,I.,SMITH,F./AUSTR.J.CHEM.10,423(1957).

NO	P (MM)	T (C)	X1	X2	Y1	Y2
1	250.73	45.00	.0470	.9530	.1444	.8556
2	275.02	45.00	.0963	.9037	.2574	.7426
3	324.25	45.00	.2207	.7793	.4417	.5583
4	348.40	45.00	.2936	.7064	.5204	.4796
5	379.88	45.00	.4011	.5989	.6139	.3861
6	399.73	45.00	.4759	.5241	.6697	.3303
7	432.95	45.00	.6125	.3875	.7614	.2386
8	453.99	45.00	.7045	.2955	.8201	.1799
9	475.39	45.00	.8081	.1919	.8805	.1195
10	495.32	45.00	.9084	.0916	.9418	.0582
11	503.96	45.00	.9519	.0481	.9699	.0301

STANDARD DEVIATIONS

1	1.00	.05	.0010		.0030
2	1.00	.05	.0010		.0030
3	1.00	.05	.0010		.0030
4	1.00	.05	.0010		.0030
5	1.00	.05	.0010		.0030
6	1.00	.05	.0010		.0030
7	1.00	.05	.0010		.0030
8	1.00	.05	.0010		.0030
9	1.00	.05	.0010		.0030
10	1.00	.05	.0010		.0030
11	1.00	.05	.0010		.0030

INITIAL PARAMETERS

P1 =   300.000
P2 =   -50.000

ITERATION = 0   SUM OF THE SQUARES IS   326.393259

PARAMETERS ARE -
300.00000
-50.00000

DELTA LINEARIZED SUM OF SQUARES = -313.06124

CHANGES IN PARAMETERS
116.95516
-61.51710

AL   SSQN

1.00000   20.03465

Example Calculation: Output

REGRESSION SUMMARY

3 ITERATIONS REQUIRED

THE SUM OF THE SQUARES OF THE WEIGHTED RESIDUALS IS   12.17879

AND THE ESTIMATED VARIANCE OF THE FIT IS   1.35320

PARAMETERS AND ESTIMATES OF THEIR STANDARD DEVIATIONS

P1 =   410.28 +/-   35.95

P2 =   -100.09 +/-   27.52

COVARIANCE MATRIX

1292.15367   -981.65836

-981.65836   757.40196

CORRELATION COEFFICIENT MATRIX

1.00000   -.99229

-.99229   1.00000

Example Calculation: Output

MEASURED VALUES, CALCULATED VALUES, AND RESIDUALS (MEAS-CALC)

NO	PM	PC	PD	TM	TC	TD	X1M	X1C	X1D	Y1M	Y1C	Y1D
1	250.73	250.00	.73	45.00	45.02	-.02	.0470	.0472	-.0002	.1444	.1451	-.0007
2	275.02	274.36	.66	45.00	45.02	-.02	.0963	.0966	-.0003	.2574	.2575	-.0001
3	324.25	324.32	-.07	45.00	45.00	.00	.2207	.2202	.0005	.4417	.4454	-.0037
4	348.40	348.71	-.31	45.00	44.99	.01	.2936	.2932	.0004	.5204	.5234	-.0030
5	379.88	380.20	-.32	45.00	44.99	.01	.4011	.4008	.0003	.6139	.6158	-.0019
6	399.73	399.88	-.15	45.00	44.99	.01	.4759	.4758	.0001	.6697	.6707	-.0010
7	432.95	432.64	.31	45.00	45.01	-.01	.6125	.6126	-.0001	.7614	.7605	.0009
8	453.99	453.18	.81	45.00	45.03	-.03	.7045	.7049	-.0004	.8201	.8172	.0029
9	475.39	474.62	.77	45.00	45.03	-.03	.8081	.8083	-.0002	.8805	.8799	.0006
10	495.32	494.49	.83	45.00	45.04	-.04	.9084	.9086	-.0002	.9418	.9416	.0002
11	503.96	502.97	.99	45.00	45.05	-.05	.9519	.9521	-.0002	.9699	.9691	.0008

ROOT MEAN SQUARED DEVIATIONS

		.62			.03			.0003			.0019

282

Example Calculation: Input

```
10  1

                    BLANK
                    BLANK
                    BLANK
ETHANOL       .2520      C2H6O      46.070     516.26     63.80     167.0    ETHNOL
     154.0               2.250               2.11        1.97              .92      ETHNO3
-.90909918893565E+02-.34658734683496E+04-.6230139174O768E-01              FC1  ETHNOL
 .20486492956207E+02 .20664221042773E-040.                                FO2  ETHNOL
ACETONITRILE             C2H3N       41.050     548.00     48.30     173.0    ACTNTR
     150.0       .2000      1.821       3.94        1.87        1.72              FO1  ACTNTR
 .41946799514893E+03-.14100072119555E+05 .18197343710789F+00              FO2  ACTNTR
-.74005889434l5E+02-.7813428l738740E-040.
1.40     1.50      1.65       T=20
ETHNOL      ACTNTR      8
VIERK,A.L./Z.ANORG.CHEM.261,283(1950).
75.0     20.0      0.050     0.111              ETHNOL      ACTNTR
77.8     20.0      0.090     0.180              ETHNOL      ACTNTR
81.0     20.0      0.190     0.289              ETHNOL      ACTNTR
81.9     20.0      0.320     0.350              FTHNOL      ACTNTR
81.5     20.0      0.500     0.400              ETHNOL      ACTNTR
80.5     20.0      0.590     0.405              ETHNOL      ACTNTR
78.0     20.0      0.680     0.450              ETHNOL      ACTNTR
72.5     20.0      0.810     0.517              ETHNOL      ACTNTR
1.0      0.05      0.001     0.003
1.0      1.0       0.0       0.0

                    BLANK
```

Example Calculation: Output

MAXIMUM LIKELIHOOD ESTIMATION OF PARAMETERS FROM VLE DATA

CONTROL PARAMETERS WERE SET AS FOLLOWS −

```
KK   = 2
LL   = 3
ILIQ=10
IVAP= 1
IST  = 1
IPRT= 2
ITMX=15
LMV  = 1
LMP  = 1
ALST=    1.00
BETA=     .25
PRCG= 1000.0
RP  =    1.50
SSTL= .00050
```

FINITE−DIFFERENCE INCREMENTS FOR VARIABLES

```
DX(1)= .5000
DX(2)= .0010
```

FINITE−DIFFERENCE INCREMENTS FOR PARAMETERS

```
DP(1)= .0010
DP(2)= .0010
DP(3)= .0010
```

Example Calculation: Output

2 COMPONENT SYSTEM

   (1)   ETHNOL   C2H6O   ETHANOL

   (2)   ACTNTR   C2H3N   ACETONITRILE

PURE COMPONENT PROPERTIES

NO	MOL WT	TC	PC	VC	VSTR	ZRA	RD	DM	L	R	Q	QP
(1)	46.070	516.26	63.80	167.0	154.0	.2520	2.250	1.69	-.41	2.11	1.97	.92
(2)	41.050	548.00	48.30	173.0	150.0	.2000	1.821	3.94	-.12	1.87	1.72	1.72

CONSTANTS FOR ZERO PRESSURE REFERENCE FUGACITY EQUATION

(1)	$-.9090991889565E+02$	$-.3465873468496E+04$	$-.6230139174075 8E-01$
	$.2048649295 6207E+02$	$.2066422104 2773E-04$	
(2)	$.4194679951 4893E+03$	$-.1410007211 9555E+05$	$.1819734371078 9E+00$
	$-.7400598894 3416E+02$	$-.7813428173974 0E-04$	

ASSOCIATION PARAMETERS

   ETHNOL   1.40

   ACTNTR   1.65

SOLVATION PARAMETERS

   ACTNTR/ETHNOL   1.50

UNIQUAC, WILSON, OR NRTL THIRD PARAMETER SET EQUAL TO 1.00

BINARY VAPOR-LIQUID EQUILIBRIUM DATA   1 ETHNOL   2 ACTNTR

    REF   VIERK,A.L./Z.ANORG.CHEM.261,283(1950).

NO	P (MM)	T (C)	X1	X2	Y1	Y2
1	75.00	20.00	.0500	.9500	.1110	.8890
2	77.80	20.00	.0900	.9100	.1800	.8200
3	81.00	20.00	.1900	.8100	.2890	.7110
4	81.90	20.00	.3200	.6800	.3500	.6500
5	81.50	20.00	.5000	.5000	.4000	.6000
6	80.50	20.00	.5900	.4100	.4050	.5950
7	78.00	20.00	.6800	.3200	.4500	.5500
8	72.50	20.00	.8100	.1900	.5170	.4830

STANDARD DEVIATICNS

1	1.00	.05	.0010		.0030
2	1.00	.05	.0010		.0030
3	1.00	.05	.0010		.0030
4	1.00	.05	.0010		.0030
5	1.00	.05	.0010		.0030
6	1.00	.05	.0010		.0030
7	1.00	.05	.0010		.0030
8	1.00	.05	.0010		.0030

INITIAL PARAMETERS

    P1 =   1.000
    P2 =   1.000
    P3 =   0.

ITERATION = 0   SUM OF THE SQUARES IS   470.737520

    PARAMETERS ARE -
    1.00000
    1.00000
    0.

DELTA LINEARIZED SUM OF SQUARES = -470.26189

CHANGES IN PARAMETERS
        .59002
        .64000
        .13004

    AL   SSON

    1.00000   10.89471

ITERATION = 1   SUM OF THE SQUARES IS   10.894710

Example Calculation: Output

REGRESSION SUMMARY

4 ITERATIONS REQUIRED

THE SUM OF THE SQUARES OF THE WEIGHTED RESIDUALS IS   .42242

AND THE ESTIMATED VARIANCE OF THE FIT IS   .84484E-01

PARAMETERS AND ESTIMATES OF THEIR STANDARD DEVIATIONS

P1 =   1.47 +/-   .02

P2 =   1.51 +/-   .02

P3 =   .04 +/-   .03

COVARIANCE MATRIX

.38714E-03   .21844E-03   .52186E-03

.21844E-03   .38106E-03   .24902E-03

.52186E-03   .24902E-03   .76477E-03

CORRELATION COEFFICIENT MATRIX

1.00000   .56872   .95909

.56872   1.00000   .46129

.95909   .46129   1.00000

Example Calculation: Output

MEASURED VALUES, CALCULATED VALUES, AND RESIDUALS (MEAS-CALC)

NO	PM	PC	PD	TM	TC	TD	X1M	X1C	X1D	Y1M	Y1C	Y1D
1	75.30	74.88	.12	20.00	20.00	-.00	.0500	.0500	-.0000	.1110	.1095	.0015
2	77.80	77.61	.19	20.00	20.00	-.00	.0900	.0900	-.0000	.1800	.1699	.0102
3	81.00	81.14	-.14	20.00	20.00	.00	.1900	.1900	.0000	.2900	.2645	.0245
4	81.90	82.15	-.25	20.00	20.00	.00	.3200	.3200	-.0000	.3500	.3301	.0199
5	81.50	81.26	.24	20.00	20.00	-.00	.5000	.5000	.0000	.4000	.3833	.0167
6	80.50	80.23	.27	20.00	20.00	-.00	.5900	.5900	.0000	.4050	.4079	-.0029
7	78.00	78.37	-.37	20.00	20.00	.00	.6800	.6800	-.0000	.4500	.4389	.0111
8	72.50	72.43	.07	20.00	20.00	-.00	.9100	.9100	.0000	.5170	.5166	.0004

ROOT MEAN SQUARED DEVIATIONS

			.22			.00			.0000			.0138

Appendix E

DESCRIPTIONS AND LISTINGS OF SUBROUTINES FOR CALCULATION OF THERMODYNAMIC PROPERTIES

The computer subroutines for calculation of vapor-phase and liquid-phase fugacity (activity) coefficients, reference fugacities, and molar enthalpies, as well as vapor-liquid and liquid-liquid equilibrium ratios, are described and listed in this Appendix. These are source routines written in American National Standard FORTRAN (FORTRAN IV), ANSI X3.9-1978, and, as such, should be compatible with most computer systems with FORTRAN IV compilers. Approximate storage requirements and CDC 6400 execution times for these subroutines are given in Appendix J.

These subroutines are all capable of treating multi-component systems with up to 20 components, and the lower level routines not usually referenced by the user (PHIS, BIJS, PURF, GAMMA, TAUS), use vectors of length 20 for compositions, activity coefficients, etc., in their argument list. The higher level routines, which may be frequently referenced by some users (VALIK, LILIK, ENTH), use vectors of length N in their argument list, where N ($\leq$ 20) is the number of components involved. This provides compatibility with user-written main programs having differing vector lengths.

Also described in this section are the labeled common storage blocks associated with the thermodynamic subroutines.

SUBROUTINE VALIK (N, ID, KEY, X, Y, T, P, K, ERR)

Purpose

VALIK calculates vapor-liquid vaporization equilibrium
ratios, K(I), for each component in a mixture of N components
(N $\leq$ 20) at specified liquid composition, vapor composition,
temperature, and pressure.

Method

Equilibrium ratios are calculated from Equation (7-5),

$$K_i = y_i/x_i = \frac{\gamma_i f_i^{OL}}{\phi_i P}$$

with the activity coefficients γ_i, standard state fugacity, f_i^{OL},
and fugacity coefficient ϕ_i, determined at the specified conditions by appropriate subroutine calls.

Arguments

Input: N   number of components (N $\leq$ 20).

ID(I) vector of indices for the components (I = 1, N).

KEY   integer for initialization control.

1: Initial calculation with new system

2: Subsequent calculation, all conditions changed

3: Subsequent calculation, only X and Y changed

4: Subsequent calculation, only T and P changed

5: Subsequent calculation, only Y, T, and P changed

6(7): Same as 2(4) but with determination of
virial coefficient derivatives for enthalpy
calculations

8: Same as 2 but with virial coefficients already
available

9(10): Same as 1(2) for use when ENTH is called
without VALIK being called.

X(I)   vector of liquid-phase mole fractions of the components (I = 1, N).

Y(I)   vector of vapor-phase mole fractions of the components (I = 1, N).

T   temperature (K)

P   pressure (bars)

Output: K(I)   vector of vaporization equilibrium ratios for all
               components (I = 1, N).

        ERR   error flag, integer variable normally zero;   ERR =
               1 indicates parameters are not available for one or
               more binary pairs in the mixture; ERR = 2 indicates
               that no solution was obtained.

Subroutine References

Calls: PHIS
              PURF
              GAMMA
Called by: FLASH
              BUDET
              BUDEP

Common Storage

None

Glossary of Principal Variable Names

GAM(I)   Vector (length 20) of activity coefficients (I = 1, N).

PHI(I)   Vector (length 20) of fugacity coefficients (I = 1, N).

SX(I)   Vector (length 20) of liquid-phase mole fractions
         (I = 1, N).

YF(I)   Vector (length 20) of vapor-phase mole fractions (I =
         1, N).

Listing

```
      SUBROUTINE VALIK(N,ID,KEY,X,Y,T,P,K,ERR)
* VALIK CALCULATES VAPORIZATION EQUILIBRIUM COEFFICIENTS ,K, FOR ALL N
* COMPONENTS (N.LE.20) WHOSE INDICES APPEAR IN VECTOR ID, GIVEN
* TEMPERATURE T(K), PRESSURE P(BAR), AND ESTIMATES OF PHASE COMPOSITION
* X AND Y (USED WITHOUT CORRECTION IN EVALUATION OF ACTIVITIES, ETC).
* CALCULATIONS ARE AN IMPLEMENTATION OF THE UNIQUAC MODEL.
* VALIK NORMALLY RETURNS ERR=0, BUT IF COMPONENT COMBINATIONS LACKING
* DATA ARE INVOLVED IT RETURNS ERR=1 AND IF NO SOLUTION IS FOUND IT
* RETURNS ERR=2.  KEY SHOULD BE 1 ON INITIAL CALL FOR A SYSTEM, 2 ON
* SUBSEQUENT CALLS WHEN ALL VARIABLES ARE CHANGED, 3 IF ONLY COMPOSITION
* IS CHANGED, AND 4 IF ONLY T (AND P) IS CHANGED.  KEY IS 6 OR 7,
* INSTEAD OF 2 OR 4, IF ENTH CALLS ARE TO BE MADE FOR SAME CONDITIONS.
      REAL X(N),Y(N),K(N),PHI(20),GAM(20),XF(20),YF(20)
      INTEGER ID(N),ERR,ER,ERG,IDF(20)
      COMMON/VAL/FIP(20)
  100 ERR=0
* CONVERT VECTORS TO DIMENSION 20 TO MATCH LOWER LEVEL SUBROUTINES
  101 DO 102 I=1,N
      XF(I)=X(I)
      YF(I)=Y(I)
  102 IDF(I)=ID(I)
```

```
*  GET VAPOR PHASE FUGACITY COEFFICIENTS, PHI
   110 CALL PHIS(N,IDF,KEY,YF,T,P,PHI,ER)
       IF(ER.GT.1) GO TO 900
       GO TO(120,120,130,120,120,120,120,120,120,120),KEY
*  GET PURE COMPONENT LIQUID FUGACITIES, FIP
   120 CALL PURF(N,IDF,T,P,FIP)
*  GET LIQUID PHASE ACTIVITY COEFFICIENTS, GAM
   130 CALL GAMMA(N,IDF,KEY,XF,T,GAM,ERG)
*  CALCULATE VAPORIZATION EQUILIBRIUM RATIOS, K
   140 DO 149 I=1,N
       K(I)=GAM(I)*FIP(I)/(PHI(I)*P)
       IF(K(I).LE.0..OR.K(I).GT.1.E19) GO TO 900
   149 CONTINUE
       ERR=ERG
       RETURN
*  ON FAILURE TO FIND PHI SET K TO ZERO, ERR TO 2.
   900 ERR=ER
   905 DO 906 I=1,N
   906 K(I)=0.
       RETURN
       END
```

SUBROUTINE LILIK (N, ID, KEY, XR, XE, T, K, GAR, GAE, ERR)

Purpose
LILIK calculates liquid-liquid equilibrium ratios, K(I), for each component in a mixture of N components (N $\leq$ 20) at specified temperature and liquid-phase compositions.

Method
Equilibrium ratios are determined from the ratio of the activity coefficients, $K_i = \gamma_i^R/\gamma_i^E$. The activity coefficients are determined at the specified conditions by appropriate subroutine calls.

Arguments
Input: N   number of components (N $\leq$ 20)

ID(I)   vector of indices for the components (I = 1,N)

KEY   integer initialization control variable (KEY = 1 . . . 10) see VALIK

XR(I)   vector of raffinate-phase mole fractions of the components (I = 1,N)

XE(I)   vector of extract-phase mole fractions of the components (I = 1,N)

T   temperature (K)

Output: K(I)   vector of equilibrium ratios for all components (I = 1,N)

GAR(I)   vector of activity coefficients for the raffinate phase (I = 1,N)

GAE(I)   vector of activity coefficients for the extract phase (I = 1,N)

ERR   error flag, integer variable normally zero; ERR = 1 indicates parameters are not available for one or more binary pairs in the mixture, ERR = 2 indicates the calculated K value is out of range.

Subroutine References
Calls: GAMMA
Called by: ELIPS

Common Storage
None

Glossary of Principal Variable Names

GX(I)   Vector (length 20) of raffinate phase activity coefficients (I = 1,N)

GY(I)   Vector (length 20) of extract phase activity coefficients
(I = 1,N)

X(I)   Vector (length 20) of raffinate phase mole fractions
(I = 1,N)

Y(I)   Vector (length 20) of extract phase mole fractions
(I = 1,N)

Listing

```
      SUBROUTINE LILIK(N,ID,KEY,XR,XE,T,K,GAR,GAE,ERR)
* LILIK CALCULATES TWO PHASE EFFECTIVE LIQUID EQUILIBRIUM RATIOS K
* FOR ALL N COMPONENTS (N.LE.20) WHOSE INDICES APPEAR IN VECTOR ID,
* GIVEN TEMPERATURE T(K) AND ESTIMATES OF PHASE COMPOSITIONS XR AND XE
* (USED WITHOUT CORRECTION TO EVALUATE ACTIVITY COEFFICIENTS GAR AND
* GAE). LILIK NORMALLY RETURNS ERR=0, BUT IF COMPONENT COMBINATIONS
* LACKING DATA ARE INVOLVED IT RETURNS ERR=1, AND IF A K IS OUT OF RANGE
* THEN ERR=2.  KEY SHOULD BE 1 ON INITIAL CALL FOR A SYSTEM, 2 (OR 6)
* ON SUBSEQUENT CALLS WHEN ALL VARIABLES ARE CHANGED, 3 IF ONLY
* COMPOSITIONS ARE CHANGED, AND 4 (OR 7) IF ONLY T IS CHANGED.
      REAL XR(N),XE(N),K(N),GAR(N),GAE(N),X(20),Y(20),GX(20),GY(20)
      INTEGER ID(N),IDF(20),ERR,ERG
  100 ERR=0
* CONVERT COMPOSITION VECTORS TO DIMENSION 20 TO MATCH LOWER LEVEL
* SUBROUTINES
  101 DO 102 I=1,N
      X(I)=XR(I)
      Y(I)=XE(I)
  102 IDF(I)=ID(I)
* GET R AND E PHASE ACTIVITY COEFFICIENTS
      CALL GAMMA(N,IDF,KEY,X,T,GX,ERG)
      CALL GAMMA(N,IDF,3,Y,T,GY,ERG)
* CONVERT ACTIVITY COEFFICIENT VECTORS TO DIMENSION N
  110 DO 119 I=1,N
      GAR(I)=GX(I)
      GAE(I)=GY(I)
      K(I)=GAR(I)/GAE(I)
      IF(K(I).LE.0..OR.K(I).GT.1.E+19) GO TO 900
  119 CONTINUE
      ERR=ERG
      RETURN
* FOR A K VALUE OUT OF RANGE SET ERR TO 2
  900 ERR=2
      DO 905 I=1,N
  905 K(I)=0.
      RETURN
      END
```

SUBROUTINE ENTH (N, ID, KEY, LEV, ZF, T, P, H, ERE)

Purpose

ENTH calculates vapor or liquid enthalpies, J/mole, (reference, ideal gas at 300 K) for a mixture of N components (N $\leq$ 20) at specified temperature, pressure, and liquid or vapor composition.

Method

Enthalpies are evaluated directly using the methods presented in Chapter 5 and Appendices A and B.

Arguments

Input:
- N   number of components (N $\leq$ 20)
- ID(I)   Vector of indices for the components (I = 1, N)
- KEY   integer initialization control variable (KEY = 1,...,10); see VALIK
- LEV   integer variable indicating type of calculation; LEV = 0, liquid enthalpy is calculated with excess enthalpy of mixing taken as zero; LEV = 1, liquid enthalpy is calculated with excess enthalpy of mixing determined from UNIQUAC equation; LEV = 2, vapor enthalpy is calculated.
- ZF(I)   vector of liquid or vapor stream mole fractions (I = 1, N)
- T   temperature (K)
- P   pressure (bars)

Output:
- H   liquid or vapor enthalpy (J/mole)
- ERE   error flag, integer variable normally zero; ERE = 1 indicates binary data are missing for one or more binary pairs.

Subroutine References

Calls:
- BIJS
- PHIS
- TAUS

Called by: FLASH

Common Storage

/PURE/

/BINARY/

/GS/

/VIRIAL/

/PS/

Glossary of Principal Variable Names

(plus see common storage descriptions)

ID(I)	Vector (length 20) of indices for the components $(I = 1,N)$.
Z(I)	Vector (length 20) of stream composition $(I = 1,N)$.
BU	Contribution from temperature dependence of UNIQUAC binary interaction parameters, here taken as 0.
ST	$\Sigma\ x_i q_i$ UNIQUAC Equation (4-15).
STP	$\Sigma\ x_i q_i'$ UNIQUAC Equation (4-15).
HIG	Ideal gas enthalpy (J/mole).
HV	Vapor enthalpy (J/mole).
SØ	Sum given by first term in brackets in Equation (A-49).
SD	Sum given by bracketed term in Equation (A-44).
SL	Sum given by second term in brackets in Equation (A-49).
HL	Liquid enthalpy (J/mole).
HC	Excess enthalpy (J/mole).

Listing

```
      SUBROUTINE ENTH(N,ID,KEY,LEV,ZF,T,P,H,ERE)
* ENTH CALCULATES VAPOR OR LIQUID ENTHALPIES (REF IDEAL GAS AT 300 K) H
* IN JOULES/GMOL FOR MIXTURES OF N COMPONENTS (N.LE.20) WHOSE INDICES
* APPEAR IN VECTOR ID, GIVEN TEMPERATURE T, PRESSURE P, AND LIQUID OR
* VAPOR COMPOSITIONS Z. ENTH RETURNS ERE=0 UNLESS BINARY DATA ARE
* MISSING FOR THE SYSTEM, IN WHICH CASE IT RETURNS ERE=1, OR NO
* SOLUTION IS FOUND IN PHIS, WHEN IT RETURNS ERE = 2. FOR LEV = 0 THE
* LIQUID ENTHALPY IS CALCULATED WITH EXCESS ENTHALPY OF MIXING TAKEN
* AS 0, FOR LEV=1 EXCESS ENTHALPY IS CALCULATED FROM THE TEMPERATURE
* DEPENDENCE OF ACTIVITY COEFFICIENTS, AND FOR LEV=2 VAPOR ENTHALPY IS
* CALCULATED.  IF VALIK HAS BEEN CALLED FOR THE SAME CONDITIONS, KEY
* SHOULD BE AS IN VALIK.  IF VALIK HAS NOT BEEN CALLED, KEY MUST BE 9.
* FOR A NEW SYSTEM, 10 OTHERWISE.
      REAL Z(20),PHI(20),ZF(N)
      INTEGER ID(20),ERE
      COMMON/PURE/NM1(100),NM2(100),TC(100),PC(100),RD(100),DM(100),
     1 A(100),C1(100),C2(100),C3(100),C4(100),C5(100),RU(100),QU(100),
     2 QP(100),D1(100),D2(100),D3(100),D4(100)
      COMMON/BINARY/ETA(5050),U(100,100)
      COMMON/GS/IER,RL(20),TH(20),TP(20),GCL(20),TAU(20,20)
      COMMON/VIRIAL/KV,B(210),BD(210),DB(210),DBD(210),BM
      COMMON/PS/PHL(20),ZI(20),C(210),VNT,TL,PL
      DATA R,CJ/8.31439,0.0988075/,TR,TR2,TRF/300.,90000.,1411./
      ERE=0
```

```
* LACKING T DEPENDENT UNIQUAC INTERACTION PARAMETER SET THEIR T
* DERIVATIVE TERM TO 0.
      BU=0.
  100 DO 104 I=1,N
* CONVERT COMPOSITION VECTOR TO DIMENSIONS 20 TO MATCH LOWER LEVEL
* SUBROUTINES
  104 Z(I)=ZF(I)
* SKIP FUGACITY CALCULATIONS IF VALIK CALLED AT SAME CONDITIONS
      IF(KEY.LT.9) GO TO 120
* SKIP VAPOR CALCULATIONS FOR LIQUID
      IF(LEV.LT.2) GO TO 110
* GET VIRIAL COEFFICIENTS IF NOT PREVIOUSLY CALCULATED
      CALL BIJS(N,ID,KEY,T,ERB)
      IF(ERB.GT.1) GO TO 900
      IF(KV.EQ.0) GO TO 110
* GET TRUE COMPOSITIONS FOR ASSOCIATING VAPORS
      CALL PHIS(N,ID,8,Z,T,P,PHI,ERR)
      IF(ERR.GT.1) GO TO 900
      GO TO 120
  110 IF(LEV.EQ.0) GO TO 120
* GET UNIQUAC INTERACTION TERMS IF EXCESS ENTHALPY IS CALCULATED FOR
* LIQUID
      CALL TAUS(N,ID,T,TAU,IER)
* SET ERE = 1 FOR BINARY DATA MISSING
      IF(IABS(IER).EQ.1) ERE=1
      ST=0.
      STP=0.
* CALCULATION OF TERMS FOR EXCESS ENTHALPY EVALUATION
  111 DO 115 I=1,N
      II=ID(I)
      TH(I)=Z(I)*QU(II)
      TP(I)=Z(I)*QP(II)
      STP=STP+TP(I)
  115 ST=ST+TH(I)
  116 DO 119 I=1,N
      TH(I)=TH(I)/ST
  119 TP(I)=TP(I)/STP
* CALCULATION OF IDEAL GAS ENTHALPY
  120 TM1=T-TR
      TM2=ALOG(T/TR)
      TM3=(T**2-TR2)/2.
      TLT=T*ALOG(T)-T-TRF
      HIG=0.
  125 DO 129 I=1,N
      II=ID(I)
  129 HIG=HIG+Z(I)*(D1(II)*TM1+D2(II)*TM2+D3(II)*TM3+D4(II)*TLT)
      IF(LEV.LT.2) GO TO 150
* CALCULATION OF ENTHALPY FOR NONASSOCIATING VAPOR
  130 IF(KV.EQ.1) GO TO 140
      HV=0.
  131 DO 135 I=1,N
      LI=(I-1)*I/2
      DO 135 J=1,I
      CMT=2.
      IF(J.EQ.I) CMT=1.
  135 HV=HV+CMT*Z(I)*Z(J)*(B(LI+J)-DB(LI+J))
  139 HV=HIG+CJ*P*HV
      H=HV
      RETURN
* CALCULATION OF ENTHALPY FOR ASSOCIATING VAPOR
  140 SD=0.
      S0=0.
      S1=0.
  141 DO 145 I=1,N
      LI=(I-1)*I/2
```

```
          S0=S0+ZI(I)*(B(LI+I)-DB(LI+I))
          DO 143 J=1,I
          ZIJ=ZI(I)*ZI(J)*C(LI+J)
          SD=SD+ZIJ*(1.-DBD(LI+J)/BD(LI+J))
  143     S1=S1+ZIJ*(B(LI+J)-DB(LI+J))
  145 CONTINUE
          HV=HIG+VNT*(CJ*P*(S0+S1)-R*T*SD)
          H=HV
          RETURN
*  CALCULATION OF LIQUID ENTHALPY
  150 TMR=T**2
          TMS=2.*T*TMR
          HL=0.
  151 DO 155 I=1,N
          II=ID(I)
  155 HL=HL+Z(I)*(-C2(II)+C3(II)*TMR+C4(II)*T+C5(II)*TMS)
          H=HIG-R*HL
*  FOR LEV =.0 SKIP EXCESS ENTHALPY CALCULATION
  160 IF(LEV.EQ.0) RETURN
*  CALCULATION OF EXCESS ENTHALPY CONTRIBUTION FOR LIQUID
          HE=0.
          HC=0.
  161 DO 169 I=1,N
          II=ID(I)
          TM1=0.
          TM2=0.
          IF(U(II,II).GT.1.E+19) GO TO 166
          DO 165 J=1,N
          JJ=ID(J)
          TM3=TP(J)*TAU(J,I)
          TM1=TM1+TM3
  165 TM2=TM2+TM3*(U(JJ,II)-2.*BU/T)
          HE=HE+QP(II)*Z(I)*TM2/TM1
          GO TO 169
  166 DO 168 J=1,N
          JJ=ID(J)
  168 TM1=TM1+TH(J)*U(JJ,II)
          HC=HC+Z(I)*TM1
  169 CONTINUE
          H=H+R*(HE+HC)
          RETURN
*  FOR FAILURE TO FIND SOLUTION IN PHIS SET ERE = 2. (H.= 0)
  900 ERE=2
          H=0.
          RETURN
          END
```

SUBROUTINE PHIS (N, ID, KEY, Y, T, P, PHI, ERR)

Purpose

PHIS calculates vapor-phase fugacity coefficients, PHI, for each component in a mixture of N components (N $\leq$ 20) at specified temperature, pressure, and vapor composition.

Method

For mixtures of nonassociating or weakly associating compounds, fugacity coefficients are evaluated using Equation (3-10b). For mixtures including one or more strongly associating compound, fugacity coefficients are calculated from Equation (3-13) with true mole fractions, z_i, determined from a first-order iterative solution of material balance and equilibrium equations. (See discussion on chemical theory in Appendix A.)

Arguments

Input: N   number of components (N $\leq$ 20)

ID(I)   vector (length 20) of indices for the components (I = 1, N).

Y(I)   vector (length 20) of vapor-phase mole fractions for the components (I = 1, N).

KEY   integer initialization control variable (KEY = 1,...,10; see VALIK).

T   temperature (K)

P   pressure (bars)

Output:

PHI(I)   vector (length 20) of calculated fugacity coefficients for the components (I = 1, N).

ERR   error flag, integer variable normally zero; if no solution is found, ERR = 2.

Subroutine References

Calls: BIJS
Called by: VALIK
                ENTH

Common Storage

/PS/

/VIRIAL/

Glossary of Principal Variable Names

(plus see common storage descriptions)

SS(I)   Vector (length 20) of summations given in Equation
         (3-10b)

TL   Temperature (K) at previous call of PHIS.

PL   Pressure (bars) at previous call of PHIS.

ZT,ZJ   Temporary true vapor composition estimates.

K̄O   True vapor composition estimate flag (0 = a priori;
         1 = previous program call).

Z̄O(I)   Vector (length 20) of true vapor compositions from pre-
         vious iteration (I = 1, N).

SI(I)   Vector of sums given by Equation (A-36).

RM   Sum given by denominator in right-hand side of Equation
         (A-45).

Listing

```
      SUBROUTINE PHIS(N,ID,KEY,Y,T,P,PHI,ERR)
* PHIS CALCULATES VAPOR PHASE FUGACITY COEFFICIENTS, PHI, FOR ALL N
* COMPONENTS (N.LE.20) WHOSE INDICES APPEAR IN VECTOR ID, GIVEN
* TEMPERATURE T(K), PRESSURE P(BAR), AND VAPOR PHASE COMPOSITION Y. PHIS
* RETURNS ERR=0 UNLESS NO SOLUTIONS ARE FOUND IN WHICH CASE IT RETURNS
* ERR=2.  KEY SHOULD BE 1 FOR A NEW SYSTEM, 2 FOR ALL CONDITIONS CHANGED
* SINCE LAST CALL FOR SAME SYSTEM, 3 IF TEMPERATURE IS UNCHANGED FRO'*
* LAST CALL FOR SAME SYSTEM, 4 (7) IF ONLY TEMPERATURE HAS CHANGED, AND
* 8 IF BIJS HAS ALREADY BEEN CALLED AT SAME CONDITIONS.
      REAL Y(20),SI(20),ZO(20),PHI(20),SS(20)
      INTEGER ID(20),ERR,ERB
      COMMON/VIRIAL/KV,B(210),BD(210),DB(210),DBD(210),BM
      COMMON/PS/PHL(20),ZI(20),C(210),VNT,TL,PL
      DATA R/83.1473/
  100 ERR=0
      PRT=P/(R*T)
      GO TO(110,101,120,101,101,101,101,120,110,101),KEY
* CHECK FOR SIGNIFICANT CHANGE IN T OR P SINCE LAST CALL FOR SYSTEM
  101 IF(ABS(T-TL).LT.0.02) GO TO 103
      GO TO 110
  103 IF(ABS(P-PL).LT.0.01) GO TO 105
      GO TO 110
* RETURN IF NO CHANGE IN T, ʳ, OR Y SINCE LAST CALL.
  105 IF(KEY.EQ.4.OR.KEY.EQ.7) RETURN
      GO TO 120
* GET SECOND VIRIAL COEFFICIENTS BIJ IN /VIRIAL/
  110 CALL BIJS(N,ID,KEY,T,ERB)
      IF(ERB.GT.1) GO TO 900
* GO TO SPECIAL CALCULATION FOR ASSOCIATING GAS MIXTURES
  120 IF(KV.EQ.1) GO TO 200
* CALCULATE SECOND VIRIAL COEFFICIENT FOR GAS MIXTURE, BM
      BM=0.
  130 DO 139 I=1,N
* CALCULATE EFF SECOND VIRIAL COEFFICIENT FOR COMP I IN MIXTURE, SS(I)
```

```
      SS(I)=0.
      LI=(I-1)*I/2
      DO 133 J=1,I
  133 SS(I)=SS(I)+Y(J)*B(LI+J)
      I1=I+1
      IF(I1.GT.N) GO TO 136
      DO 135 J=I1,N
      LJ=(J-1)*J/2
  135 SS(I)=SS(I)+Y(J)*B(LJ+I)
  136 BM=BM+Y(I)*SS(I)
  139 CONTINUE
* CALCULATE VAPOR PHASE FUGACITY COEFFICIENTS, PHI(I)
  140 DO 149 I=1,N
      PHI(I)=EXP(PRT*(2.*SS(I)-BM))
* SAVE FUGACITY COEFFICIENTS FOR USE AT SIMILAR CONDITIONS
      PHL(I)=PHI(I)
  149 CONTINUE
* SAVE CONDITIONS AT WHICH PHIS CALCULATED
      TL=T
      PL=P
      RETURN
* SPECIAL CALCULATION FOR ASSOCIATING GAS MIXTURES
  200 KO=0
      GO TO(203,201,201,201,201,201,201,203,203,201),KEY
* IF PREVIOUS PHI VALUES AVAILABLE USE TO GET FIRST ESTIMATES OF ACTUAL
* VAPOR COMPOSITION
  201 DO 202 I=1,N
      LI=(I-1)*I/2
  202 ZI(I)=PHL(I)*Y(I)*EXP(-PRT*B(LI+I))
      IF(KEY.EQ.3) GO TO 208
      KO=1
* FOR NO PREVIOUS PHI VALUES AVAILABLE (KO = 0) MAKE FIRST ESTIMATES OF
* ACTUAL VAPOR COMPOSITION
* FOR ALL CASES (EXCEPT KEY = 3) FIND VALUES OF ASSOCIATION EQUILIBRIUM
* CONSTANTS C.
  203 DO 207 I=1,N
      IF(KO.EQ.0) ZI(I)=Y(I)
      LI=(I-1)*I/2
      DO 206 J=1,I
      LJ=(J-1)*J/2
      C(LI+J)=-2.*PRT*BD(LI+J)*EXP(PRT*(B(LI+I)+B(LJ+J)-B(LI+J)))
      IF(C(LI+J).LT.0.) GO TO 900
      IF(J.EQ.I) GO TO 205
      IF(KO.EQ.1) GO TO 206
* INITIAL ESTIMATES OF ZI(I)
      IF(C(LI+J).LE.0.5) GO TO 206
      IF(Y(J).LT.Y(I)) GO TO 204
      ZT=Y(I)/(C(LI+J)*Y(J)+1.)
      ZJ=Y(J)/(C(LI+J)*ZT+1.)
      IF(ZT.LT.ZI(I)) ZI(I)=ZT
      IF(ZJ.LT.ZI(J)) ZI(J)=ZJ
      GO TO 206
  204 ZJ=Y(J)/(C(LI+J)*Y(I)+1.)
      ZT=Y(I)/(C(LI+J)*ZJ+1.)
      IF(ZT.LT.ZI(I)) ZI(I)=ZT
      IF(ZJ.LT.ZI(J)) ZI(J)=ZJ
      GO TO 206
  205 C(LI+J)=C(LI+J)/2.
      IF(KO.EQ.1) GO TO 206
      IF(C(LI+J).LE.0.5) GO TO 206
      ZT=(SQRT(1.+8.*C(LI+J)*Y(I))-1.)/(4.*C(LI+J))
      IF(ZT.LT.ZI(I)) ZI(I)=ZT
  206 CONTINUE
  207 CONTINUE
```

301

```
* START ITERATIVE CALCULATION OF ACTUAL VAPOR COMPOSITION, ZI(I)
* STORE FIRST ITERATION VALUES
  208 DO 209 I=1,N
  209 ZO(I)=ZI(I)
      IT=0
  210 IT=IT+1
      IF(IT.GT.20) GO TO 900
      RM=1.
  220 DO 229 I=1,N
      SI(I)=0.
* DAMP ITERATION 20 PERCENT
      ZI(I)=.2*ZO(I)+.8*ZI(I)
      ZO(I)=ZI(I)
      LI=(I-1)*I/2
      DO 221 J=1,I
  221 SI(I)=SI(I)+C(LI+J)*ZI(J)
      DO 223 J=I,N
      LJ=(J-1)*J/2
  223 SI(I)=SI(I)+C(LJ+I)*ZI(J)
      RM=RM+ZI(I)*SI(I)/2.
  229 CONTINUE
  230 DO 235 I=1,N
  235 ZI(I)=RM*Y(I)/(1.+SI(I))
      DO 239 I=1,N
      IF(Y(I).LT.1.E-09) GO TO 239
* CHECK CONVERGENCE OF EACH ZI(I)
      IF(ABS((ZI(I)-ZO(I))/Y(I)).GT.0.005) GO TO 210
  239 CONTINUE
* CALCULATE VAPOR PHASE FUGACITY COEFFICIENTS FOR ACTUAL COMPOSITION OF
* ASSOCIATING VAPOR.
  240 DO 249 I=1,N
      LI=(I-1)*I/2
      PHI(I)=RM*EXP(PRT*B(LI+I))/(1.+SI(I))
      IF(KEY.EQ.8) GO TO 249
* SAVE FUGACITY COEFFICIENTS FOR USE AT SIMILAR CONDITIONS
      PHL(I)=PHI(I)
  249 CONTINUE
* CALCULATE TOTAL MOLS OF ASSOCIATING VAPOR PER MOL STOICHIOMETRIC VAPOR
  250 VNT=1./RM
      TL=T
      PL=P
      RETURN
* ERROR RETURN FOR FAILURE OF ITERATION FOR ZI(I) TO CONVERGE.
  900 ERR=2
      DO 901 I=1,N
      PHL(I)=1.
  901 PHI(I)=1.
      RETURN
      END
```

SUBROUTINE BIJS (N, ID, KEY, T, ERB)

Purpose

BIJS calculated second virial coefficients for pure compoments and all binary pairs in a mixture of N components (N $\leq$ 20) at specified temperature. These coefficients are placed in common storage /VIRIAL/.

Method

Second virial coefficients are calculated using the equations for the Hayden-O'Connell correlation (see Appendix A). If the mixture includes organic acids, the equations of Hayden and O'Connell yield equilibrium constants for all possible dimerization reactions.

Arguments

Input: N   number of components (N $\leq$ 20)
        ID(I) vector (length 20) of indices for the components
              (I = 1, N)
         KEY   integer initialization control variable (KEY =
              1,...,10); see VALIK.
           T   temperature (K)

Output: ERB   error flag, integer variable normally zero; if any
              anomalies are detected (e.g. negative molecular
              size or energy parameters), ERB = 2.

Subroutine References

Called by: PHIS
             ENTH

Common Storage

/PURE/
/BINARY/
/VIRIAL/
/BS/

Glossary of Principal Variable Names

(plus see common storage descriptions):

E(L)   Vector (length 210) of energy terms E_{ij} given by the
        term in parenthesis in Equation (A-15).

303

ET(L)   Vector (length 210) of η_{ij}, association or solvation parameters.

G(L)   Vector (length 210) of μ^*_{ij}, reduced (and modified) dipole moment, Equation (A-11).

H(L)   Vector (length 210) of n_{ij} or ΔH_{ij}.

L $L = (I-1)\,I/2 + J$ for $(I = 1,20$ and $J = 1,I)$.

S(L)   Vector (length 210) of b_{0ij}, Equation (A-10)

TS(L)   Vector (length 210) of ε_{ij}/k, and $(\varepsilon_{ij}/k)^*$, characteristic energy, Equation (A-17)(A-28).

W(L)   Vector (length 210) of ω'_{ij}, nonpolar acentric factor, Equation (A-16).

Z(L)   Vector (length 210) of A_{ij}, Equation (A-12) or ζ_{ij}.

BN, BP, T1, T2, T3, TK, TA
        Temporary storage.

Listing

```
      SUBROUTINE BIJS(N,ID,KEY,T,ERB)
* BIJS CALCULATES SECOND VIRIAL COEFFICIENTS, BIJ, FOR ALL PAIRS OF N
* COMPONENTS (N.LE.20) WHOSE INDICES APPEAR IN VECTOR ID, FOR
* TEMPERATURE T(K). COEFFICIENTS ARE RETURNED IN COMMON STORAGE/VIRIAL
* WITH B(I,J)=B(L), L=(I-1)*I/2+J. IF CARBOXYLIC ACIDS ARE PRESENT
* KV (IN COMMON/VIRIAL) IS SET TO 1 (OTHERWISE 0), AND B0 IS RETURNED
* IN B, BT IN BD. IF ANY ANOMALIES ARE DETECTED IN CALCULATION ERB IS
* SET TO 2 (OTHERWISE 0). TEMPERATURE INDEPENDENT PARAMETERS ARE
* EVALUATED ONLY IF KEY = 1 OR 9, TEMPERATURE DERIVATIVES OF
* COEFFICIENTS (MULTIPLIED BY TEMPERATURE) ARE FOUND AND RETURNED IN
* DB(L) (AND DBD(L)) IN COMMON/VIRIAL IF KEY IS 6 OR LARGER.
      INTEGER ID(20),ERB
      COMMON/PURE/NM1(100),NM2(100),TC(100),PC(100),RD(100),DM(100),
     1 A(100),C1(100),C2(100),C3(100),C4(100),C5(100),RU(100),QU(100),
     2 QP(100),D1(100),D2(100),D3(100),D4(100)
      COMMON/BINARY/ETA(5050),U(100,100)
      COMMON/VIRIAL/KV,B(210),BD(210),DB(210),DBD(210),BM
      COMMON/BS/G(210),TS(210),S(210),Z(210),H(210),E(210),W(210),
     1 ET(210)
      DATA B1,B2,B3/1.2618,7243.8,1.7941E07/,CN1,CN2,CN3,CN4/0.94,-1.47,
     1 -0.85,1.015/,CP1,CP2,CP3,CP4/-0.75,3.,-2.1,-2.1/,CA1,CA2,CH1,CH2/
     2 -0.3,-0.05,1.99,0.2/,CW1,CW2,CW3/0.006026,0.02096,-0.001366/,
     3 CS1,CK1,CK2,CE1,CE2,CE3,CE4,CE5/2.4507,0.7,0.6,650.,300.,4.27,
     4 42800.,22400./,CD1,CD2,CD3/0.748,0.91,0.4/,E3/0.3333333/
 100 ERB=0
* CALCULATE TEMPERATURE-INDEPENDENT PARAMETERS ONLY FOR NEW SYSTEM
      GO TO (109,200,200,200,200,200,200,200,109,200),KEY
* RESET ASSOCIATING VAPOR FLAG
 109 KV=0
* CALCULATE TEMPERATURE-INDEPENDENT PARAMETERS FOR PURE COMPONENTS
 110 DO 119 I=1,N
* IDENTIFY COMPONENT
      II=ID(I)
      L=(I+1)*I/2
* NONPOLAR ACENTRIC PARAMETER
      W(L)=CW1*RD(II)+CW2*RD(II)**2+CW3*RD(II)**3
```

```
* MOLECULAR SIZE PARAMETER (CUBED)
      S(L)=(CS1-W(L))**3*TC(II)/PC(II)
      IF(S(L).LT.0.) GO TO 900
      III=(II+1)*II/2
      ET(L)=ETA(III)
      IF(ET(L).GE.4.4999) KV=1
* ENERGY PARAMETER
      TS(L)=TC(II)*(CO1+CO2*W(L)-CO3*ET(L)/(2.+20.*W(L)))
      IF(TS(L).LT.0.) GO TO 900
      IF(DM(II).LT.1.45) GO TO 117
* MODIFICATION OF PARAMETERS FOR LARGE DIPOLE MOMENTS
      H(L)=16.+400.*W(L)
      T1=H(L)/(H(L)-6.)
      T2=3./(H(L)-6.)
      TK=2.882-(1.882*W(L)/(0.03+W(L)))
      Z(L)=B3*DM(II)**4/(TS(L)*S(L)**2*TC(II)*TK)
      IF(Z(L).LT.-1.) GO TO 900
* MODIFIED MOLECULAR SIZE PARAMETER (CUBED)
      S(L)=S(L)*(1.+T2*Z(L))
* MODIFIED ENERGY PARAMETER
      TS(L)=TS(L)*(1.-T1*Z(L)+T1*(T1+1.)*Z(L)**2/2.)
* REDUCED DIPOLE MOMENT
  117 G(L)=B2*DM(II)**2/(TS(L)*S(L))
  119 CONTINUE
      IF(N.EQ.1) GO TO 130
* CALCULATE TEMPERATURE-INDEPENDENT PARAMETERS FOR COMPONENT PAIRS
  120 DO 129 I=2,N
      II=ID(I)
      LI=(I+1)*I/2
      I1=I-1
      DO 129 J=1,I1
      JJ=ID(J)
      LJ=(J+1)*J/2
      L=(I-1)*I/2+J
* CROSS NONPOLAR ACENTRIC PARAMETER
      W(L)=(W(LI)+W(LJ))/2.
* CROSS MOLECULAR SIZE PARAMETER
      S(L)=SQRT(S(LI)*S(LJ))
* CROSS ENERGY PARAMETER
      TS(L)=CK1*SQRT(TS(LI)*TS(LJ))+CK2/(1./TS(LI)+1./TS(LJ))
      IF(DM(II).LT.1.E-19) GO TO 123
      IF(DM(JJ).GT.1.E-19) GO TO 124
      IF(DM(II).LT.2.5) GO TO 124
      Z(L)=DM(II)**2*(TS(LJ)**2*S(LJ))**E3/(TS(L)*S(LI))
      GO TO 125
  123 IF(DM(JJ).LT.2.5) GO TO 124
* MODIFICATION OF PARAMETERS IN POLAR-NONPOLAR PAIRS
      Z(L)=DM(JJ)**2*(TS(LI)**2*S(LI))**E3/(TS(L)*S(LJ))
      GO TO 125
  124 Z(L)=0.
      GO TO 126
  125 H(L)=16.+400.*W(L)
      T1=H(L)/(H(L)-6.)
      T2=3./(H(L)-6.)
* MODIFIED CROSS MOLECULAR SIZE PARAMETER
      S(L)=S(L)*(1.-T2*Z(L))
* MODIFIED CROSS ENERGY PARAMETER
      TS(L)=TS(L)*(1.+T1*Z(L))
* CROSS REDUCED DIPOLE MOMENT
  126 G(L)=B2*DM(II)*DM(JJ)/(TS(L)*S(L))
* DETERMINE EFFECTIVE ASSOCIATION/SOLVATION PARAMETER
      IJ=(II-1)*II/2+JJ
      IF(JJ.GT.II) IJ=(JJ-1)*JJ/2+II
      IF(ABS(ETA(IJ)).LT.1.E-19) GO TO 127
      ET(L)=ETA(IJ)
```

305

```
      IF(ET(L).GE.4.4999) KV=1
      GO TO 129
  127 ET(L)=0.
      IF(ABS(ET(LI)-ET(LJ)).LT.1.E-19) ET(L)=ET(LI)
  129 CONTINUE
*  CALCULATE TEMPERATURE-INDEPENDENT TERMS IN VIRIAL COEFFICIENTS FOR
*  PURE COMPONENTS AND PAIRS.
  130 DO 139 I=1,N
      DO 139 J=1,I
      L=(I-1)*I/2+J
      S(L)=B1*S(L)
      H(L)=CH1+CH2*G(L)**2
      Z(L)=CA1+CA2*G(L)
*  DETERMINE MODIFIED REDUCED DIPOLE PARAMETER
      IF(G(L).LT.0.04) GO TO 135
      IF(G(L).GE.0.25) GO TO 134
      G(L)=0.
      GO TO 135
  134 G(L)=G(L)-0.25
  135 IF(ET(L).GE.4.5) GO TO 137
      IF(ET(L).LT.1.E-19) GO TO 139
*  ENERGY TERM FOR NONASSOCIATING PAIR
      E(L)=CE1/(TS(L)+CE2)-CE3
      GO TO 139
*  ENERGY TERM FOR ASSOCIATING PAIR
  137 E(L)=CE4/(TS(L)+CE5)-CE3
  139 CONTINUE
*  CALCULATE TEMPERATURE-DEPENDENT TERMS AND VIRIAL COEFFICIENTS
  200 DO 209 I=1,N
      DO 209 J=1,I
      L=(I-1)*I/2+J
      TA=T/TS(L)
      T1=1./(1./TA-1.6*W(L))
      T2=T1*T1
      T3=T2*T1
*  NONPOLAR FREE CONTRIBUTION
      BN=CN1+CN2/T1+CN3/T2+CN4/T3
      IF(G(L).GT.1.E-19) GO TO 201
      BP=0.
      GO TO 202
*  POLAR FREE CONTRIBUTION
  201 BP=(CP1+CP2/T1+CP3/T2+CP4/T3)*G(L)
*  TOTAL FREE CONTRIBUTION TO VIRIAL COEFFICIENT
  202 B(L)=S(L)*(BN+BP)
*  METASTABLE PLUS BOUND CONTRIBUTIONS
      BN=Z(L)*EXP(H(L)/TA)
      IF(ET(L).LT.1.E-19) GO TO 204
*  CHEMICAL CONTRIBUTION
      BP=EXP(ET(L)*E(L))-EXP(ET(L)*(1500./T+E(L)))
      GO TO 205
  204 BP=0.
*  METASTABLE, BOUND, AND CHEMICAL CONTRIBUTIONS TO VIRIAL COEFFICIENT
  205 BD(L)=S(L)*(BN+BP)
      IF(KEY.GT.1.AND.KEY.LE.5) GO TO 208
*  CALCULATION OF T DERIVATIVES OF VIRIAL COEFFICIENTS (ALL MULTIPLIED
*  BY T)
      DBN=-CN2-2.*CN3/T1-3.*CN4/T2
      DBP=(-CP2-2.*CP3/T1-3.*CP4/T2)*G(L)
*  DERIVATIVE OF TOTAL FREE CONTRIBUTION TO VIRIAL COEFFICIENT
      DB(L)=S(L)*(DBN+DBP)/TA
      DBN=-H(L)*BN/TA
      IF(ET(L).LT.1.E-19) GO TO 206
      DBP=1500.*ET(L)*EXP(ET(L)*(1500./T+E(L)))/T
      GO TO 207
```

```
   206 DBP=0.
*  DERIVATIVE OF METASTABLE, BOUND, AND CHEMICAL CONTRIBUTIONS TO VIRIAL
*  COEFFICIENT
   207 DBD(L)=S(L)*(DBN+DBP)
*  CALCULATION OF TOTAL VIRIAL COEFFICIENT FOR CASES WITHOUT ASSOCIATING
*  VAPORS.
       IF(KV.EQ.0) DB(L)=DB(L)+DBD(L)
   208 IF(KV.EQ.0) B(L)=B(L)+BD(L)
   209 CONTINUE
       RETURN
*  ERROR RETURN FOR FAILURE TO FIND VALID VIRIAL COEFFICIENTS.
   900 ERB=2
       NL=(N+1)*N/2
       DO 902 L=1,NL
       B(L)=0.
   902 BD(L)=0.
       RETURN
       END
```

SUBROUTINE PURF (N, ID, T, P, FIP)

Purpose

PURF calculates pure liquid standard-state fugacities at zero pressure, pure-component saturated liquid molar volume $(cm^3/mole)$, and pure-component liquid standard-state fugacities at system pressure. Pure-component hypothetical liquid reference fugacities are calculated for noncondensable components. Liquid molar volumes for noncondensable components are taken as zero. All of the above calculations are done at the specified system temperature.

Method

Standard-state fugacities at zero pressure are evaluated using the Equation (A-2) for both condensable and noncondensable components. The Rackett Equation (B-2) is evaluated to determine the liquid molar volumes as a function of temperature. Standard-state fugacities at system temperature and pressure are given by the product of the standard-state fugacity at zero pressure and the Poynting correction shown in Equation (4-1). Double precision is advisable.

Arguments

Input: N   number of components ($N \leq 20$)

ID(I)   vector (length 20) of indices for the components
(I = 1,N)

T   temperature (K)

P   pressure (bars)

Output: FIP(I)   vector (length 20) of standard-state fugacity
(bars) (I = 1,N)

Subroutine References

Called by: VALIK

Common Storage

/PURE/

Glossary of Principal Variable Names

FO(I) Vector (length 20) of pure-component liquid standard-state
fugacities at zero pressure or hypothetical liquid standard-

state fugacity for a noncondensable component (bars) (I = 1,N).

VIP(I) Vector (length 20) of saturated liquid molar volumes (cm^3/ mole) for condensable components; for noncondensable components VIP(I) = 0; (I = 1,N).

TAU = 1.60 + 0.00693/(T_r - 0.655).

(Also see common-block storage descriptions.)

Listing

```
      SUBROUTINE PURF(N,ID,T,P,FIP)
* PURF CALCULATES PURE COMPONENT LIQUID FUGACITIES, FIP, AT SYSTEM
* TEMPERATURE T(K) AND PRESSURE P(BAR) FOR ALL N COMPONENTS (N.LE.20)
* WHOSE INDICES APPEAR IN VECTOR ID. FUGACITIES OF HYPOTHETICAL LIQUID
* PHASES ARE CALCULATED FOR NONCONDENSABLE COMPONENTS.
      REAL FIP(20),F0(20),VIP(20)
      INTEGER ID(20)
      COMMON/PURE/NM1(100),NM2(100),TC(100),PC(100),RD(100),DM(100),
     1  A(100),C1(100),C2(100),C3(100),C4(100),C5(100),RU(100),QU(100),
     2  QP(100),D1(100),D2(100),D3(100),D4(100)
      DATA R,CA,CB,CC,E/83.1473,1.60,0.655,0.00693026,0.28571429/
  100 RT=R*T
      AT=ALOG(T)
      T2=T*T
  101 DO 109 I=1,N
* IDENTIFY COMPONENT
      II=ID(I)
* GET PURE COMPONENT 0-PRESSURE FUGACITIES, F0.
      F0(I)=EXP(C1(II)+C2(II)/T+C3(II)*T+C4(II)*AT+C5(II)*T2)
* GET PURE COMPONENT LIQUID MOLAR VOLUMES, VIP
      TR=T/TC(II)
      IF(TR.GT.0.75) GO TO 105
      TAU=1.+(1.-TR)**E
      GO TO 107
  105 TAU=CA+CC/(TR-CB)
  107 VIP(I)=R*TC(II)*A(II)**TAU/PC(II)
  109 CONTINUE
* CALCULATE PURE COMPONENT LIQUID FUGACITIES AT P
  110 DO 119 I=1,N
      FIP(I)=F0(I)*EXP(VIP(I)*P/RT)
  119 CONTINUE
      RETURN
      END
```

<u>SUBROUTINE GAMMA (N, ID, KEY, X, T, GAM, ERG)</u>

Purpose

GAMMA calculates activity coefficients for N components ($N \leq 20$) at system temperature. For noncondensable components effective infinite-dilution activity coefficients are calculated.

Method

Activity coefficients for condensable components are calculated with the UNIQUAC Equation (4-15), and infinite-dilution activity coefficients for noncondensable components are calculated with Equation (4-22).

Arguments

Input:
    N   number of components ($N \leq 20$)

    ID(I)   vector (length 20) of indices for the components ($I = 1$, N)

    KEY   integer initialization control variable (KEY = 1,...,10); see VALIK.

    X(I)   vector (length 20) of liquid-phase mole fractions for the components ($I = 1$, N).

    T   temperature (K)

Output:

    GAM(I)   vector (length 20) of calculated activity coefficients for the components ($I = 1$, N)

    ERG   Error flag, integer variable normally zero; ERG = 1 indicates binary data are missing.

Subroutine References

Calls: TAUS

Called by: VALIK

                LILIK

Common Storage

/PURE/

/BINARY/

/GS/

<u>Glossary of Principal Variable Names</u>
(plus see common storage descriptions)

GRL $\ln \gamma_i$ (residual) for a condensable component.

PT(I) $\sum \theta_k' \tau_{ki}$ vector of summations given in Equation (4-15).

PTS(I) $\displaystyle\sum \frac{\theta_j' \tau_{ij}}{\sum \theta_k' \tau_{kj}}$ vector of summations given in Equation (4-15).

SL $\sum x_i \ell_i$ in Equation (4-15); noncondensable free basis

SP $\sum x_i \gamma_i$ UNIQUAC Equation (4-15).

SS   Summation of mole fractions over condensable components
       only.

ST $\sum x_i q_i$ UNIQUAC Equation (4-15).

STP $\sum x_i q_i'$ UNIQUAC Equation (4-15).

<u>Listing</u>

```
      SUBROUTINE GAMMA(N,ID,KEY,X,T,GAM,ERG)
* GAMMA CALCULATES LIQUID PHASE ACTIVITY COEFFICIENTS, GAM, FOR ALL N
* COMPONENTS(N.LE.20) WHOSE INDICES APPEAR IN VECTOR ID, GIVEN
* TEMPERATURE T(K) AND LIQUID COMPOSITION X, USING THE UNIQUAC MODEL.
* FOR NONCONDENSABLE COMPONENTS (U(I,I) SET TO 1.E+20) AND UNSYMMETRIC
* CONVENTION IS USED TO DERIVE EFFECTIVE ACTIVITY COEFFICIENTS. GAMMA
* RETURNS ERG=0 UNLESS BINARY DATA ARE MISSING FOR THE SYSTEM, IN WHICH
* CASE IT RETURNS ERG=1. KEY SHOULD BE 1 FOR A NEW SYSTEM, 3 FOR T
* UNCHANGED, AND 4 OR 5 FOR X UNCHANGED.
      REAL X(20),GAM(20),PT(20),PTS(20)
      INTEGER ID(20),ERG
      COMMON/PURE/NM1(100),NM2(100),TC(100),PC(100),RD(100),DM(100),
     1 A(100),C1(100),C2(100),C3(100),C4(100),C5(100),RU(100),QU(100),
     2 QP(100),D1(100),D2(100),D3(100),D4(100)
      COMMON/BINARY/ETA(5050),U(100,100)
      COMMON/GS/IER,RL(20),TH(20),TP(20),GCL(20),TAU(20,20)
      DATA Z/10./
* SKIP SYSTEM INITIALIZATION ON SUBSEQUENT CALCULATIONS
  100 GO TO(110,120,120,130,130,120,130,120,110,120),KEY
  110 ERG=0
* CALCULATE COMPOSITION INDEPENDENT TERMS
  111 DO 119 I=1,N
      II=ID(I)
  119 RL(I)=Z*(RU(II)-QU(II))/2.-RU(II)+1.
* CALCULATE SEGMENT AND AREA FRACTIONS FOR COMPONENTS IN MIXTURE
  120 SP=1.E-30
      ST=1.E-30
      STP=1.E-30
      SS=0.
      SL=0.
  121 DO 125 I=1,N
      II=ID(I)
      TH(I)=X(I)*QU(II)
      TP(I)=X(I)*QP(II)
      SP=SP+X(I)*RU(II)
      ST=ST+TH(I)
      STP=STP+TP(I)
* SKIP FOR NONCONDENSABLE COMPONENTS
```

```
      IF(U(II,II).GT.1.E+19) GO TO 125
      SS=SS+X(I)
      SL=SL+X(I)*RL(I)
  125 CONTINUE
  126 DO 129 I=1,N
      II=ID(I)
      TH(I)=TH(I)/ST
      TP(I)=TP(I)/STP
      IF(U(II,II).GT.1.E+19) GO TO 128
* CALCULATE COMBINATORIAL COMTRIBUTION TO EXCESS FREE ENERGY
  127 GCL(I)=RL(I)-RJ(II)*SL/SP+ALOG(RU(II)*SS/SP)+Z*QU(II)*ALOG(QU(II)
     1    *SP/(RU(II)*ST))/2.
      GO TO 129
  128 GCL(I)=0.
  129 CONTINUE
      IF(KEY.EQ.3) GO TO 140
* GET UNIQUAC BINARY INTERACTION PARAMETER TERMS
  130 CALL TAUS(N,ID,T,TAU,IER)
* CALCULATE RESIDUAL CONTRIBUTION TO EXCESS FREE ENERGY
  140 DO 141 I=1,N
  141 PTS(I)=0.
  142 DO 149 I=1,N
      PT(I)=1.E-30
      DO 143 J=1,N
  143 PT(I)=PT(I)+TP(J)*TAU(J,I)
      DO 145 J=1,N
  145 PTS(J)=PTS(J)+TP(I)*TAU(J,I)/PT(I)
  149 CONTINUE
  150 DO 159 I=1,N
      II=ID(I)
      IF(U(II,II).GT.1.E+19) GO TO 155
* RESIDUAL FREE ENERGY FOR CONDENSABLE COMPONENTS
      GRL=QP(II)*(1.-ALOG(PT(I))-PTS(I))
      GO TO 158
  155 GRL=0.
      DO 156 J=1,N
      JJ=ID(J)
* RESIDUAL FREE ENERGY FOR NONCONDENSABLE COMPONENTS
  156 GRL=GRL+TH(J)*(U(II,JJ)+U(JJ,II)/T)
* CALCULATE ACTIVITY COEFFICIENT
  158 GAM(I)=EXP(GCL(I)+GRL)
  159 CONTINUE
      IF(IABS(IER).EQ.1) ERG=1
      RETURN
      END
      SUBROUTINE TAUS(N,ID,T,TAU,IER)
* TAUS CALCULATES TEMPERATURE DEPENDENT INTERACTION COEFFICIENTS TAU FOR
* USE IN SUBROUTINE GAMMA.  IF SYSTEM DATA ARE MISSING (SOME REQUIRED
* ENTRY IN MATRIX U IN COMMON/BINARY IS ZERO) CORRESPONDING TAU IS
* SET TO 1   AND IER IS RETURNED AS +/- 1.  FOR NONCONDENSABLES PRESENT
* IER IS -2 OR -1 (OTHERWISE 0).
      REAL TAU(20,20)
      INTEGER ID(20)
      COMMON/BINARY/ETA(5050),U(100,100)
  100 IER=0
  110 DO 119 I=1,N
      II=ID(I)
* CHECK IF ANY COMPONENT IS A NONCONDENSABLE AND FLAG IER
      IF(U(II,II).GT.1.E+19) IER=ISIGN(IER**2-2,-1)
      DO 119 J=1,N
      IF(J.EQ.I) GO TO 115
      JJ=ID(J)
* CHECK IF BINARY PAIR ARE BOTH NONCONDENSABLES.
```

312

SUBROUTINE TAUS (N, ID, T, TAU, IER)

Purpose

TAUS calculates temperature dependent UNIQUAC binary interaction parameters, τ_{ij}, for use in subroutine GAMMA and ENTH.

Method

τ_{ij} are evaluated from Equations (4-9) and (4-10) for all possible binary pairs in a mixture of N components ($N \leq 20$).

Arguments

Input: N   number of components ($N \leq 20$).

ID(I)   vector (length 20) of indices for the components
         (I = 1, N).

T   temperature (I)

Output:

TAU(I,J) Array (dimension 20 x 20) of UNIQUAC interaction
          parameters, τ_{ij} (I,J = 1,N).

IER   integer control variable, zero if all binary data
      are available; IER = ±1 indicates some binary data
      were not found; IER <0 indicates noncondensables
      are present in the mixture.

Subroutine References

Called by: GAMMA
              ENTH

Common Storage

/BINARY/

Glossary of Principal Variable Names

(See common storage descriptions.)

Listing

```
      IF(U(II,II).GT.1.E+19.AND.U(JJ,JJ).GT.1.E+19) GO TO 115
* CHECK IF BINARY DATA ARE MISSING
      IF(ABS(U(II,JJ)).LT.1.E-19) GO TO 112
* CHECK IF EITHER COMPONENT IN BINARY PAIR IS A NONCONDENSABLE
      IF((U(II,II)+U(JJ,JJ)).GT.1.E+19) GO TO 115
* CALCULATE INTERACTION TERM
```

```
      TAU(I,J)=EXP(-U(II,JJ)/T)
      GO TO 119
  112 IER=ISIGN(1,IER)
*  SET INTERACTION TERM EQUAL TO UNITY FOR PAIR WITH MISSING DATA
  115 TAU(I,J)=1.
  119 CONTINUE
      RETURN
      END
```

/BINARY/

ETA(L)   vector (length 5050) of η_{ij}, association and
         solvation parameters

U(I,J)   array (dimensions 100,100) of UNIQUAC binary
         interaction parameters (set to 0 if not available;
         U_{ii} is 0 for condensable components and 10^{20} for
         noncondensable components).

/BS/

G(L)   vector (length 210) of μ_{ij}^{*}, reduced dipole moment
         Equation (A-11)

TS(L)   vector (length 210) of ε_{ij}/k and $(\varepsilon_{ij}/k)^{*}$, charac-
         teristic energy, Equation (A-17) and (A-25)

S(L)   vector (length 210) of b_{oij}, Equation (A-10)

Z(L)   vector (length 210) of A_{ij}, Equation (A-12), or
         ξ_{ij}, Equations (A-21) and (A-29)

H(L)   vector (length 210) of Δh_{ij}, Equation (A-13)

E(L)   vector (length 210) of E_{ij}, Equation (A-15)

W(L)   vector (length 210) of ω_{ij}, nonpolar acentric
         factor, Equation (A-16)

ET(L)   vector (length 210) of η_{ij}, association and sol-
         vation parameters

/GS/

IER   integer error flag, zero if all binary data are
         available; IER = ±1 indicates some binary data
         were not found; IER < 0 indicates noncondensables
         are present in the mixture

RL(I)   vector (length 20) of ℓ_i, Equation (4-15)

TH(I)   vector (length 20) of θ_i, Equation (4-15)

TP(I)   vector (length 20) of θ_i', Equation (4-15)

GCL(I)   vector (length 20) of $\ell n\ \gamma_{i\,(combinatorial)}$, Equa-
         tion (4-15)

TAU(I,J) array (dimensions 20,20) of UNIQUAC interaction
         parameters, τ_{ij}

/PS/

PHL(I)   vector (length 20) of last calculated fugacity
         coefficients (I = 1,N)

ZI(I)   vector (length 20) of true mole fractions (I = 1,N)

C(L)	vector (length 210) of mole fraction equilibrium constants, Equation (A-33)
VNT	ratio of true number of moles to stoichiometric number of moles, Equation (A-45)
TL	temperature at previous call (K)
PL	pressure at previous call (bar)

/PURE/

NM1(I)	vector (length 100) of first 10 characters of the names of all components (I = 1,M, where M is the number of components in the data base).
NM2(I)	vector (length 100) of last 10 characters of the names of all components (I = 1,M)
TC(I)	vector (length 100) of critical temperatures (K) of all components (I = 1,M)
PC(I)	vector (length 100) of critical pressures (bar) of all components (I = 1,M)
RD(I)	vector (length 100) of the mean radii of gyration (Angstrom) of all components (I = 1,M)
DM(I)	vector (length 100) of the dipole moments (D) of all components (I = 1,M)
A(I)	vector (length 100) of the Rackett parameters for all components (I = 1,M). (Set to 10^{-20} for non-condensables)
C1(I) C2(I) C3(I) C4(I) C5(I)	vectors (length 100) of constants for the zero-pressure reference fugacity equation [Equation (B-2)] for all components (I = 1,M)
RU(I)	vector (length 100) of UNIQUAC parameters, r_i, for all components (I = 1,M)
QU(I)	vector (length 100) of UNIQUAC parameters, q_i, for all components (I = 1,M)
QP(I)	vector (length 100) of UNIQUAC parameters, q_i', for all components (I = 1,M)
D1(I) D2(I) D3(I) D4(I)	vectors (length 100) of constants for the ideal-gas heat capacity equation [Equation (5-1)] for all components (I - 1,M)

/VAL/

FIP(I)	vector (length 20) of liquid reference fugacities (I = 1,N)

/VIRIAL/

KV   integer key; KV = 0 indicates no organic acids are present; KV = 1 indicates one or more organic acids are present

B(L)   vector (length 210) of B_{ij}, Equation (A-1) [or B_{ij}^F, Equation (A-2) if KV = 1]

BD(L)   vector (length 210) of B_{ij}^D, Equation (A-2)

DB(L)   vector (length 210) of $T(dB_{ij}/dT)$ [or $T(dB_{ij}^F/dT)$ if KV = 1]

DBD(L)   vector (length 210) of $T(dB_{ij}^D/dT)$

BM   second virial coefficient of the mixture

Appendix F

DESCRIPTIONS AND LISTINGS OF SUBROUTINES FOR CALCULATION
OF VAPOR-LIQUID EQUILIBRIUM SEPARATIONS

The computer subroutines for calculation of vapor-liquid
equilibrium separations, including determination of bubble-point
and dew-point temperatures and pressures, are described and listed
in this Appendix. These are source routines written in American
National Standard FORTRAN (FORTRAN IV), ANSI X3.9-1978, and, as
such, should be compatible with most computer systems with FORTRAN
IV compilers. Approximate storage requirements for these subroutines are given in Appendix J; their execution times are
strongly dependent on the separations being calculated but can be
estimated (CDC 6400) from the times given for the thermodynamic
subroutines they call (essentially all computation effort is in
these thermodynamic subroutines).

These subroutines are capable of treating multicomponent
systems with up to 20 components. For compatibility with diverse,
user-written main programs, they employ vectors of length N, where
N ($\leq$ 20) is the number of components involved, in their argument
lists.

SUBROUTINE FLASH (TYPE,N,ID,KEY,VF,XF,YF,TL,TV,PF,A,X,Y,T,P,K,ERF)

Purpose

FLASH determines the equilibrium vapor and liquid compositions resulting from either an isothermal or adiabatic equilibrium flash vaporization for a mixture of N components ($N \leq 20$).
The subroutine allows for presence of separate vapor and liquid
feed streams for adaption to countercurrent staged processes.
The temperature and composition of each feed stream and the stream
ratios are specified along with a common feed pressure (significant only for the vapor stream) and the flash pressure. For an
isothermal flash the flash temperature is also specified. Resulting vapor and liquid compositions, phase ratios, vaporization
equilibrium ratios, and, for an adiabatic flash, flash temperature
are returned.

Method

A step-limited Newton-Raphson iteration, applied to the
Rachford-Rice objective function, is used to solve for A, the
vapor to feed mole ratio, for an isothermal flash. For an adiabatic flash, an enthalpy balance is included in a two-dimensional
Newton-Raphson iteration to yield both A and T. Details are given
in Chapter 7.

Arguments

Input: TYPE   indicates type of flash; 1 = isothermal, 2 =
               adiabatic

        N   number of components ($N \leq 20$)

        ID(I) vector of indices for the components involved
               (I = 1,N)

        KEY   integer initialization control variable; set
               KEY = 1 on initial call for new system, otherwise
               set KEY = 2

        VF   fraction of the total feed that is vapor (mole
               basis)

        XF(I) composition vector for liquid portion of feed
               (mole fraction) (I = 1,N)

        YF(I) composition vector for vapor portion of feed
               (mole fraction) (I = 1,N)

        TL   temperature of liquid feed stream (K)

        TV   temperature of vapor feed stream (K)

PF   pressure of vapor feed stream (bar)

A   estimate of vapor to feed mole ratio, if known; otherwise set A=0 to activate default initial estimate.

X(I)   vector of estimated equilibrium liquid composition (mole fraction) if known (I = 1, N); otherwise can be any vector not summing to 1.

Y(I)   vector of estimated equilibrium the vapor composition (mole fraction) if known (I = 1, N); otherwise can be any vector not summing to 1.

T   temperature (K) of isothermal flash; for adiabatic flash, estimate of flash temperature if known, otherwise set to 0 to activate default initial estimate.

P   pressure (bars) of flash.

Output: A   calculated vapor to feed mole ratio

X(I)   vector of calculated liquid mole fractions (I = 1, N)

Y(I)   vector of calculated vapor mole fractions (I = 1,N)

T   calculated flash temperature (°K) for adiabatic flash

K(I)   vector of calculated vaporization equilibrium ratios (I = 1, N)

ERF   error flag, integer variable normally zero; ERF = 1 indicates parameters are not available for one or more binary pairs in the mixture; ERF = 2 indicates no solution was obtained; ERF = 3 or 4 indicates the specified flash temperature is less than the bubble-point temperature or greater than the dew-point temperature respectively; ERF = 5 indicates bad input arguments.

Subroutine References
Calls: BUDET
          ENTH
          VALIK

Common Storage
None

320

Glossary of Principal Variable Names

DET   Determinant of Jacobian partial derivative matrix whose elements are given by the partial derivatives of the Rachford-Rice objective function (7-13) and the enthalpy balance (7-14) with respect to temperature and vapor-feed ratio.

DA   Change in the vapor-feed ratio from one iteration to the next

DFA   Partial derivative of the Rachford-Rice objective function (7-13) with respect to the vapor-feed ratio.

DFT   Partial derivative of the Rachford-Rice objective function (7-13) with respect to temperature.

DGA   Partial derivative of the enthalpy balance equation (7-14) with respect to the vapor-feed ratio.

DGT   Partial derivative of the enthalpy balance equation (7-14) with respect to the temperature.

DT   Change in the temperature from one iteration to the next.

EPS   Convergence tolerance.

F   Rachford-Rice objective function (7-13)

G   Enthalpy balance criterion function (7-14).

HG   Enthalpy of combined liquid and vapor feed streams, J/mole.

HL   Enthalpy of exiting liquid stream, J/mole.

HLF   Enthalpy of liquid feed stream, J/mole.

HV   Enthalpy of exiting vapor stream, J/mole.

HVF   Enthalpy of vapor feed stream, J/mole.

IT   Integer iteration counter.

KD   Integer control variable for step-limiting.

KV   Integer control variable for combining liquid and vapor feed streams.

SL   Scaler used for step-limiting, or damping, the Newton-Raphson iteration.

Z(I)   Vector (length 20) of overall feed composition resulting from combined liquid and vapor feeds (I = 1,N).

V(I)   Vector (length 20) of unnormalized liquid compositions (I = 1,N)

W(I)   Vector (length 20) of unnormalized vapor compositions (I = 1,N).

Listing

```
          SUBROUTINE FLASH(TYPE,N,ID,KEY,VF,XF,YF,TL,TV,PF,A,X,Y,T,P,K,ERF)
*  FLASH CONDUCTS ISOTHERMAL (TYPE=1) OR ADIABATIC (TYPE=2) EQUILIBRIUM
*  FLASH VAPORIZATION CALCULATIONS AT GIVEN PRESSURE P(BAR) FOR THE
*  SYSTEM OF N COMPONENTS (N.LE.20) WHOSE INDICES APPEAR IN VECTOR ID.
*  THE SUBROUTINE ACCEPTS BOTH A LIQUID FEED OF COMPOSITION XF AT
*  TEMPERATURE TL(K) AND A VAPOR FEED OF COMPOSITION YF AT TV(K) AND
*  PRESSURE PF(BAR), WITH THE VAPOR FRACTION OF THE FEED BEING VF (MOL
*  BASIS).  FOR AN ISOTHERMAL FLASH THE TEMPERATURE T(K) MUST ALSO BE
*  SUPPLIED.  THE SUBROUTINE DETERMINES THE V/F RATIO A, THE LIQUID AND
*  VAPOR PHASE COMPOSITIONS X AND Y, AND, FOR AN ADIABATIC FLASH, THE
*  TEMPERATURE T(K).   THE EQUILIBRIUM RATIOS K ARE ALSO PROVIDED.  IT
*  NORMALLY RETURNS ERF=0, BUT IF COMPONENT COMBINATIONS LACKING DATA ARE
*  INVOLVED IT RETURNS ERF=1, AND IF NO SOLUTION IS FOUND IT RETURNS ERF
*  =2.  FOR FLASH T.LT.TB OR T.GT.TD FLASH RETURNS ERF=3 OR 4
*  RESPECTIVELY, AND FOR BAD INPUT DATA IT RETURNS ERF=5.
*  IF ESTIMATES ARE AVAILABLE FOR A,X,Y,(AND T) THEY CAN BE ENTERED
*  IN THESE VARIABLES - OTHERWISE THESE VARIABLES SHOULD BE SET TO ZERO.
*  KEY SHOULD BE 1 ON INITIAL CALL FOR A NEW SYSTEM AND 2 OTHERWISE.
          REAL XF(N),YF(N),X(N),Y(N),K(N),Z(20),U(20),W(20),K1
          INTEGER ID(N),TYPE,ERF,ER,EB
          DATA EPS/0.001/
      100 ERF=0
          KEE=KEY
          KV=1
*  CHECK IF FEED IS LIQUID (KV=1), VAPOR (KV=3), OR BOTH (KV=2).
          IF(VF.GT.0.001) KV=2
          IF(VF.GT.0.999) KV=3
          SX=0.
          SY=0.
          SXF=0.
          SYF=0.
*  SUM FEED AND PHASE MOLE FRACTIONS
          DO 109 I=1,N
          SX=SX+X(I)
          SY=SY+Y(I)
          SXF=SXF+XF(I)
      109 SYF=SYF+YF(I)
*  CHECK THAT SUM OF FEED MOL FRACTIONS IS UNITY
          IF(KV.LE.2.AND.ABS(SXF-1.).GT.0.01) GO TO 903
          IF(KV.GE.2.AND.ABS(SYF-1.).GT.0.01) GO TO 903
*  CHECK THAT FLASH PRESSURE IS WITHIN LIMITS.
          IF(P.LT.1.E-6.OR.P.GT.100.) GO TO 903
      110 GO TO (112,114,116),KV
*  FOR LIQUID FEED USE XF AS ESTIMATE FOR X AND Y IF NOT GIVEN AND SET
*  TOTAL FEED Z TO XF
      112 DO 113 I=1,N
          IF(ABS(SX-1.).GT.0.01) X(I)=XF(I)
          IF(ABS(SY-1.).GT.0.01) Y(I)=XF(I)
      113 Z(I)=XF(I)
          GO TO 120
*  FOR MIXED FEED USE XF AS ESTIMATE FOR X AND YF FOR Y IF NOT GIVEN AND
*  FIND TOTAL FEED Z.
      114 DO 115 I=1,N
          IF(ABS(SX-1.).GT.0.01) X(I)=XF(I)
          IF(ABS(SY-1.).GT.0.01) Y(I)=YF(I)
      115 Z(I)=(1.-VF)*XF(I)+VF*YF(I)
          GO TO 120
*  FOR VAPOR FEED USE YF AS ESTIMATE FOR X AND Y IF NOT GIVEN AND SET
*  TOTAL FEED Z TO YF
      116 DO 117 I=1,N
          IF(ABS(SX-1.).GT.0.01) X(I)=YF(I)
          IF(ABS(SY-1.).GT.0.01) Y(I)=YF(I)
      117 Z(I)=YF(I)
*  INITIALIZE LIQUID AND VAPOR PRODUCT COMPOSITIONS
      120 DO 121 I=1,N
```

```
      U(I)=XF(I)
  121 W(I)=YF(I)
* FIND INITIAL ESTIMATE FOR A (IF NOT GIVEN) FOR ISOTHERMAL FLASH
      TB=TL
      IF(T.GT.200..AND.T.LT.600.) TB=T
      TD=TB
      CALL BUDET(1,N,ID,KEE,Z,W,TB,P,K,EB)
      IF(EB.GT.1) GO TO 900
      CALL BUDET(2,N,ID,2,U,Z,TD,P,K,ER)
      IF(ER.GT.1) GO TO 900
      IF(TYPE.EQ.2) GO TO 125
      IF(EB.LT.-2) TB=(T-TD*VF)/(1.1-VF)
      IF(T.LT.TB) GO TO 901
      IF(T.GT.TD) GO TO 902
      IF(A.LT.0.001.OR.A.GT.0.999) A=(T-TB)/(TD-TB)
      GO TO 150
* FIND FEED ENTHALPY FOR ADIABATIC FLASH
  125 HLF=0.
      HVF=0.
      IF(PF.LT.1.E-6.OR.PF.GT.100.) GO TO 903
      IF(KV.GT.2) GO TO 126
      IF(TL.LT.200..OR.TL.GT.600.) GO TO 903
      CALL ENTH(N,ID,10,0,XF,TL,PF,HLF,ER)
      IF(ER.GT.1) GO TO 900
  126 IF(KV.LT.2) GO TO 127
      IF(TV.LT.200..OR.TV.GT.600.) GO TO 903
      CALL ENTH(N,ID,10,2,YF,TV,PF,HVF,ER)
      IF(ER.GT.1) GO TO 900
  127 HF=(1.-VF)*HLF+VF*HVF
* FIND INITIAL ESTIMATES FOR A AND T (IF NOT GIVEN) FOR ADIABATIC FLASH.
      IF(EB.LT.-2) TB=200.
      IF(A.GT.0.001.AND.A.LT.0.999) GO TO 128
      CALL ENTH(N,ID,10,0,Z,TB,P,HL,ER)
      IF(ER.GT.1) GO TO 900
      CALL ENTH(N,ID,10,2,Z,TD,P,HV,ER)
      IF(ER.GT.1) GO TO 900
      IF(HF.LT.HL) GO TO 901
      IF(HF.GT.HV) GO TO 902
      A=(HF-HL)/(HV-HL)
  128 IF(T.GT.TB.AND.T.LT.TD) GO TO 150
      T=TB+A*(TD-TB)
* INITIALIZE ITERATIVE SOLUTION
  150 IT=0
      SL=1.
      KEE=2
      IF(TYPE.EQ.2) KEE=6
      KD=0
      TN=T
      AN=A
      GO TO 250
* CONDUCT NEWTON-RAPHSON ITERATION (200 SERIES STATEMENTS).
  200 IT=IT+1
      IF(IT.EQ.1) GO TO 205
      IF(TYPE.EQ.2) GO TO 203
      IF(ABS(DFA).GE.0.1) GO TO 203
      EPF=EPS*10.*ABS(DFA)
      IF(ABS(F).LE.EPF) GO TO 300
      GO TO 204
* CHECK CONVERGENCE OF OBJECTIVE FUNCTION
  203 IF(FV.LE.EPS) GO TO 300
* EXIT WITHOUT SOLUTION AFTER 20 ITERATIONS
  204 IF(IT.GT.20) GO TO 900
  205 FO=FV
      KD=0
```

323

```
        IF(TYPE.NE.2) GO TO 280
*  DETERMINE KS AND ENTHALPIES AT 0.2 K T INCREASE FOR T DERIVATIVES
*  (ADIABATIC)
   210 CALL VALIK(N,ID,7,X,Y,T+0.2,P,K,ER)
        IF(ER.GT.1) GO TO 900
        CALL ENTH(N,ID,7,0,X,T+0.2,P,HL,ER)
        IF(ER.GT.1) GO TO 900
        CALL ENTH(N,ID,7,2,Y,T+0.2,P,HV,ER)
        IF(ER.GT.1) GO TO 900
        FP=0.
*  UPDATE PHASE COMPOSITIONS (ADIABATIC)
   211 DO 214 I=1,N
        X(I)=U(I)/SX
        Y(I)=W(I)/SY
        K1=K(I)-1.
   214 FP=FP+K1*Z(I)/(K1*A+1.)
        GP=(A*(HV-HL)+HL)/HF-1.
*  FIND TEMPERATURE DERIVATIVES BY FINITE DIFFERENCE (ADIABATIC)
        DFT=(FP-F)*5.
        DGT=(GP-G)*5.
*  SOLVE 2 DIMENSIONAL NEWTON-RAPHSON ITERATION FOR A AND T CORRECTIONS
*  (ADIABATIC)
   220 DET=DFA*DGT-DFT*DGA
        DA=(F*DGT-G*DFT)/DET
        DT=(G*DFA-F*DGA)/DET
   230 AN=A-SL*DA
*  LIMIT A TO RANGE 0 - 1
        IF(AN.LT.0.) AN=0.01
        IF(AN.GT.1.) AN=0.99
        IF(TYPE.NE.2) GO TO 239
        TN=T-SL*DT
*  LIMIT T TO RANGE TB - TD (ADIABATIC)
        IF(TN.LT.TB) GO TO 235
        IF(TN.GT.TD) GO TO 237
        GO TO 250
*  CORRECT Y FOR T SET TO TB
   235 IF(EB.LT.-2) GO TO 903
        CALL BUDET(1,N,ID,2,Z,Y,TB,P,K,EB)
        IF(EB.GT.1) GO TO 900
        TN=TB
        GO TO 251
*  CORRECT X FOR T SET TO TD
   237 CALL BUDET(2,N,ID,2,X,Z,TD,P,K,ER)
        IF(ER.GT.1) GO TO 900
        TN=TD
        GO TO 251
   239 IF(KD.EQ.1) GO TO 251
*  GET NEW K VALUES
   250 CALL VALIK(N,ID,KEE,X,Y,TN,P,K,ER)
        IF(ER.GT.1) GO TO 900
   251 SX=0.
        SY=0.
        DFA=0.
*  FIND EQUILIBRIUM OBJECTIVE FUNCTION F AND UNNORMALIZED COMPOSITIONS
   252 DO 254 I=1,N
        K1=K(I)-1.
        U(I)=Z(I)/(K1*AN+1.)
        W(I)=K(I)*U(I)
        SX=SX+U(I)
        SY=SY+W(I)
*  FIND DERIVATIVE OF F WRT A
        DFA=DFA-(W(I)-U(I))**2/Z(I)
   254 CONTINUE
```

```
      F=SY-SX
      IF(TYPE.NE.2) GO TO 260
* FIND ENTHALPY OBJECTIVE FUNCTION G (ADIABATIC)
  255 CALL ENTH(N,ID,KEE,0,X,TN,P,HL,ER)
      IF(ER.GT.1) GO TO 900
      CALL ENTH(N,ID,KEE,2,Y,TN,P,HV,ER)
      IF(ER.GT.1) GO TO 900
      G=(AN*(HV-HL)+HL)/HF-1.
* FIND DERIVATIVE OF G WRT A (ADIABATIC)
      DGA=(HV-HL)/HF
* FIND NORM OF OBJECTIVE FUNCTION AND CHECK FOR DECREASE
  260 FV=ABS(F)
      IF(TYPE.EQ.2) FV=SQRT((F*F+G*G)/2.)
      IF(IT.EQ.0) GO TO 200
      IF(IT.LE.2) GO TO 270
      IF(KD.EQ.2) GO TO 270
      IF(FV.LE.FO) GO TO 270
* APPLY STEP-LIMITING PROCEDURE TO DECREASE OBJECTIVE FUNCTION
  265 KD=1
      SL=0.7*SL
* CHECK FOR FAILURE OF STEP-LIMITING
      IF(SL.LT.0.20) GO TO 268
      GO TO 230
* ABANDON STEP-LIMITING AND PROCEED
  268 AN=A
      TN=T
      KD=2
      GO TO 250
* PROCEED TO NEXT ITERATION
  270 A=AN
      T=TN
      SL=1.
      GO TO 200
* FIND NEWTON-RAPHSON CORRECTION TO A (ISOTHERMAL)
  280 DA=F/DFA
* UPDATE PHASE COMPOSITIONS (ISOTHERMAL)
  281 DO 284 I=1,N
      X(I)=U(I)/SX
  284 Y(I)=W(I)/SY
      GO TO 230
* FOR CONVERGED ITERATION GET FINAL NORMALIZED PHASE COMPOSITIONS
  300 DO 304 I=1,N
      X(I)=U(I)/SX
  304 Y(I)=W(I)/SY
      ERF=ER
      RETURN
* ON FAILURE TO CONVERGE ITERATION SET A TO 0 (T TO 0 ADIABATIC) AND
* ERF TO 2.
  900 ERF=2
      GO TO 905
* FOR T LESS THAN TB SET ERF TO 3 (A TO 0)
  901 ERF=3
      GO TO 905
* FOR T GREATER THAN TD SET ERF TO 4 (A TO 1)
  902 ERF=4
      A=1.
      GO TO 906
* FOR BAD INPUT DATA SET ERF TO 5 (A TO 0)
  903 ERF=5
  905 A=0.
  906 IF(TYPE.EQ.2) T=0.
      RETURN
      END
```

SURBOUTINE BUDET (TYPE,N,ID,KEY,X,Y,T,P,K,ERR)

Purpose
   BUDET calculates the bubble-point temperature or dew-point
temperature for a mixture of N components (N ≤ 20) at specified
pressure and liquid or vapor composition. The subroutine also
furnishes the composition of the incipient vapor or liquid and
the vaporization equilibrium ratios.

Method
   Bubble-point temperature or dew-point temperatures are calculated iteratively by applying the Newton-Raphson iteration to
the objective functions given by Equations (7-23) or (7-24)
respectively.

Arguments
Input: TYPE indicates type of calculation; 1 = bubble-point
              temperature, 2 = dew-point temperature.
           N number of components (N ≤ 20).
       ID(I) vector of indices for the components (I = 1,N).
         KEY integer initialization control variable; set
              KEY = 1 on initial call for a new system; other-
              wise set KEY = 2.
        X(I) vector of liquid mole functions for a bubble-point
              temperature calculation (I = 1,N). An estimate
              of incipient liquid composition can be provided
              here for a dew-point calculation.
        Y(I) vector of vapor mole fractions for a dew-point
              temperature calculation (I = 1,N). An estimate of
              incipient vapor composition can be provided here
              for a bubble-point calculation.
           T initial estimate of bubble or dew-point temperature
              (°K). Otherwise set to 0 to activate default ini-
              tial estimate.
           P pressure, bars.

Output: X(I) vector of calculated liquid-phase mole fractions at
              equilibrium with the specified vapor phase (I=1,N).
        Y(I) vector of calculated vapor-phase mole fractions
              at equilibrium with the specified liquid phase
              (I = 1, N).

326

T   calculated bubble or dew-point temperature (°K).

K(I)   vector of calculated vaporization equilibrium ratios
(I = 1, N).

ERR   error flag, integer variable normally zero; ERR = 1
indicates parameters are not available for one or
more binary pairs in the mixture, ERR = 2 indicates
no solution was found; ERR = 5 indicates input arguments are inconsistent or are out of range of
the correlations used here.

Subroutine References

Calls: VALIK

Called by: FLASH

Common Storage

None

Glossary of Principal Variable Names

CN(I) Vector (length 20) of intermediate, unconverged values of
the vapor or liquid mole fractions (I = 1, N).

FO   Current value of the objective function given by either
(7-23) or (7-24)

F1   Value of the objective function [(7-23) or (7-24)] at
$T + \Delta T$; used for finite difference approximation of the
derivative.

EPS   Convergence tolerance.

S   Sum of the calculated vapor or liquid mole fractions at
temperature T.

SS   Sum of the calculated vapor or liquid mole fractions at
$T + \Delta T$.

Listing

```
      SUBROUTINE BUDET(TYPE,N,ID,KEY,X,Y,T,P,K,ERR)
*  BUDET CALCULATES BUBBLE (TYPE=1) OR DEW (TYPE=2) POINT TEMPERATURE
*  T(K) FOR GIVEN PRESSURE P(BAR) AND FEED COMPOSITION X (OR Y) FOR
*  THE SYSTEM OF N COMPONENTS (N.LE.20) WHOSE INDICES APPEAR IN ID.
*  IT RETURNS T AND INCIPIENT PHASE COMPOSITION Y (OR X), UTILIZING AN
*  INITIAL ESTIMATE OF T AND Y (OR X) IF SUPPLIED (NE.0). THE EQUILIBRIUM
*  RATIOS K ARE ALSO PROVIDED BY THE SUBROUTINE. THE PROGRAM NORMALLY
*  RETURNS ERR=0, BUT IF COMPONENT COMBINATIONS LACKING DATA ARE INVOLVED
*  IT RETURNS ERR=1, AND IF NO SOLUTION IS FOUND IT RETURNS ERR=2.
*  FOR BAD OR OUT OF RANGE INPUT DATA THE PROGRAM RETURNS ERR=5, AND FOR
*  SYSTEMS WITH BP BELOW 200 K (WITH NONCONDENSABLES) ERR=-5.  KEY
*  SHOULD BE 1 ON INITIAL CALL FOR A NEW SYSTEM AND 2 OTHERWISE.
      REAL X(N),Y(N),K(N),CN(20)
      INTEGER ID(N),TYPE,ERR,ER
      DATA EPS/0.001/
  100 ERR=0
```

```
*  CHECK FOR VALID PRESSURE
       IF(P.LT.1.E-6.OR.P.GT.100.) GO TO 903
       KEE=KEY
       S=0.
       SS=0.
*  CHECK FOR VALID FEED COMPOSITIONS AND FOR ESTIMATE OF INCIPIENT PHASE
*  COMPOSITION
   101 DO 109 I=1,N
       S=S+X(I)
   109 SS=SS+Y(I)
       IF(TYPE.EQ.1.AND.ABS(S-1.).GT.0.01) GO TO 903
       IF(TYPE.EQ.2.AND.ABS(SS-1.).GT.0.01) GO TO 903
   110 IF(TYPE.EQ.1.AND.ABS(SS-1.).GT.0.01) GO TO 114
       IF(TYPE.EQ.2.AND.ABS(S-1.).GT.0.01) GO TO 118
       GO TO 120
*  FOR NO ESTIMATE OF INCIPIENT VAPOR COMPOSITION SET EQUAL TO FEED
   114 DO 115 I=1,N
   115 Y(I)=X(I)
       GO TO 120
*  FOR NO ESTIMATE OF INCIPIENT LIQUID COMPOSITION SET EQUAL TO FEED
   118 DO 119 I=1,N
   119 X(I)=Y(I)
   120 IT=0
*  FOR NO ESTIMATE OF TEMPERATURE SET TO 400 K
   200 IF(T.LT.200..OR.T.GT.600.) T=400.
*  CONDUCT ITERATION STEP
   210 IT=IT+1
       IF(IT.GT.10) GO TO 900
*  GET K VALUES
   220 CALL VALIK(N,ID,KEE,X,Y,T,P,K,ER)
       IF(ER.GT.1) GO TO 900
       S=0.
*  CALCULATE SUM OF KX (BP) OR Y/K (DP)
   221 DO 229 I=1,N
       IF(TYPE.EQ.2) GO TO 225
       CN(I)=K(I)*X(I)
       GO TO 229
   225 CN(I)=Y(I)/K(I)
   229 S=S+CN(I)
   230 F0=ALOG(S)
*  CHECK CONVERGENCE
       IF(ABS(F0).LE.EPS) GO TO 290
*  GET K VALUES AT T+1 FOR FINITE DIFFERENCE DERIVATIVE
       CALL VALIK(N,ID,4,X,Y,T+1.,P,K,ER)
       IF(ER.GT.1) GO TO 900
       SS=0.
       IF(TYPE.EQ.2) GO TO 235
```

```
* CALCULATE NEW VAPOR COMPOSITION FOR BP
  231 DO 234 I=1,N
      Y(I)=CN(I)/S
  234 SS=SS+K(I)*X(I)
      GO TO 240
* CALCULATE NEW LIQUID COMPOSITION FOR DP
  235 DO 239 I=1,N
      X(I)=CN(I)/S
  239 SS=SS+Y(I)/K(I)
  240 F1=ALOG(SS)
* DETERMINE NEW NEWTON-RAPHSON TEMPERATURE ITERATE
      T=(F1-F0)*T/(F1-T*F0/(T+1.))
* CHECK FOR T IN RANGE FOR POSSIBLE CONVERGENCE
      IF(T.GT.700.) GO TO 900
      IF(T.GT.100.) GO TO 245
      IF(ER.LT.0) GO TO 901
      GO TO 900
  245 KEE=2
      IF(TYPE.EQ.1) KEE=5
      GO TO 210
* GET NORMALIZED INCIPIENT PHASE COMPOSITIONS
  290 DO 299 I=1,N
      IF(TYPE.EQ.2) GO TO 295
      Y(I)=CN(I)/S
      GO TO 299
  295 X(I)=CN(I)/S
  299 CONTINUE
* CHECK FOR T IN RANGE OF THERMO SUBROUTINES
      IF(T.GT.600.) GO TO 903
      IF(T.GT.200.) GO TO 199
      IF(ER.LT.0) GO TO 901
      GO TO 903
* SET ERR RETURN FOR MISSING BINARY DATA
  199 IF(IABS(ER).EQ.1) ERR=1
      RETURN
* ON FAILURE TO CONVERGE SET T TO 0 AND ERR TO 2
  900 ERR=2
      GO TO 905
* ON TB LESS THAN 200 K SET T TO 0 AND ERR TO -5
  901 ERR=-5
      GO TO 905
* FOR BAD INPUT DATA (OR TB/TD OUT OF RANGE) SET T TO 0 AND ERR TO 5
  903 ERR=5
  905 T=0.
      RETURN
      END
```

SUBROUTINE BUDEP (TYPE, N, ID, KEY, X, Y, T, P, K, ERR)

Purpose

BUDEP calculates the bubble-point pressure or the dew-point pressure for a mixture of N components (N ≤ 20) at specified temperature and liquid or vapor composition. The subroutine also furnishes the composition of the incipient vapor or liquid and the vaporization equilibrium ratios.

Method

Bubble-point and dew-point pressures are calculated using a first-order iteration procedure described by the footnote to Equation (7-25).

Arguments

Input: TYPE indicates type of calculation; 1 = bubble-point pressure calculation; 2 = dew-point pressure calculation.

        N number of components (N ≤ 20).

     ID(I) vector of indices for the components (I = 1, N).

     KEY integer initialization control variable; set KEY = 1 on initial call for a new system; otherwise set KEY = 2.

     X(I) vector of liquid mole fractions for a bubble-point pressure calculation (I = 1, N). An estimate of incipient liquid composition can be provided here for a dew-point calculation.

     Y(I) vector of vapor-phase mole fractions for a dew-point pressure calculation (I = 1,N). An estimate of incipient vapor composition can be provided here for a bubble-point calculation.

       T temperature (K).

       P initial estimate of bubble or dew-point pressure (bars) if available; otherwise set to 0 to activate default initial estimate.

Output: X(I) vector of calculated liquid-phase mole fractions at equilibrium with the specified vapor composition (I = 1, N).

     Y(I) vector of calculated vapor-phase mole fractions at equilibrium with the specified liquid phase

$(I = 1, N)$.

P calculated bubble or dew-point pressure (bars).

K(I) vector of calculated vaporization equilibrium ratios $(I = 1, N)$.

ERR error flag, integer variable normally zero; ERR = 1 indicates parameters are not available for one or more binary pairs in the mixture; ERR = 2 indicates no solution was found; ERR = 5 indicates input arguments are inconsistent or are out of the range of the correlations used here.

Subroutine References
Calls: VALIK

Common Storage
None

Glossary of Principal Variable Names

CN(I) Vector (length 20) of intermediate (unconverged) values of the vapor or liquid phase mole fractions $(I = 1, N)$.

EPS Convergence tolerance.

S Sum of the calculated vapor or liquid phase mole fractions at pressure P.

Listing

```
      SUBROUTINE BUDEP(TYPE,N,ID,KEY,X,Y,T,P,K,ERR)
* BUDEP CALCULATES BUBBLE (TYPE=1) OR DEW (TYPE=2) POINT PRESSURE
* P(BAR) FOR GIVEN TEMPERATURE T(K) AND FEED COMPOSITION X (OR Y)
* FOR THE SYSTEM OF N COMPONENTS (N.LE.20) WHOSE INDICES APPEAR IN ID.
* IT RETURNS P AND INCIPIENT PHASE COMPOSITION Y (OR X), UTILIZING AN
* INITIAL ESTIMATE OF P AND Y (OR X) IF SUPPLIED (NE.0). THE EQUILIBRIUM
* RATIOS K ARE ALSO PROVIDED BY THE SUBROUTINE.  THE PROGRAM NORMALLY
* RETURNS ERR=0, BUT IF COMPONENT COMBINATIONS LACKING DATA ARE INVOLVED
* IT RETURNS ERR=1, AND IF NO SOLUTION IS FOUND IT RETURNS ERR=2.
* FOR BAD OR OUT OF RANGE INPUT DATA THE PROGRAM RETURNS ERR=5.  KEY
* SHOULD BE 1 ON INITIAL CALL FOR A NEW SYSTEM AND 2 OTHERWISE.
      REAL X(N),Y(N),K(N),CN(20)
      INTEGER ID(N),TYPE,ERR,ER
      DATA EPS/0.001/
  100 ERR=0
* CHECK FOR VALID TEMPERATURE
      IF(T.LT.200..OR.T.GT.600.) GO TO 903
      KEE=KEY
      S=0.
      SS=0.
* CHECK FOR VALID FEED COMPOSITIONS AND FOR ESTIMATE OF INCIPIENT PHASE
* COMPOSITION
  101 DO 109 I=1,N
      S=S+X(I)
```

331

```
      109 SS=SS+Y(I)
          IF(TYPE.EQ.1.AND.ABS(S-1.).GT.0.01) GO TO 903
          IF(TYPE.EQ.2.AND.ABS(SS-1.).GT.0.01) GO TO 903
      110 IF(TYPE.EQ.1.AND.ABS(SS-1.).GT.0.01) GO TO 114
          IF(TYPE.EQ.2.AND.ABS(S-1.).GT.0.01) GO TO 118
          GO TO 120
* FOR NO ESTIMATE OF INCIPIENT VAPOR COMPOSITION SET EQUAL TO FEED.
      114 DO 115 I=1,N
      115 Y(I)=X(I)
          GO TO 120
* FOR NO ESTIMATE OF INCIPIENT LIQUID COMPOSITION SET EQUAL TO FEED
      118 DO 119 I=1,N
      119 X(I)=Y(I)
      120 IT=0
* FOR NO ESTIMATE OF PRESSURE SET TO 1 BAR
      200 IF(P.LT.0.001.OR.P.GT.50.) P=1.
* CONDUCT ITERATION STEP
      210 IT=IT+1
          IF(IT.GT.10) GO TO 900
* GET K VALUES
      220 CALL VALIK(N,ID,KEE,X,Y,T,P,K,ER)
          IF(ER.GT.1) GO TO 900
          S=0.
* CALCULATE SUM OF KX (BP) OR Y/K (DP)
      221 DO 229 I=1,N
          IF(TYPE.EQ.2) GO TO 225
          CN(I)=K(I)*X(I)
          GO TO 229
      225 CN(I)=Y(I)/K(I)
      229 S=S+CN(I)
      230 IF(TYPE.EQ.2) GO TO 235
* CALCULATE NEW VAPOR COMPOSITION FOR BP
      231 DO 234 I=1,N
      234 Y(I)=CN(I)/S
* CORRECT P FOR BP
          P=P*S
          GO TO 240
* CALCULATE NEW LIQUID COMPOSITION FOR DP
      235 DO 239 I=1,N
      239 X(I)=CN(I)/S
* CORRECT P FOR DP
          P=P/S
* CHECK FOR P OUT OF RANGE FOR CONVERGENCE
      240 IF(P.LT.1.E-7.OR.P.GT.150.) GO TO 900
          KEE=2
          IF(TYPE.EQ.1) KEE=5
* CHECK CONVERGENCE
          IF(ABS(S-1.).GT.EPS) GO TO 210
* CHECK FOR P IN RANGE OF THERMO SUBROUTINES
          IF(P.LT.1.E-6.OR.P.GT.100.) GO TO 903
* SET ERR RETURN FOR MISSING BINARY DATA
          IF(IABS(ER).EQ.1) ERR=1
          RETURN
* ON FAILURE TO CONVERGE SET P TO 0 AND ERR TO 2
      900 ERR=2
          GO TO 905
* FOR BAD INPUT DATA (OR PB/PD OUT OF RANGE) SET P TO 0 AND ERR TO 5
      903 ERR=5
      905 P=0.
          RETURN
          END
```

Appendix G

DESCRIPTION AND LISTING OF SUBROUTINE FOR CALCULATION OF
LIQUID-LIQUID EQUILIBRIUM SEPARATIONS

The computer subroutine for calculation of liquid-liquid
equilibrium separations is described and listed in this Appendix.
This is a source routine written in American National Standard
FORTRAN (FORTRAN IV), ANSI X3.9-1978, and, as such, should be
compatible with most computer systems with FORTRAN IV compilers.
The approximate storage requirements for this subroutine are given
in Appendix J; the execution time is strongly dependent on the
separation being calculated but can be estimated (CDC 6400) from
the times given for the thermodynamic subroutines it calls
(essentially all computation effort is in these thermodynamic subroutines).

This subroutine is capable of treating multicomponent systems
with up to 20 components. For compatibility with diverse, user-
written main programs, it employs vectors of length N, where N
($\leq$ 20) is the number of components involved, in its argument list.

333

<u>SUBROUTINE ELIPS</u> (N,ID,KEY,IR,IE,Z,T,A,XR,XE,K,ERR)

Purpose

ELIPS calculates equilibrium phase compositions for a partially miscible liquid system of N components (N $\leq$ 20). The temperature, T, and overall mole fractions, $Z(I)$, of the system must be specified.

Method

Liquid phase compositions and phase ratios are calculated by Newton-Raphson iteration for given K values obtained from LILIK. K values are corrected by a linearly accelerated iteration over the phase compositions until a solution is obtained or until it is determined that calculations are too near the plait point for resolution.

Arguments

Input:
- N — number of components (N $\leq$ 20)
- ID(I) — vector of indices for the components (I = 1,N)
- KEY — integer initialization control variable (KEY = 1 . . . 10) see VALIK
- IR — integer number identifying I for the solvent in the raffinate phase
- IE — integer number identifying I for the solvent in the extract phase
- Z(I) — vector of feed mole fractions (I = 1,N)
- T — system temperature (K)

Output:
- A — fraction of system in extract phase
- XR(I) — vector of calculated raffinate phase mole fractions (I = 1,N)
- XE(I) — vector of calculated extract phase mole fractions (I = 1,N)
- K(I) — vector of equilibrium ratios (I = 1,N)
- ERR — error flag, integer variable normally zero; ERR = 1 indicates parameters are not available for one or more binary pairs in the mixture; ERR = 2 if convergence is not achieved; ERR = 7 if Z(I) is too near the plait point for resolution.

334

Calls: LILIK

Common Storage

None

Glossary of Principal Variable Names

AA   Next to last iterative value of extract-feed ratio.

AN   Temporary new value for extract-feed ratio.

AO   Last iterative value of extract-feed ratio.

DA   Change in extract-feed ratio from one iteration to the next.

DF   Partial derivative of Rachford-Rice objective function with respect to extract-feed ratio.

DI   Fractional change in extract-feed ratio during last two iterations.

ESS   Convergence tolerance.

EX(I)   Vector (length 20) of last iteration values of component mole fractions in extract phase (I = 1,N).

F   Rachford-Rice objective function for liquid-liquid separation

IT   Integer iteration counter.

KAC   Integer control variable for acceleration of phase compositions at alternate iterations.

KEE   Temporary storage of control variable (see KEY).

KP   Temporary storage of equilibrium ratio for raffinate solvent.

KS   Temporary storage of equilibrium ratio for extract solvent.

PPI   Plait point vicinity criterion variable.

RX(I)   Vector (length 20) of last iteration values of component mole fractions in raffinate phase (I = 1,N).

SE   Sum of calculated extract mole fractions.

SL   Previous value of the sum of the absolute values of the deviations in component fugacities.

SR   Sum of calculated raffinate mole fractions.

SZ   Sum of given feed mole fractions.

WG,WK   Weighting functions for Wegstein acceleration.

Listing

```
      SUBROUTINE ELIPS(N,ID,KEY,IR,IE,Z,T,A,XR,XE,K,ERR)
* ELIPS CALCULATES CONJUGATE PHASE COMPOSITIONS XR AND XE FOR PARTIALLY
* MISCIBLE N COMPONENT LIQUID SYSTEMS (N.LE.20) OF OVERALL COMPOSITION
* Z AT GIVEN TEMPERATURE T(K).  INDICES OF THE COMPONENTS INVOLVED
* SHOULD BE IN THE VECTOR ID.  IR AND IE ARE THE COMPONENT NRS OF THE
* R AND E PHASE SOLVENTS (SHOULD BE SET TO 0 IF NO SUCH SOLVENTS
* DESIGNATED).  THE FRACTION OF THE SYSTEM GOING INTO THE E PHASE IS
* RETURNED IN A, WITH A BEING 0 OR 1 (AND XE=XR) IF Z IS IN A SINGLE
* PHASE REGION.  ELIPS NORMALLY RETURNS ERR=0, BUT IF COMPONENT
* COMBINATIONS LACKING DATA ARE INVOLVED IT RETURNS ERR=1, IF
* CONVERGENCE IS NOT ACHIEVED ERR=2, AND IF Z IS TOO NEAR THE PLAIT
* POINT FOR RESOLUTION ERR=7.  KEY SHOULD BE 1 (OR 9) FOR THE FIRST
* CALL WITH NEW SYSTEMS, OTHERWISE 2 (OR 3-8,10).
      REAL Z(N),XR(N),XE(N),K(N),GAR(20),GAE(20),K1(20),KS,KP,K2,
     1 RX(20),EX(20)
      INTEGER ID(N),ERR,ERL
  100 ERR=0
      KEE=KEY
      IF(IR.EQ.0.OR.IE.EQ.0) GO TO 101
      GO TO(101,130,130,130,130,130,130,130,101,130),KEY
* FOR NEW SYSTEMS WITHOUT IR, IE SPECIFIED, FIND IR, IE AS LEAST SOLUBLE
* PAIR
  101 KS=1.
      KP=1.
  110 DO 115 I=1,N
      XR(I)=0.
  115 XE(I)=0.
      J1=1
      IF(IR.NE.0) GO TO 120
      IF(IE.NE.0) GO TO 121
  116 DO 119 J=2,N
      XR(J1)=0.
      XE(J1)=0.
      J1=J-1
      XR(J)=0.98
      XE(J)=0.02
      XR(J-1)=0.
      XE(J-1)=0.
      IF(Z(J).LT.0.10) GO TO 119
      DO 118 I=1,J1
      XE(I)=0.98
      XR(I)=0.02
      IF(I.GT.1) XE(I-1)=0.
      IF(I.GT.1) XR(I-1)=0.
      IF(Z(I).LT.0.10) GO TO 118
      CALL LILIK(N,ID,KEE,XR,XE,T,K,GAR,GAE,ERL)
      IF(ERL.GT.1) GO TO 900
      KEE=3
      IF(K(I).LE.KS) GO TO 117
      KS=K(I)
      IS=I
  117 IF(1./K(J).LE.KS) GO TO 118
      KS=1./K(J)
      IS=J
  118 CONTINUE
  119 CONTINUE
      XR(N)=0.
      XE(N)=0.
      XR(J1)=0.
      XE(J1)=0.
      GO TO 125
  120 IF(IE.NE.0) GO TO 130
      IS=IR
      GO TO 125
```

```
    121 IS=IE
    125 XE(IS)=0.98
        XR(IS)=0.02
    126 DO 129 J=1,N
        IF(J.EQ.IS) GO TO 129
        XR(J)=0.98
        XE(J)=0.02
        IF(J.EQ.(IS+1)) GO TO 128
        IF(J.GT.1) XR(J-1)=0.
        IF(J.GT.1) XE(J-1)=0.
    127 IF(Z(J).LT.0.10) GO TO 129
        CALL LILIK(N,ID,KEE,XR,XE,T,K,GAR,GAE,ERL)
        IF(ERL.GT.1) GO TO 900
        KEE=3
        IF(K(J).GE.KP) GO TO 129
        KP=K(J)
        IP=J
        GO TO 129
    128 IF(J.GT.2) XR(J-2)=0.
        IF(J.GT.2) XE(J-2)=0.
        GO TO 127
    129 CONTINUE
        IE=IP
        IF(IR.NE.IS) IE=IS
        IF(IR.NE.IS) IR=IP
* INITIALIZE R AND E PHASE COMPOSITIONS
    130 DO 131 I=1,N
        XR(I)=0.
    131 XE(I)=0.
        XR(IR)=0.98
        XE(IR)=0.02
        XR(IE)=0.02
        XE(IE)=0.98
* GET INITIAL ESTIMATES FOR K VALUES
    135 CALL LILIK(N,ID,KEE,XR,XE,T,K,GAR,GAE,ERL)
        IF(ERL.GT.1) GO TO 900
        SZ=0.
    136 DO 137 I=1,N
        SZ=SZ+Z(I)
    137 K1(I)=K(I)-1.
        IF(ABS(SZ-1.).GT.0.01) GO TO 903
* GET INITIAL ESTIMATE FOR A
        A=Z(IE)/(Z(IE)+Z(IR))
        A=-A/K1(IR)+(A-1.)/K1(IE)
        IF(A.LT.0.) A=0.
        IF(A.GT.1.) A=1.
        AO=A
        IT=0
        KAC=0
* CONDUCT ITERATION OVER PHASE COMPOSITION (OUTER LOOP)
    200 IT=IT+1
        IF(IT.GT.50) GO TO 900
        AA=AO
        AO=A
* CONDUCT NEWTON-RAPHSON ITERATION FOR A AT FIXED K VALUES
    300 DO 319 M=1,10
        F=0.
        DF=0.
    301 DO 309 I=1,N
        K2=A*K1(I)+1.
        F=F+Z(I)*K1(I)/K2
    309 DF=DF-Z(I)*(K1(I)/K2)**2
        IF(ABS(F).LT.1.E-05) GO TO 210
```

337

```
      DA=F/DF
  310 AN=A-DA
* LIMIT A TO RANGE 0 - 1 IF IT FALLS OUTSIDE POLES OF F.
      IF(AN.LT.-1./K1(IR)) GO TO 312
      AN=A+(1./K1(IR)+A)/2.
      GO TO 319
  312 IF(AN.GT.-1./K1(IE)) GO TO 319
      AN=A+(1./K1(IE)+A)/2.
  319 A=AN
      GO TO 900
  210 A=AN
      SR=0.
      SE=0.
* FIND NEW R AND E PHASE COMPOSITIONS
  211 DO 215 I=1,N
      XR(I)=Z(I)/(A*K1(I)+1.)
      XE(I)=K(I)*XR(I)
      SR=SR+XR(I)
  215 SE=SE+XE(I)
  220 DO 225 I=1,N
      XR(I)=XR(I)/SR
  225 XE(I)=XE(I)/SE
* AT ALTERNATE ITERATIONS AFTER 3 ACCELERATE PHASE COMPOSITIONS BY
* WEGSTEIN METHOD BASED ON SUM OF DEVIATIONS OF COMPONENT FUGACITIES.
  230 IF(IT.LT.3) GO TO 250
      IF(IT.LE.5.AND.(A.LT.0..OR.A.GT.1.)) GO TO 250
      IF(KAC.GE.1) GO TO 239
      IF(SS.GT.SL) GO TO 250
      IF(SS.GT.0.2) GO TO 250
      KAC=1
      WG=SS/(SL-SS)
      WK=1.+WG
      SR=0.
      SE=0.
      DO 235 I=1,N
      XR(I)=WK*XR(I)-WG*RX(I)
* ALLOW NO NEGATIVE MOL FRACTIONS
      IF(XR(I).LT.0.) XR(I)=0.
      XE(I)=WK*XE(I)-WG*EX(I)
      IF(XE(I).LT.0.) XE(I)=0.
      SR=SR+XR(I)
  235 SE=SE+XE(I)
* NORMALIZE ACCELERATED COMPOSITIONS (TO ALLOW FOR XR(I) OR XE(I) SET
* TO 0)
  236 DO 237 I=1,N
      XR(I)=XR(I)/SR
  237 XE(I)=XE(I)/SE
      GO TO 250
  239 KAC=0
* GET NEW K AND GAMMA VALUES
  250 CALL LILIK(N,ID,3,XR,XE,T,K,GAR,GAE,ERL)
      IF(ERL.GT.1) GO TO 900
      ESS=1.E-03
      IF(IT.LE.5) GO TO 252
* CHECK FOR VICINITY OF A PLAIT POINT
      PPI=K(IR)/K(IE)+K(IE)/K(IR)
      IF(PPI.GT.10..OR.SS.GT.0.05) GO TO 251
* EXIT IF TOO NEAR PLAIT POINT
      IF(IT.GE.20) GO TO 290
      IF(PPI.GT.7.) GO TO 251
* CHECK IF CALCULATION NEAR PLAIT POINT IS PROBABLY IN SINGLE PHASE
* REGION--IF SO CONTINUE.
      DE=AO
      IF(DE.GT.0.5) DE=AO-1.
```

```
      DI=(AA-A)/DE
      IF(DI.LT.0.1) GO TO 290
  251 IF(PPI.LT.20) ESS=2.E-04
  252 SL=SS
      SS=0.
  253 DO 255 I=1,N
      K1(I)=K(I)-1.
* CALCULATE OBJECTIVE FUNCTION
  255 SS=SS+ABS(GAE(I)*XE(I)-GAR(I)*XR(I))
* CHECK CONVERGENCE
      IF(SS.LE.ESS) GO TO 190
      IF(A.GE.0..AND.A.LE.1.) GO TO 260
      IF(IT.LT.3.OR.SS.GT.0.20) GO TO 260
      IF(IT.LT.5.AND.SS.GT.0.05) GO TO 260
* CHECK IF A MOVING AWAY FROM 0 - 1 REGION
      IF((ABS(A)-ABS(AO)).GT.0.) GO TO 195
* SAVE LAST PHASE COMPOSITIONS FOR USE IN ACCELERATION
  260 DO 265 I=1,N
      RX(I)=XR(I)
  265 EX(I)=XE(I)
      GO TO 200
* FEED IN VICINITY OF PLAIT POINT--CHECK IF IN TWO PHASE REGION
  290 IF(A.GE.0..AND.A.LE.1.) GO TO 905
      GO TO 195
* DO NOT ALLOW CONVERGENCE ON ACCELERATED ITERATION
  190 IF(KAC.EQ.1) GO TO 260
* CONVERGED SOLUTION--CHECK IF IN TWO PHASE REGION
  191 IF(A.LT.0..OR.A.GT.1.) GO TO 195
      ERR=ERL
      RETURN
* FEED OUTSIDE TWO PHASE REGION
  195 DO 196 I=1,N
      XR(I)=Z(I)
  196 XE(I)=Z(I)
      IF(A.LT.0.) A=0.
      IF(A.GT.1.) A=1.
      ERR=ERL
      RETURN
* ON FAILURE TO CONVERGE SET A TO -1 AND ERR TO 2
  900 ERR=2
      GO TO 910
* FOR BAD INPUT DATA SET A TO -1 AND ERR TO 5
  903 ERR=5
      GO TO 910
* FOR FEED TOO NEAR PLAIT POINT SET A TO -1 AND ERR TO 7
  905 ERR=7
  910 DO 911 I=1,N
      XR(I)=Z(I)
  911 XE(I)=Z(I)
      A=-1.
      RETURN
      END
```

Appendix H

DESCRIPTIONS AND LISTINGS OF SUBROUTINES FOR LOADING AND CHANGING PARAMETERS

The parameters characterizing pure components and their binary interactions are stored in labeled common blocks /PURE/ and /BINARY/ for a maximum of 100 components (see Appendix E). Pure component parameters for 92 components, and as many binary interaction parameters as have been established, are cited in Appendix C. These parameters can be loaded from formated cards, or other input file containing card images, by subroutine PARIN.

The addition of components to this set of 92, the change of a few parameter values for existing components, or the inclusion of additional UNIQUAC binary interaction parameters, as they may become available, is best accomplished by adding or changing cards in the input deck containing the parameters. The formats of these cards are discussed in the subroutine PARIN description. Where many parameters, especially the binary association and solvation parameters (η_{ij}), are to be changed for an existing component, or where such a component is to be replaced by another, it may be more convenient to use the subroutine PARCH provided for this purpose. This subroutine can also be used to add data for new components to the common storage blocks.

The subroutines PARIN and PARCH are source routines written in American National Standard FORTRAN (FORTRAN IV), ANSI X3.9-1978, and should be compatible with most computer systems where input can be taken from logical unit 3.

340

SUBROUTINE PARIN (M, ERIN)

Purpose

PARIN loads values of pure component and binary parameters
from formatted card images into labeled common blocks /PURE/ and
/BINARY/ for a maximum of 100 components.

Method

PARIN first loads all pure component data by reading two
records per component. The total number of components, M, in the
library or data deck must be known beforehand. Next the associ-
ation/solvation parameters are input for M components. Finally
all the established UNIQUAC binary interaction parameters (or
noncondensable-condensable interaction parameters) are read.

Arguments

Input: M   number of components in library

Output: ERIN   integer error flag
                0 = no discrepancy
                5 = discrepancy detected in input data file

Input Data

Formatted card images arranged as follows:

Set of 2 cards for each component (2M total); component
    index J

    Card 1: J, NM1(J), NM2(J), TC(J), PC(J), A(J), RD(J),
           DM(J), RU(J), QU(J), QP(J)
           FORMAT(I3, 2A10, 2F7.2, F5.4, F6.4, 4F5.2)

    Card 2: J, C1(J), C2(J), C3(J), C4(J), C5(J)
           FORMAT(I3, 7X, 5E14.8)

    Card 3: J, D1(J), D2(J), D3(J), D4(J)
           FORMAT(I3, 7X, 4E13.7)

    (continue until a total of 2M cards has been read)

Blank card

Set of $(M_{18} + 1)(M - 9M_{18})$ cards containing binary asso-
        ciation/solvation parameters

    Card 1: $\eta_{1,1}$ FORMAT(18F4.2)
    Card 2: $\eta_{2,1}$ $\eta_{2,2}$

      $\vdots$
      $\vdots$

341

Card 18: $n_{18,1}$ $n_{18,2}$ $\cdots$ $n_{18,18}$
Card 19: $n_{19,1}$ $n_{19,2}$ $\cdots$ $n_{19,18}$
Card 20: $n_{19,19}$
Card 21: $n_{20,1}$ $n_{20,2}$ $\cdots$ $n_{20,18}$
Card 22: $n_{20,19}$ $n_{20,20}$

$\vdots$

Card 53: $n_{36,1}$ $n_{36,2}$ $\cdots$ $n_{36,18}$
Card 54: $n_{36,19}$ $n_{36,20}$ $\cdots$ $n_{36,36}$
Card 55: $n_{37,1}$ $n_{37,2}$ $\cdots$ $n_{37,18}$
Card 56: $n_{37,19}$ $n_{37,20}$ $\cdots$ $n_{37,36}$
Card 57: $n_{37,37}$

$\vdots$

Card 107: $n_{54,19}$ $n_{54,20}$ $\cdots$ $n_{54,36}$
Card 108: $n_{54,37}$ $n_{54,38}$ $\cdots$ $n_{54,54}$
etc.

Blank card

Set of cards for UNIQUAC binary interaction parameters
up to M(M-1)/2 cards); component indices I
and J

Card 1: I, J, U(I,J), U(J,I)   FORMAT(2I5, 2F10.2)
Card 2: (same as card 1 but for a new IJ binary)

$\vdots$

etc.

Blank card

Subroutine References
None

Common Storage
/PURE/
/BINARY/

Glossary of Principal Variable Names
See labeled common storage description for /PURE/ and /BINARY/
(Appendix E).

Listing

```
      SUBROUTINE PARIN(M,ERIN)
* PARIN READS PURE COMPONENT AND BINARY PARAMETERS FROM FORMATTED CARDS,
* OR OTHER FILES CONTAINING EQUIVALENT RECORDS, INTO COMMON
* STORAGE BLOCKS /PURE/ AND /BINARY/ FOR A LIBRARY OF M (LE.100)
* COMPONENTS.  INPUT IS TAKEN FROM LOGICAL UNIT 3.  PARIN RETURNS ERIN=0
* UNLESS A DISCREPANCY IS DETECTED IN THE INPUT FILE, IN WHICH CASE IT
* RETURNS ERIN = 5.
      INTEGER ERIN
      COMMON/PURE/NM1(100),NM2(100),TC(100),PC(100),RD(100),DM(100),
     1    A(100),C1(100),C2(100),C3(100),C4(100),C5(100),RU(100),QU(100),
     2    QP(100),D1(100),D2(100),D3(100),D4(100)
      COMMON/BINARY/ETA(5050),U(100,100)
  100 ERIN=0
      IF(M.GT.100) GO TO 900
* READ IN PURE COMPONENT PARAMETERS
      DO 109 I=1,M
* FIRST CARD FOR PURE COMPONENT
      READ(3,01) J,NM1(J),NM2(J),TC(J),PC(J),A(J),RD(J),DM(J),RU(J),
     1    QU(J),QP(J)
   01 FORMAT(I3,2A10,2F7.2,F5.4,F6.4,4F5.2)
* CHECK CARD SEQUENCE
      IF(J.NE.I) GO TO 900
* SECOND CARD FOR PURE COMPONENT
      READ(3,02) J,C1(J),C2(J),C3(J),C4(J),C5(J)
   02 FORMAT(I3,7X,5E14.8)
      IF(J.NE.I) GO TO 900
* THIRD CARD FOR PURE COMPONENT
      READ(3,03) J,D1(J),D2(J),D3(J),D4(J)
   03 FORMAT(I3,7X,4E13.7)
      IF (J.NE.I) GO TO 900
  109 CONTINUE
* CHECK FOR REQUIRED BLANK CARD SEPARATOR
      READ(3,01) J
      IF(J.NE.0) GO TO 900
* READ IN BINARY ASSOCIATION PARAMETERS ETA
  110 DO 119 I=1,M
      I1=(I-1)*I/2+1
      I2=(I-1)*I/2+I
      READ(3,04) (ETA(IJ),IJ=I1,I2)
   04 FORMAT(18F4.2)
  119 CONTINUE
* CHECK FOR REQUIRED BLANK CARD SEPARATOR
      READ(3,04) E
      IF(ABS(E).GT.1.E-19) GO TO 900
* INITIALLY ZERO UNIQUAC BINARY INTERACTION PARAMETERS
  120 DO 121 I=1,M
      DO 121 J=1,M
  121 U(I,J)=0.
* READ IN UNIQUAC BINARY PARAMETERS
  125 READ(3,05) I,J,UIJ,UJI
   05 FORMAT(2I5,2F10.2)
* TERMINATE READ ON BLANK FINAL CARD
      IF(I.EQ.0) GO TO 130
      U(I,J)=UIJ
      U(J,I)=UJI
      GO TO 125
  130 DO 139 I=1,M
* SET U(I,I) TO 1E+20 FOR NONCONDENSABLE I
      IF(A(I).LT.1.E-19) U(I,I)=1.E+20
  139 CONTINUE
      RETURN
* ERROR RETURN FOR DISCREPANCY IN INPUT DATA FILE
  900 ERIN=5
      RETURN
      END
```

SUBROUTINE PARCH (M, N, ERP)

Purpose

PARCH changes all parameters for N components in the common storage blocks /PURE/ and /BINARY/ either by replacing a previous component and/or its parameters or by adding components, thus increasing the library of M ($\leq$ 100) components.

Method

PARCH reads update information from formatted cards, or other input file containing card images. Input having existing component indices replace old data with those indices. Otherwise the new component data is added to the library, with the restriction that the total number of components is always less than or equal to 100.

Arguments

Input: M starting number of components in library

             N number of components being added or changed

Output: M final number of components in library

      ERIN integer error flag

           0 = no discrepancy

           5 = discrepancy detected in input data file

Input Data

Formatted card images arranged as follows:

Set of cards for each component added and/or changed; component index J

1st pure component card: J, NM1(J), NM2(J), TC(J), PC(J), A(J), RD(J), DM(J), RU(J), QU(J), QP(J) FORMAT(I3, 2A10, 2F7.2, F5.4, F6.4, 4F5.2)

2nd pure component card: J, C1(J), C2(J), C3(J), C4(J), C5(J)   FORMAT(I3, 7X, 5E14.8)

3rd pure component card: J, D1(J), D2(J), D3(J), D4(J)   FORMAT(I3, 7X, 4E13.7)

$(M - 1)_{18} + 1$ binary association/solvation parameter cards:

ET(IJ)   18 per card   FORMAT(18F4.2)

giving M ηs for new component J with all components in library in sequence.

344

Series of cards (up to M - l cards) for UNIQUAC binary
interaction parameters: I, J, U(I,J),
U(J,I)   FORMAT(2I5, 2F10.2)   giving all
known interaction parameters.
    Blank card

Subroutine References
    None

Common Storage
/PURE/

/BINARY/

Glossary of Principal Variable Names
See labeled common storage description for /PURE/ and /BINARY/
(Appendix E).

Listing

```
      SUBROUTINE PARCH(M,N,ERP)
* PARCH CHANGES THE PARAMETERS FOR N COMPONENTS IN THE COMMON STORAGE
* BLOCKS /PURE/ AND /BINARY/, EITHER BY REPLACING A PREVIOUS COMPONENT
* AND/OR ITS PARAMETERS OR BY ADDING COMPONENTS. INCREASING THE LIBRARY
* OF M (LE.100) COMPONENTS. FORMATTED INPUT (CARDS OR EQUIVALENT) IS
* TAKEN FROM LOGICAL UNIT 3.  PARCH RETURNS ERP=0 UNLESS A DISCREPRANCY
* IS DETECTED IN THE INPUT FILE. IN WHICH CASE IT RETURNS ERP=5.
      REAL ET(18)
      INTEGER ERP
      COMMON/PURE/NM1(100),NM2(100),TC(100),PC(100),RD(100),DM(100),
     1  A(100),C1(100),C2(100),C3(100),C4(100),C5(100),RU(100),QU(100),
     2  QP(100),D1(100),D2(100),D3(100),D4(100)
      COMMON/BINARY/ETA(5050),U(100,100)
  100 ERP=0
      IF(M.GT.100) GO TO 900
      IN=0
* START INPUT FOR ONE ADDED/REPLACED COMPONENT
  110 IN=IN+1
      IF(IN.GT.N) RETURN
* FIRST CARD FOR PURE COMPONENT
      READ(3,01) J,NM1(J),NM2(J),TC(J),PC(J),A(J),RD(J),DM(J),RU(J),
     1  QU(J),QP(J)
   01 FORMAT(I3,2A10,2F7.2,F5.4,F6.4,4F5.2)
* CARD/S MISSING
      IF(J.EQ.0) GO TO 900
      JS=J
      IF(J.LE.M) GO TO 112
* ADDED COMPONENT
      M=M+1
      IF(M.GT.100) GO TO 900
* WRONG INDEX FOR ADDED COMPONENT
      IF(J.NE.M) GO TO 900
* SECOND CARD FOR PURE COMPONENT
  112 READ(3,02) J,C1(J),C2(J),C3(J),C4(J),C5(J)
   02 FORMAT(I3,7X,5E14.8)
```

```
*  INDICES DO NOT MATCH
      IF(J.NE.JS) GO TO 900
*  THIRD CARD FOR PURE COMPONENT
      READ(3,03) J,D1(J),D2(J),D3(J),D4(J)
   03 FORMAT(I3,7X,4E13.7)
      IF(J.NE.JS) GO TO 900
  120 ME=(M-1)/18+1
*  READ IN BINARY ASSOCIATION PARAMETERS ETA
      DO 129 L=1,ME
      READ(3,04) (ET(K),K=1,18)
   04 FORMAT(18F4.2)
      DO 124 K=1,18
      I=(L-1)*18+K
      IF(I.GT.M) GO TO 130
      IJ=(I-1)*I/2+J
      IF(J.GT.I) IJ=(J-1)*J/2+I
  124 ETA(IJ)=ET(K)
  129 CONTINUE
*  INITIALLY ZERO UNIQUAC BINARY PARAMETERS FOR NEW COMPONENT
  130 DO 131 I=1,M
      U(J,I)=0.
  131 U(I,J)=0.
      IF(A(J).LT.1.E-19) U(J,J)=1.E+20
*  READ IN UNIQUAC BINARY PARAMETERS FOR NEW COMPONENT
  135 READ(3,05) I,J,UIJ,UJI
   05 FORMAT(2I5,2F10.2)
*  TERMINATE READ ON BLANK FINAL CARD
      IF(I.EQ.0) GO TO 110
      IF(J.NE.JS) GO TO 900
      U(I,J)=UIJ
      U(J,I)=UJI
      GO TO 135
*  ERROR RETURN FOR DISCREPANCY IN INPUT DATA FILE
  900 ERP=5
      RETURN
      END
```

Appendix I

DRIVER PROGRAMS FOR VAPOR-LIQUID AND LIQUID-LIQUID
EQUILIBRIUM SEPARATION CALCULATIONS

Examples of main programs calling subroutines FLASH and
ELIPS for vapor-liquid and liquid-liquid separation calculations,
respectively, are described in this Appendix. These are intended
only to illustrate the use of the subroutines and to provide a
means of quickly evaluating their performance on systems of
interest. It is expected that most users will write their own
main prograns utilizing FLASH and ELIPS, and the other subroutines
presented in this monograph, to suit the requirements of their
separation calculations.

The programs DRFLA for vapor-liquid and DRELI for liquid-
liquid calculations are written in FORTRAN IV source language for
the CDC 6400 of the Computer Center, University of California,
Berkeley. Minor modifications, mostly with regard to input and
output, will be required for implementation on most other computer
systems.

347

PROGRAM DRFLA

Purpose
Illustrates use of subroutine FLASH for vapor-liquid
equilibrium separation calculations for up to 10 components and
of subroutine PARIN for parameter loading.

Subroutine References
Calls: PARIN

            FLASH

            BUDET

Input (see Glossary)
(Following input deck for PARIN, if read from cards.)   For each
case to be calculated:

      Card 1a: N, L, T, P, (ID(I), F(I), I = 1,N($\leq$ 5))

                FORMAT(2I5, F10.2, F10.3, 5(I4, F6.3))

     [Card 1b: (ID(I), F(I), I = 6,N) for N > 5 FORMAT(5(I4,

                F6.3))]

      Card 2a: TF, PF, VF, TV, (V(I), I = 1,N($\leq$ 5))

                FORMAT(F10.2, 2F10.3, F10.2, 5F8.3)

     [Card 2b: (V(I), I = 6,N) for N > 5 FORMAT(5F8.3)]

     Blank card

Output
130 columns of printout giving type of flash; components;
liquid and vapor feed compositions; temperature and pressure;
vapor fraction of feed; bubble and dew-point temperature of
the feed at flash pressure; flash pressure and temperature;
fractional vaporization occurring in the flash; and liquid
and vapor product mole fractions.

Listing

```
      PROGRAM DRFLA(INPUT,OUTPUT,TAPE3,TAPE7=INPUT,TAPE4=OUTPUT)
*  DRIVER PROGRAM FOR SUBROUTINES FLASH AND BUDET FOR SYSTEMS OF UP TO
*  10 COMPONENTS WITH VAPOR AND LIQUID FEED STREAMS.
      REAL Z(10),X(10),Y(10),K(10),V(10),BD(2),F(10)
      INTEGER ID(10),ER
      COMMON/PURE/NM1(100),NM2(100),TC(100),PC(100),RD(100),DM(100),
     1  A(100),C1(100),C2(100),C3(100),C4(100),C5(100),RU(100),QU(100),
     2  QP(100),D1(100),D2(100),D3(100),D4(100)
      COMMON/BINARY/ETA(5050),U(100,100)
100 CALL PARIN(92,ER)
      IF(ER.GT.0) GO TO 900
```

```
  200 READ(7,02) N,L,T,P,(ID(I),F(I),I=1,N)
   02 FORMAT(I5,I5,F10.2,F10.3,(5(I4,F6.3)))
      IF(N.EQ.0) STOP
      READ(7,03) TF,PF,VF,TV,(V(I),I=1,N)
   03 FORMAT(F10.2,2F10.3,F10.2,(5F8.3))
      IF(L.EQ.1) WRITE(4,11)
      IF(L.EQ.2) WRITE(4,12)
      WRITE(4,13)
   11 FORMAT(////48X,*FEED*,26X,*TB/TD*,15X,*ISOTHERMAL FLASH*)
   12 FORMAT(////48X,*FEED*,26X,*TB/TD*,15X,*ADIABATIC FLASH*)
   13 FORMAT(1X,*INDEX*,5X,*COMPONENT*,10X,*XF*,4X,*TF(K)   PF(BAR)*,
      1   2X,*VFRACT    YF*,4X,*TV(K)*,6X,* (K) *,5X,*P(BAR)    T(K)*,
      2   4X,*V/F*,6X,*X*,7X,*Y*/)
      DO 207 I=1,2
  207 BD(I)=0.
      DO 209 I=1,N
      Z(I)=(1.-VF)*F(I)+VF*V(I)
      X(I)=0.
  209 Y(I)=0.
  210 CALL BUDET(1,N,ID,1,Z,Y,BD(1),P,K,ER)
      CALL BUDET(2,N,ID,2,X,Z,BD(2),P,K,ER)
  220 DO 221 I=1,N
      X(I)=0.
  221 Y(I)=0.
      Q=0.
      CALL FLASH(L,N,ID,2,VF,F,V,TF,TV,PF,Q,X,Y,T,P,K,ER)
      IF(ER.GT.0) WRITE(4,15) ER
   15 FORMAT(/* ERROR IN FLASH*,I5/)
      II=ID(1)
      WRITE(4,16) II,NM1(II),NM2(II),F(1),TF,PF,VF,V(1),TV,BD(1),P,T,Q,
      1   X(1),Y(1)
      II=ID(2)
      WRITE(4,17) II,NM1(II),NM2(II),F(2),V(2),BD(2),X(2),Y(2)
   16 FORMAT(I3,2X,2A10,F8.3,F8.1,F7.1,F9.3,F7.3,F8.1,F12.2,F9.1,F8.1,
      1   3F8.4)
   17 FORMAT(I3,2X,2A10,F8.3,24X,F7.3,8X,F12.2,25X,2F8.4)
   18 FORMAT(I3,2X,2A10,F8.3,24X,F7.3,45X,2F8.4)
      IF(N.LT.3) GO TO 230
      DO 228 I=3,N
      II=ID(I)
      WRITE(4,18) II,NM1(II),NM2(II),F(I),V(I),X(I),Y(I)
  228 CONTINUE
  230 GO TO 200
  900 WRITE(4,19)
   19 FORMAT(/* ERROR IN PARAMETER INPUT DECK*/)
      STOP
      END
```

PROGRAM DRELI

Purpose

Illustrates use of subroutine ELIPS for liquid-liquid
equilibrium separation calculations for up to 10 components and
of subroutine PARIN for parameter loading.

Subroutine References

Calls: PARIN

               ELIPS

Input (see Glossary)

(Following input deck for PARIN, if read from cards.) For each
system to be calculated, 1 card:

    N, T, (ID(I), I = 1,N) FORMAT(I2, 8X, F10.2, 10I5)

For each case to be calculated for the system, 1 or 2 cards:

    IR, IE, (Z(I), I = 1,N($\leq$ 7)) FORMAT(2I5, 7F10.4)

    (Z(I), I = 8, N(> 7)) FORMAT(3F10.4)

    Blank card

Output

80 columns of printout giving number of components in
liquid-liquid system; system temperature; component indices;
component names; feed composition; raffinate and extract
compositions; K values; and ratio of total moles in extract
phase to total moles in raffinate phase.

Listing

```
      PROGRAM DRELI(INPUT,OUTPUT,TAPE3,TAPE7=INPUT,TAPE4=OUTPUT)
* DRIVER PROGRAM FOR SUBROUTINE ELIPS FOR SYSTEMS OF UP TO 10 COMPONENTS
      REAL Z(10),X(10),Y(10),K(10)
      INTEGER ID(10),ER
      COMMON/PURE/NM1(100),NM2(100),TC(100),PC(100),RD(100),DM(100),
     1  A(100),C1(100),C2(100),C3(100),C4(100),C5(100),RU(100),OU(100),
     2  QP(100),D1(100),D2(100),D3(100),D4(100)
      COMMON/BINARY/ETA(5050),U(100,100)
100   CALL PARIN(92,ER)
      IF(ER.GT.0) GO TO 900
200   READ(7,02) N,T,(ID(I),I=1,N)
 02   FORMAT(I2,8X,F10.2,10I5)
      IF(N.EQ.0) STOP
      KEY=1
210   READ(7,04) IR,IE,(Z(I),I=1,N)
 04   FORMAT(2I5,(7F10.4))
      SZ=0.
      DO 212 I=1,N
```

```
212 SZ=SZ+Z(I)
    IF(SZ.LT.0.999) GO TO 200
    WRITE(4,11) N,T
 11 FORMAT(///30H LIQUID/LIQUID EQUILIBRIUM FOR,I3,23H COMPONENT SYSTE
   1M AT T=,F4.0)
    WRITE(4,12)
 12 FORMAT(/1X,*INDEX*,5X,*COMPONENT*,15X,*FEED*,6X,*R PHASE*,5X,
   1  *E PHASE*,8X,*K*/)
    CALL ELIPS(N,ID,KEY,IR,IE,Z,T,V,X,Y,K,IER)
    IF(IER.GT.0) WRITE(4,13) IER
 13 FORMAT(/9H ***ERROR,I2,3H**/)
220 DO 229 I=1,N
    II=ID(I)
    WRITE(4,14) II,NM1(II),NM2(II),Z(I),X(I),Y(I),K(I)
 14 FORMAT(2X,I3,2X,2A10,3F12.4,1PE14.2)
229 CONTINUE
    WRITE(4,15) V
 15 FORMAT(/3X,*E/R =*,1PE9.2/////)
    GO TO 210
900 WRITE(4,19)
 19 FORMAT(/* ERROR IN PARAMETER INPUT DECK*/)
    STOP
    END
```

Appendix J

EXECUTION TIME AND STORAGE REQUIREMENTS OF THERMODYNAMIC SUBROUTINES

The computer storage requirements (floating point machine words) for the labeled common storage blocks and, approximately, for the principal computer subroutines are given in Table J-1. Also included in this table are some average execution times for the thermodynamic subroutines measured for the CDC 6400 of the Computer Center, University of California, Berkeley.

The program storage requirements will depend somewhat on the computer and FORTRAN compiler involved. The execution times can be corrected approximately to those for other computer systems by use of factors based upon bench-mark programs representative of floating point manipulations. For example, execution times on a CDC 6600 would be less by a factor of roughly 4 than those given in the table and on a CDC 7600 less by a factor of roughly 24.

Execution times for the higher level subroutines FLASH and ELIPS will be highly dependent on the problems involved. The times required per iteration can be estimated from times for lower level subroutines and the descriptions given for FLASH and ELIPS. Computation times for two specific cases calculated with FLASH and one case claculated with ELIPS are included in Table J-1 to show approximate magnitudes required.

352

TABLE J-1: Computer Storage and Execution Time Requirements for Thermodynamic Subroutines

Subroutine or Common Block	Storage (Decimal Words)	Conditions System Type*	Key**	Execution Time (ms) on CDC 6400 for: N=2	N=3	N=4	N=6
/PURE/	1900						
/BINARY/	15050						
/VIRIAL/	842						
/BS/	1680						
/GS/	481						
/VAL/	20						
/PS/	253						
PURF	153	–	–	2.7		4.7	6.6
TAUS	113	–	–	1.2		3.9	6.5
BIJS	625	–	1	3.0		9.5	20.0
		–	2-5	1.5		4.1	7.6
		–	6,7	2.0		5.6	11.4
GAMMA	312	–	1	3.5	5.5	8.5	13.5
		–	2+	2.4	3.8	7.2	12.2
PHIS	546	Non-associating	1	4.0		11.5	22.5
		Non-associating	2-5	2.4		5.8	10.4
		Non-associating	6,7	3.0		7.4	13.8
		Associating	1	8.5		22.5	45.0
		Associating	2-5	5.5		14.1	27.3
		Associating	6,7	5.9		17.0	32.6
VALIK	217	Non-associating	Average	8.		20.	34.
		Associating	Average	17.		36.	62.
LILIK	196	–	Average		8.		
ENTH	407	–	All	1.4		1.7	2.5
FLASH	1009	(Table 7-1, Cases 1&6)			(~75.)	(~170.)	(~300.)
BUDET	521						
BUDEP	421						
ELIPS	1443	(Table 7-2, Case 1)					
PARIN	195						

* System type refers to whether vapor phase contains strongly associating (dimerizing) components or not.

** Key = 1 represents an initial calculation for a new system; Key 2-5 are subsequent calculations not differing significantly in time requirements; Key = 6,7 require temperature derivatives of virial coefficients.